Understanding Token Ring Protocols and Standards

For a complete listing of the *Artech House Telecommunications Library*,
turn to the back of this book.

Understanding Token Ring Protocols and Standards

James T. Carlo
Robert D. Love
Michael S. Siegel
Kenneth T. Wilson

Artech House
Boston • London

Library of Congress Cataloging-in-Publication Data
   Understanding token ring protocols and standards / James T. Carlo... [et al.].
      p. cm. — (Artech House telecommunications library)
   Includes bibliographical references and index.
   ISBN 0-89006-458-X (alk. paper)
   1. IBM Token-Ring Network (Local area network system). 2. Computer network
protocols. 3. Ring networks (Computer networks)—Standards. I. Carlo, James T.
II. Series.
TK5105.8.I24U53   1998   98-30074
004.6'8dc21   CIP

British Library Cataloguing in Publication Data
   Understanding token ring protocols and standards. — (Artech House
telecommunications library)
1. IBM Token-Ring network (Local area network system) 2. Ring networks
(Computer networks)
1. Carlo, James T.
004.6'5'0218

   ISBN 0-89006-458-X

Cover design by Lynda Fishbourne

© 1998 ARTECH HOUSE, INC.
685 Canton Street
Norwood, MA 02062

International Standard Book Number: 0-89006-458-X
Cataloging-In-Publication: 98-30074

10 9 8 7 6 5 4 3 2 1

Contents

Preface

This book explains the architecture of Token Ring from the unique perspective of the authors' long-term participation in the development of Token Ring technology and the IEEE 802.5 Token Ring Standards. The book is written as a supplement for and a companion document to the IEEE 802.5 Token Ring Standards, providing full details of Token Ring architecture and a discussion of why certain decisions were made. The book is a comprehensive reference for network administrators, Token Ring product developers, and graduate level work in Token Ring. The primary goal of this book is architectural, for the authors believe that the Token Ring architecture must be understood to better utilize the technology and optimize future developments in Token Ring networking. Technical how-to and troubleshooting books often provide specific step-by-step procedures without explaining why those procedures are effective. This book will supplement, rather than replace, those practical texts, providing insights and deeper understanding.

The IEEE 802.5 Standards were written to specify the performance of Token Ring adapters, concentrators, hubs, switches, and repeaters in sufficient detail to reasonably assure their interoperation. The base Token Ring Standard was substantially revised and rewritten during the period between 1991 and 1995. The Dedicated Token Ring Supplement to the base standard, developed between 1994 and 1997, allows migration of Token Ring from a shared network to switched full-duplex operation. This switched protocol preserves the integrity of the Token Ring cabling infrastructure while providing dedicated bandwidth with a reliable point-to-point network architecture. More importantly, Dedicated Token Ring provides for the evolution of a Token Ring network from Classic, shared operation to full-duplex operation, gradually, based on user requirements. This text explores in depth the details of both Classic and Dedicated Token Ring.

In some cases the parameters specified in the IEEE 802.5 Standards were based on subtle operational characteristics of Token Ring. In others, they were the results of extensive committee discussion and compromise to allow existing implementations to remain in conformance. Still other issues were decided philosophically, weighing the alternatives of a specification that was easier to follow versus a more inclusive design that left room for future innovation at the cost of present complexity. In all cases, there was extensive testing and analysis to provide the highest level of assurance that proper operation would result from implementation of the standards as written. The Token Ring Standards provide an excellent basis for evaluating conformance of specific implementations. They do not, however, provide tutorial assistance as to why particular features were chosen. During the standards' development, many details and insights were expressed but not recorded. It is our hope to keep these insights on Token Ring alive by providing important details on rationales and critical decisions made. The Standards include features that have not been implemented in commercially available products. This text will stimulate vendors and users of Token Ring to better understand these latent capabilities.

It is suggested that this book be read jointly with the IEEE 802.5 Token Ring Standards for both Classic and Dedicated Token Ring operation. Detailed state diagrams and tables appearing in the Standards are not repeated here. The Standards will be required for this level of detail. However, the insights and implications of those state diagrams and tables were beyond the scope of the Standards and are a contribution that this work attempts to provide.

Chapter 1 is a general introduction to Token Ring. It includes insights into the driving motivation for the development of Token Ring and a historical perspective focusing on the standards development process.

Chapter 2 focuses on the cabling plant, which is the physical underpinning of any Token Ring network. Cable types are defined and topological configurations are discussed, including the detailed wiring rules for Classic Token Ring operation and their simplification using Dedicated Token Ring. Chapter 3 continues the description of the physical layer with a discussion of jitter accumulation. Jitter buildup in Token Ring was a major concern in 1992, and we felt the need to provide technical and historical background on this issue and in so doing highlight how interoperability issues can be resolved through the Standards consensus process.

Chapter 4 probes the details of Classic Token Ring operation. Because the Standard provides excellent coverage of the media access control (MAC) state machines and tables, that topic is only briefly described here. Reference to the Standards will be critical for both this chapter and the next. Chapter 5 provides equivalent detail on Dedicated Token Ring. Its architecture and configurations are explained along with a historical perspectives of specific decisions and choices made in development of this standard.

Chapter 6 discusses ongoing developments in the world of Token Ring, specifically high-speed operation. It provides basic information about the technology as well as guidance on when and how to apply it.

The five appendixes that follow provide important supplemental information not covered elsewhere in the text. Three of the appendixes are independently written papers providing critical supplemental material. Appendix A is an extract from Dr. Werner Bux's classic paper, "Token-Ring-Local-Area Networks and Their Performance," first published in 1989. The extract includes his entire discussion of Token Ring performance and his extensive bibliography of important Token Ring reference works. Appendix B is John Morency's "Investing in Token Ring: A Guide for Network Managers," first posted on the Alliance for Strategic Token Ring Advancement and Leadership (ASTRAL) Web site in 1996. This paper provides critical insights into the costs associated with network complexity and with migrating from one network type to another. Network administrators need to understand both the concepts presented in this paper and their implications. They will prove vital in carrying out the job of providing maximum networking capability at minimum cost over an extended time frame with evolving network protocols, topologies, and applications.

Appendix C, "Migration Issues and Strategies for Token Ring" by Ebbesen, Cowtan, Hakimi, and Love, published in 1997, shows in step-by-step progression how to fully exploit Dedicated Token Ring for the growing performance requirements of today's networks. Included in this progression is a smooth migration path from Token Ring to asynchronous transfer mode (ATM), available for those who see ATM as their strategic networking direction.

We hope this book will increase the readers' appreciation for Token Ring, for the Token Ring Standards, and for the effort involved in its development. We also believe that other network architects would profit from reading this book in applying critical features of Token Ring to other networks. Finally, the book should provide encouragement to Token Ring developers and vendors to better understand the richness of the Token Ring protocol and why Token Ring has been successful in the marketplace.

Acknowledgments

This book is the culmination of more than 15 years of Token Ring technology and standards development in which the authors feel privileged to have been participants. That history includes:

- Early theoretical papers and hardware;
- The Z-ring developed in IBM's Zurich laboratory;
- The intense development efforts at IBM's Research Triangle Park, North Carolina, facility and Texas Instruments' Texas and U.K. facilities;
- The years of standards development with participants from dozens of companies.

Token Ring has been a collaborative effort, integrating diverse ideas and approaches. The result is a vital technology, Token Ring, entrusted with the task of carrying the critical data of thousands of business establishments, including most of the world's top corporations. There is no way to properly credit or thank the hundreds of professionals who have made important contributions to this great technology or the dozens of corporations that have thrown their support and resources into the development and refinement of Token Ring. We must, however, give special thanks to IBM and Texas Instruments for their 15+ years of belief, trust, and support for Token Ring and to IEEE's LAN/MAN Standards Committee for creating the forum that allowed for the development of the Token Ring Standards. In addition, special thanks must be given to Bob Donnan, the first chair of the IEEE 802.5 Token Ring Working Group, who spent over 10 years at the helm, skillfully and diplomatically shepherding the fledgling standard through the U.S. and international standards committees, and especially for his role as mentor, showing us how to carry on where he left off.

We give special thanks to our wives, Muffy, Ann, and Trish, and to our families for their understanding and patience during the preparation of this book.

James T. Carlo, Robert D. Love, Michael S. Siegel, Kenneth T. Wilson

1

Introduction

1.1 Why IBM Chose Token Ring

Token Ring was introduced after Ethernet was already developing in the *local area network* (LAN) marketplace. Token Ring's development as a standard slightly lagged Ethernet's. So why develop and standardize a new LAN type? Wasn't Ethernet's capability enough? Wouldn't standardizing a single LAN type be beneficial to the whole data communications industry rather than introducing a competing technology?

Answering these questions requires an understanding of the different requirements that Token Ring aimed to satisfy. In the late 1970s IBM was looking at their customer's requirements for future networking. These customers included leading banks and insurance companies, whose economic livelihood depended on reliable data transport. Large computers were satisfying that requirement with applications running under IBM-developed *Systems Network Architecture* (SNA). Message delivery was highly reliable, and available bandwidth was efficiently utilized. Redundancy and reliability were built into every aspect of the system. These applications needed to be migrated to a LAN and that LAN would have to exhibit most or all of the vital reliability characteristics available on the mainframe computer.

From this point of view Ethernet was an inferior technology because the *Carrier Sense Multiple Access with Collision Detect* (CSMA/CD) protocol was chaotic and uncontrolled. Stations ready to transmit a message onto the LAN would listen to see if a message was in progress. If the medium was quiet, they would initiate their messages while continuing to listen, monitoring the LAN for another message. If another station also began transmitting, then both messages would be lost in what was termed a *collision,* and both stations would

1

stop transmitting, or *back off*, and try again after waiting for a random time period. This protocol worked very well when LAN traffic was light but bogged down in heavy traffic, when much of the usable bandwidth was wasted in collisions. Nor was this the only problem that Ethernet had from the viewpoint of the highly controlled world of large systems. It was impossible to identify which stations were on a LAN. Any station on the LAN was undetectable until it transmitted a signal. Further, if the station were to malfunction, it could take down the entire LAN, with no easy way to determine where the fault was, except by "walking the LAN," with testers and probes. Network management for Ethernet LANs was almost nonexistent. As a last straw, the bridging protocol used for Ethernet was *transparent bridging*, a technique that precluded designing the LAN with multiple paths and parallel paths to ease congestion while adding redundancy. IBM had two choices. It could start with Ethernet and convince its loyal followers to transform that technology into one that was appropriate for handling corporations' vital network traffic, or it could develop its own LAN type along with a cadre of like-minded corporations. The latter task was far more doable, and Token Ring was the technology chosen by IBM.[1]

The initial requirements for Token Ring included:

- Efficient use of bandwidth;
- Excellent network management including the abilities to know who is on the network and where errors are being generated and to identify faulty stations, including those responsible for generating soft errors;
- Compatibility with SNA;
- Priority mechanisms in place for traffic requiring fast access;
- Fairness in accessing the ring;
- Good cabling capabilities for campus LANs (when using repeaters);
- Excellent *reliability, availability, serviceability* (RAS) features including ensuring the LAN rarely goes down, minimizing outage times, and providing "work around" solutions for any single faults until replacement or repair parts are available.

To better understand these requirements, let's examine the implications of the strict requirements IBM placed on RAS for Token Ring. These RAS

---

1. It must be pointed out that these early problems of the Ethernet have largely been overcome with the introduction of 10BASE-T star-wired HUBS using the star-wired architecture of Token Ring.

features defined a methodology to design and verify that the Token Ring system would result in a high-utilization communications system.

The first part of the RAS policy, reliability, was aimed at building a networking system with a low failure rate. The semiconductor technology in the early 1980s achieved low failure rates by virtue of very large scale integration (VLSI) processing where the improved reliability of semiconductor processing and the reduction of chip count reduced system failures. The reliability goal of IBM for a Token Ring network of 100 stations was no more than two failures per year. In addition to this low system failure rate, data reliability was paramount and this integrity was insured by a frame check sequence (FCS) protection on all data packets sent over the network and the use by most early implementations of parity on all adapter data transfer buses.

The second part of the RAS policy, availability, dictated that the communications system had a high percentage of operational time versus inoperable down time. Stations inserting into the Token Ring network had to undergo self-test before insertion to insure that the adapter was operational and would not break an operational network. Failures of specific stations, cables, or concentrators had to be readily diagnosed and, where possible, the failing station had to be removed from the ring, leaving the rest of the communications system operational.

The third part of the RAS policy, serviceability, required that faults in the network be accurately diagnosed and that positive indication of these faults be available to the network manager. A main objection to early networks was that in the event of failure, the network would become inoperable, but no indication of the fault could be found. Serviceability insured that the system, when an error did occur, would record the error indication for field service technicians and in most cases recommend which component should be replaced. In order to achieve a significant level of fault isolation at the system level, the concept of a Token Ring fault domain (8802-5:1995, clause 4.1.6.1) was architected to enable all Token Ring stations to participate harmoniously in the fault diagnostic process at the *media access control* (MAC) level. Ring beaconing is a key component of this isolation process that drove a considerable portion of the functionality (and complexity) for Token Ring.

1.1.1 Token Ring "One-Page" Tutorial

Before describing the two fundamental measurements for RAS, it is necessary that the reader have a short understanding of Token Ring operation and basic implementation. This section will provide a short summary of Token Ring operation and frame format. Readers who are familiar with Token Ring can skip this section.

The Token Ring is a communications technique in which a group of stations are connected in a loop and a token (three-byte entity) is used to control ring access. Figure 1.1 illustrates a Token Ring operation with four stations on the ring and shows the four basic steps in token circulation, transmitting, receiving, and stripping.

Each station transmits data to its downstream neighbor and receives data from its upstream neighbor. The token, which is initially released by one station on the ring, termed the *Active Monitor* (AM), circulates from station to station around the ring, with each station "repeating" the data (Step 1). A station wishing to transmit data to another station must wait for a token to be received, and after detecting the start of the token, it changes the token to a frame and transmits the frame as shown in Figure 1.1 (Station B in Step 2). The frame circulates around the ring and is repeated by each station. The station recognizing the destination address of the frame (Station D in Step 3) receives the frame, in addition to continuing to repeat the frame to the next station on the ring. Finally, the original transmitting station removes the frame from the ring and releases a token (Station B in Step 4).

The bit pattern for a token and a frame are illustrated in Figure 1.2. The detailed fields within the frame and token will be described in Chapter 4.

As seen in Figure 1.2, the start of the token and the frame are the same. The token bit in the access control field octet specifies whether the data following is for a frame (token bit = 1) or token (token bit = 0). Thus, a station wishing to transmit waits for a starting delimiter and the access control field, and if the token bit is zero, it changes this bit to a one and places the frame on the wire. If the token bit is one, then the station must continue to repeat (because a frame is being circulated on the ring). The station may set a reservation bit in the access control frame utilizing the priority protocol in Token Ring as described in Chapter 4. The Token Ring frame contains a frame status field at the end of the frame that is used by a receiving station's MAC to indicate that the frame has been received and copied. This capability is utilized by network management to determine if the frame was received and copied by the destination station on the ring to aid fault isolation.

1.1.2 Typical Token Ring Adapter Components

The illustration in Figure 1.3 shows the major blocks for a typical Token Ring adapter.

The *physical* (PHY) layer is responsible for transmitting data and recovering received data on the ring. The PHY contains a *phase locked loop* (PLL) that is used for recovering the clock in the received data. The *protocol handler* (PH) performs the basic protocol functions for the ring, such as token

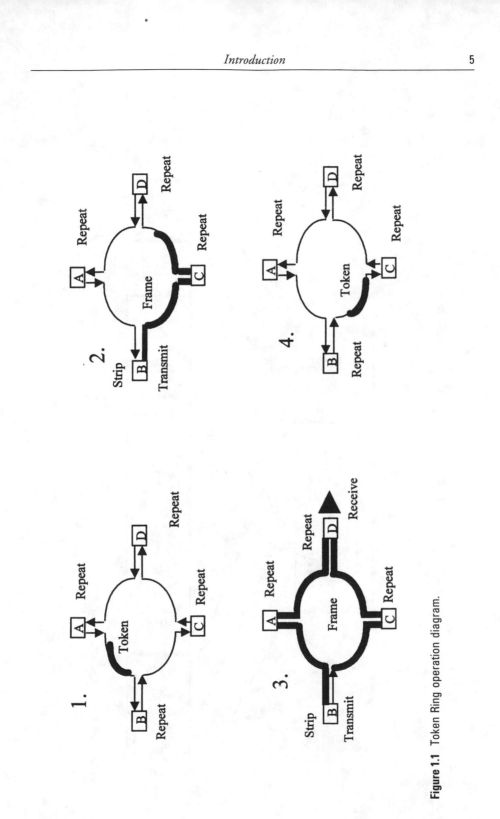

Figure 1.1 Token Ring operation diagram.

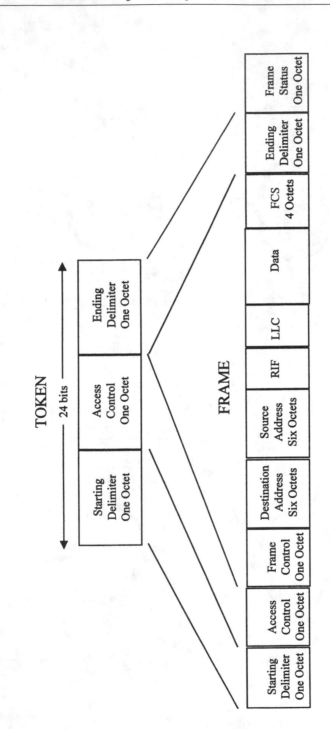

Figure 1.2 Token and frame byte patterns.

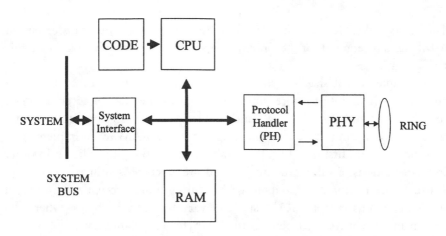

Figure 1.3 Token Ring adapter (typical).

delineation and token control, FCS checking, and address recognition. These MAC facilities are time critical and thus handled directly by hardware logic in most Token Ring implementations. The repeat path from the ring through the adapter is designed to be as short as possible (typically one to three symbols or bit delays) to keep the effective physical ring transmission path as small as possible for best performance. Note that during station transmission, the repeat path is bypassed so that the incoming data (which was transmitted) is stripped from the ring.

The adapter central processing unit (CPU) provides for data and frame processing. All frames to and from the ring are stored in the adapter buffer random access memory (RAM), where they are parsed by the CPU to determine whether they should be passed to the system or processed directly by the adapter. MAC protocol processing is handled by the CPU to reduce demands on the system processor (which might be a personal computer). In some cases a frame response is required so that the CPU software sets up a frame for transmission by the adapter without interference from the system processor. In other cases, the system is informed of a specific event on the ring for management purposes. Typical code sizes to support the MAC protocol, self-test diagnostics, and interface to the system bus and system processor for command processing was about 16K bytes.

The system interface provides the control and handshaking between the system and the adapter across the system bus. In high-speed adapters, the system interface is a *direct memory access* (DMA) controller providing both address and data for all system/adapter transactions. In other adapters, the system interface provides a slave interface that allows programmed input/output (I/O) or memory reads/writes for system/adapter transactions. For a

16-Mbit/s Token Ring, with a 16-bit wide system bus, a net transfer rate of 4 Mbytes/s is required for the simultaneous transmit/receive interaction across the system bus.

Adapter RAM size is a trade-off between minimizing RAM for lower cost versus the ability to store multiple frames in the adapter to compensate for a busy system bus during peak ring load frame traffic. In some implementations, RAM is minimized, thus allowing the entire RAM to be implemented as a short first in, first out (FIFO) contained within the CPU silicon. In other implementations, a full store and forward architecture is utilized, where the entire frame is stored in adapter RAM during frame reception. In actual implementations, adapter RAM size can vary between 64 bytes (where the system bus is always available) up to 2 Mbytes for adapters with peak load control where system bus availability cannot be guaranteed.

1.1.3 Error Detection and Fault Isolation

The RAS policy aims toward localizing faults to the adapter as distinct from the complete system, cabling, and other stations on the ring. A major part of serviceability and availability was based upon a concept of *error detection/fault isolation* (EDFI) that was developed by IBM in the 1970s [1]. EDFI provides a system of measurable metrics to prove fault localization by defining error indicators that localize the failure. In EDFI, errors are defined as incorrect actions of the system and are the results of a fault. Note that multiple error indicators may result from a single fault. The faults may be hard faults, such as the shorting of a transistor, or soft faults, such as data recovery error induced by noise or an intermittent cable short. In general, the intermittent faults are found to be the dominant class, typically creating errors five times more frequent than hard faults. Error detection is accomplished if the error caused by the failure activates an error checker, which is an indication that the error occurred. Examples of error checkers are FCS checksums, background code diagnostics, and parity errors.

The error checkers sense a fault and the protocol then causes appropriate action in the network (such as ring station removal and beaconing). In addition, management is notified of the error indication so that localization of the fault may occur. Most of the error checkers are specified in the Token Ring Standards so that multivendor Token Ring adapters may work in harmony to aid diagnosis and localization of a failing adapter. A few error checkers that are specific to the hardware implementation (such as internal adapter parity) monitor for errors that are only reported to the adapter system.

1.1.4   Adapter Self-Test

Another major part of serviceability and availability is the self-test of the adapter, which prevents faulty adapters from inserting into the ring and provides a set of field diagnostics for the adapter. These diagnostic tests consisted of three parts.

- *Bring-up diagnostics* (BUD). This BUD code (3K bytes in the TI chipset) achieved a high degree of fault diagnostics coverage by placing the adapter PHY in a loop-back mode and transmitting valid and invalid frames and checking the responses. This enabled a check of the MAC hardware functions including address compare, token handling state machines, and FCS generation. This BUD code also verified each instruction of the CPU instruction set, adapter RAM (by writing hexadecimal patterns 0000, 0101, A5A5, 5A5A, and FFFF into all memory locations), and adapter RAM or read-only memory (ROM) code by calculating a checksum for the code and then comparing it to a stored checksum. The BUD code was also run by the CPU during inactivity as background diagnostics to continuously verify system operation.
- *Initialization.* When the adapter was initialized by the system processor, the DMA interface between the system and adapter was verified by DMA writes/reads and compares.
- *Lobe test.* Before final insertion into an active ring, an extensive test of the lobe media was performed by transmitting lobe MAC test frames between the adapter and the concentrator (in loop-back mode) over the lobe cable. The duplicate address-checking state machines were also tested during this process.

Figure 1.4 illustrates the three self-test loop-back configurations described previously. A self-test coverage of 92.5% for the entire adapter was achieved on the initial chip-set implementation by Texas Instruments (TI) released in 1985. This coverage was measured analytically by partitioning the adapter into its major components and determining the coverage of each component based on transistor counts for that component. Coverage was also verified experimentally by inducing failures in an implementation of the final system in a hardware model (consisting of greater than 500 separate integrated circuits), illustrated later in this chapter in Figure 1.10, and by inducing faults and determining whether the faults were detected by the self-test. Token Ring implementations from all vendors have maintained this high degree of self-test coverage in order to enable high reliability of the communications system.

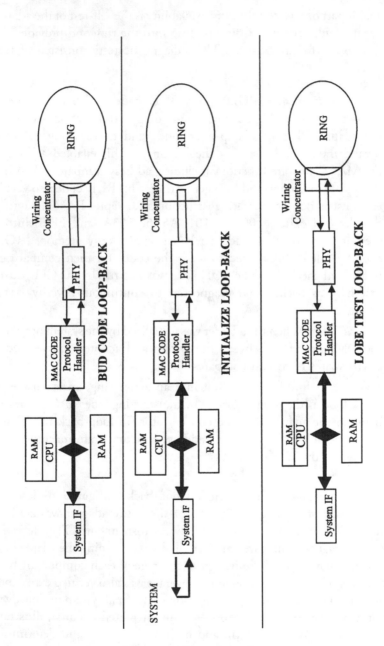

Figure 1.4 Adapter self-test loop-back configurations.

1.1.5 Token Ring Wiring Architecture

The Token Ring star-wired wiring architecture was a key feature of the RAS policy. During early network loops in the 1970s, cabling proved to be the major trouble area. With the stations wired in a physical loop arrangement, while minimizing cabling distances, the loop proved to be unstable and impossible to maintain. Breaks in the cabling could not be located easily and the entire network would become inoperable due to a single fault as shown in Figure 1.5. This single cabling break, besides shutting the entire network down, was often difficult to isolate because of poor documentation on specific cabling location.

The solution to this problem was to utilize a star-wired ring architecture as shown in Figure 1.6 for small node count rings and in Figure 1.7 for large node count rings. This star-wired system with phantom drive in most cases avoids the shut-down of the entire ring if one adapter is faulty.

In the star-wired Token Ring, each station utilizes a small dc current (phantom drive) to indicate an insertion request to the concentrator port (called the trunk coupling unit in 8802-5:1995). Token Ring concentrators typically

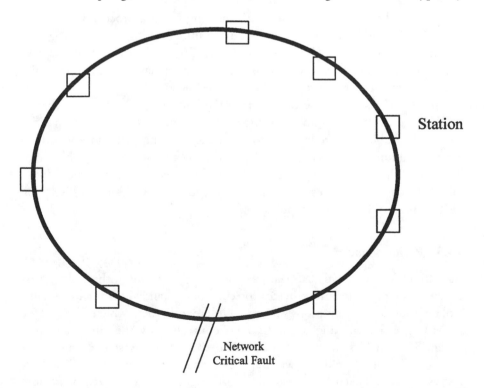

Figure 1.5 Looped ring-based wiring architecture.

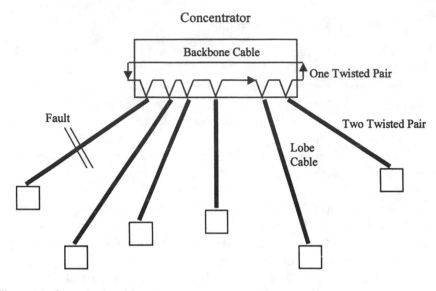

Figure 1.6 Star-wired architecture.

use a reed relay to insert a station into the ring using a process described in Chapter 2. While there were concerns during initial testing with the reliability of a mechanical reed relay, throughout Token Ring's history the authors have been seldom aware of a failure in this unit (failures have been reported in a semiconductor electronic replacement for the relay). If the phantom drive is not present the station lobe cable is in the initialize loop-back mode (Figure 1.4), permitting a self-test of the lobe cable. When the dc phantom drive current is enabled, the resistance of each cable is monitored and a wire fault error condition is indicated at the adapter due to a cable break or cable short. The use of the phantom drive and wire fault has been essential in the delivery of the Token Ring's robust architecture. Chapter 2 provides a full discussion of Token Ring wiring and topology.

1.1.6 Performance

In addition to the broad RAS features mentioned previously, the Token Ring token-passing architecture provides for controlled bandwidth for high-traffic Token Ring communications systems. Unlike Ethernet, which is based on CSMA/CD, Token Ring transmission is controlled by the circulation of a token. This enables the total bandwidth to be effectively shared among the stations wishing to transmit without loss in bandwidth due to collisions or retries. For the Token Ring system, as load increases, the utilized bandwidth also increases up to the ring data rate. Token Ring achieves full bandwidth of

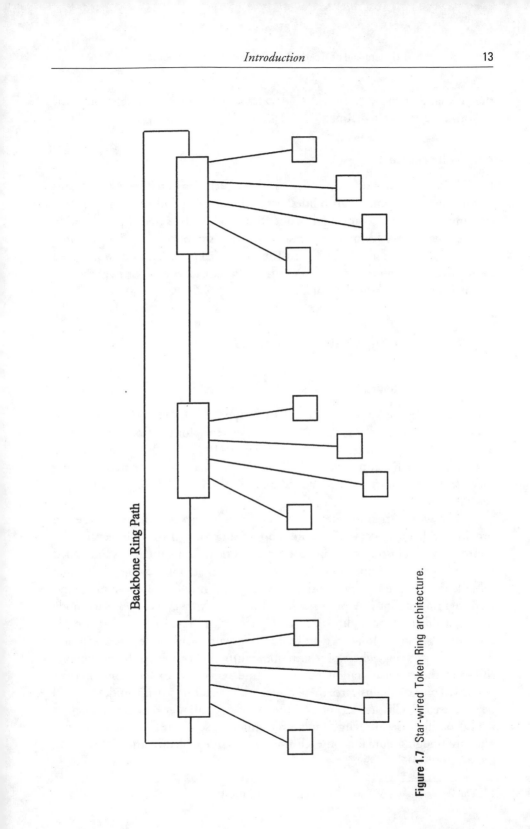

Figure 1.7 Star-wired Token Ring architecture.

the media. Appendix A provides an extract of a key paper describing the performance for Token Ring.

1.1.7 Management

In addition to providing fault isolation, the Token Ring architecture contains feature-rich management capabilities that enable system administrators to monitor ring activity and count and maintain a list of active nodes participating on the ring. These features are present in every station on the Token Ring. The total RAS capabilities for Token Ring has been achieved as demonstrated by the over ten million active nodes in the marketplace today with very robust performance and excellent reliability.

1.2 History of Token Ring

1.2.1 Earliest Papers

Token Ring, as we know it today, was developed by IBM workers in Zurich Laboratories in the late-1970s, and the best reference for the first experimental Token Ring is contained in the classic paper (1983) by Bux et al. [2]. To understand which features were critical to IBM at that time, one needs to note that the largest networks were aimed at linking 10 to 100 bank tellers in a central bank.

The wiring architecture was a key focus of development in order to provide a reliable system. The store loop,[2] using the wiring architecture of a "pure loop" as illustrated in Figure 1.6, was difficult to maintain. While called a "ring," the serial connection of a large number of stations, while minimizing cable length, had too high a cost associated with installing, maintaining, and reconfiguring stations. A pure star-wired system, with every workplace node tied to a single point as illustrated in Figure 1.6, resulted in too much cable for outlying stations in a large office building with many nodes. The solution was a star-wired ring, where concentration points were connected by a backbone ring and these concentration points became star-wired to locations in their vicinity. Figure 1.7 illustrates these options; the detailed wiring guidelines will be discussed in Chapter 2. For an extremely large number of stations, such as would be the case in a large multilevel building, separating the stations into separate bridging domains on each floor was also enabled in the Token Ring architecture [3].

---

2. The store loop was an early IBM product in the late-1970s.

1.2.2 Token Ring at 4 Mbit/s (1980–1985)

Token Ring was initially introduced into the market with a 4-Mbit/s ring speed. The original standard called for a ring speed of 1 Mbit/s and 4 Mbit/s. While the 1-Mbps is present in the first drafts of Token Ring Standard, to the author's determination, no product was ever sold at that speed. The major concepts of Token Ring were developed by IBM and tested by IBM and TI. The IEEE 802.5 Standard was primarily written by IBM contributors under the original IEEE 802.5 Token Ring Working Group chair, Bob Donnan.

At the start of the commercial development of Token Ring, IBM identified a "silicon partner," TI, who would develop commercial Token Ring chipsets to enable multiple vendor interoperability, and a system partner, Ungermann Bass, who would provide system products in addition to IBM's products. TI began Token Ring silicon development in 1982. During this development program, IBM and TI worked together with teams of people worldwide (Houston, Dallas, Bedford-UK, Raleigh) throughout the program through 1985. The management of the program was based on a concept of *ratification, verification, and validation* (RVV).

Figure 1.8 illustrates the RVV process as applied to the Token Ring program. A functional specification, primarily an architectural specification, was the basic reference text for the Token Ring design. This specification has become the basis for the Token Ring standard. From the architectural viewpoint, one specific design implementation was developed by TI that specified the system partitioning between hardware (silicon chips) and software (microcode). The implementation specification was used to design the silicon devices as well as build a TTL hardware model. Ratification efforts were aimed at proving that the network met user requirements as provided by the architecture. Verification efforts were aimed at proving that the implementations of the Token Ring protocol would operate in a ring environment. Validation efforts were aimed at proving that the different implementations (a TTL hardware model, software models, and actual silicon) were equivalent. During the program development, 617 silicon hardware test cases were written to validate that the silicon met the implementation specification, and 4,879 system-level test cases were written to verify that the system (hardware and software) operated in the ring.

The block diagram for a Token Ring adapter using the TI chipset [4] in 1985 is illustrated in Figure 1.9. The Token Ring implementation utilized a CPU coprocessor to carry out the Token Ring MAC frame management functions and processors. In the initial TI implementation, this processor was based on the TI9900 16-bit microprocessor. In IBM's initial implementation, a proprietary processor was utilized. The use of non-marketwide standard

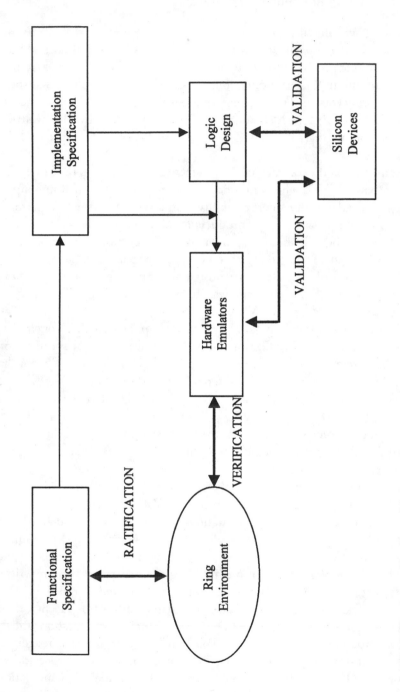

Figure 1.8 Ratification–verification–validation process.

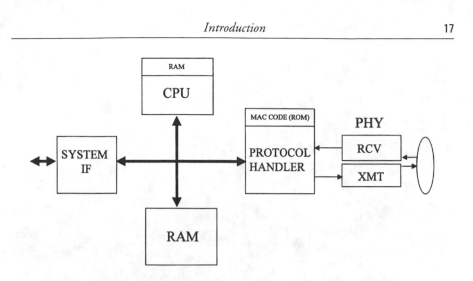

Figure 1.9 Early Token Ring system block diagram.

processors increased the level of commitment required for code modification, and thus few companies were able to deploy their unique code, which could have resulted in interoperability problems.

This MAC code, which consisted of 16K bytes, was included as part of the PH silicon. The reason for including the microcode inside the PH was to closely couple hardware and software MAC functions. A different PH implementing another MAC protocol might have a different MAC code load associated with it. Intermediate data storage area and MAC code tables were stored in RAM, of which 2.75K was initially included as part of the CPU to allow for minimum chipset adapter with no expansion RAM. Additional facilities for RAM expansion were provided to allow greater intermediate data storage, and nearly all early implementations typically included 16K bytes of additional RAM.

A key development task of the RVV program was the building of a transistor-transistor logic (TTL) hardware model based on the silicon design circuits. This hardware model was developed by a team of over 80 engineers working from silicon design schematics. The rationale for this model was to enable early test case development, the development of the MAC microcode, and early ratification of design concepts prior to silicon availability. Figure 1.10 illustrates the TTL model developed for the initial Token Ring chipset. This model consisted of nine wire wrap cards (12 in by 16 in) with a total of approximately 1,500 TTL integrated circuit packages. As each of the silicon chips became available, a silicon-based board was built to replace the specific hardware model cards, so that finally, a single board containing the five original silicon chips was constructed and validated.

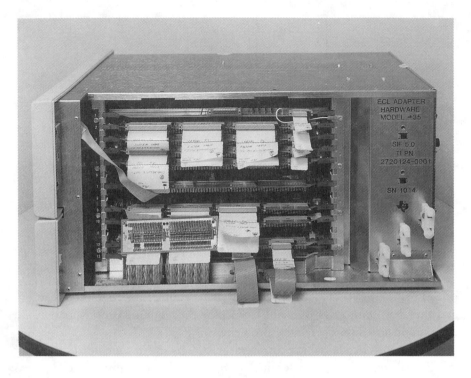

Figure 1.10 TTL hardware model of Token Ring chipset.

MAC microcode was a major development task during the Token Ring chipset development. This microcode, developed by IBM, was written in assembly language and resulted in 16K bytes of code including BUD code (3K bytes). The 16K-byte size level was dictated by the ROM code space in the PH silicon and several times proved the maxim that "code size will grow to the allowable limit." The adapter shipped in 1985 had 18 bytes of unused code space resulting in a figure of 99.9% code space utilization. The microcode was rigorously protected by both IBM and TI with the belief that if the code was not modified then interoperability was assured. Even today, over 10 years after the initial product shipments, this code continues to be carefully controlled, and in 10 years, interoperability problems due to microcode errors have been few and far between. The downside of close control of code was in the possible loss of flexibility for multiple vendors to modify and improve upon the features in Token Ring. The decision to choose non-market-standard processors and tightly control code has enabled high stability of Token Ring but possibly reduced the Token Ring market penetration.

The introduction of the cabling plant to enable customers to install cabling to support their future networking needs was made by IBM in May

1994. This cabling plant, based on shielded twisted-pair cabling (IBM Type 1), enabled customers to wire buildings in anticipation of Token Ring products. Table 1.1 shows the initial cable types introduced by IBM. In 1985, AT&T announced a competing *unshielded twisted-pair* (UTP) wiring scheme, based on four pairs that would become the predecessor to the UTP building wiring guides and eventually supplant IBM's *shielded twisted pair* (STP) [5].

Along with the IBM cabling announcement, a hermaphroditic connector shown in Figure 1.11 was introduced. This connector (as its name denotes) allowed connection without needing two connector types (a jack side and a plug side). Token Ring will always be identified with this connector, and this is the tell-tale indication of IBM Type 1 cabling worldwide.

Table 1.1
Token Ring Early Cabling Options

Type	Date	Characteristics
Type 1	1984	STP (2 pairs, 22 gauge) for horizontal wiring
Type 2	1984	STP (2 pairs, 22 gauge) plus UTP (4 pairs, 22 gauge for telephone) for horizontal wiring
Type 5	1984	2 pair 100/140 optical fiber for backbone and campus wiring
Type 6	1984	STP (2 pairs, 26 gauge) for short interconnections
Type 8	1984	Shielded under carpet cable (2 pairs, 26 gauge) for open office wiring
Type 3	1985	UTP (4 pairs) for horizontal wiring

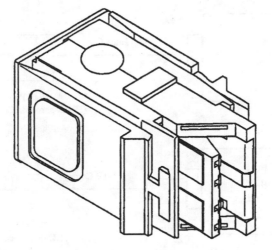

Figure 1.11 Hermaphroditic connector.

The initial announcement of Token Ring products occurred on October 15, 1985, with the joint announcement by IBM and TI in New York City. The announcement from IBM identified multiple products from IBM for their major channel controllers and the PC card. TI announced silicon availability, and several other companies announced products using this silicon. IBM announced networking software based upon SNA and NetBIOS, with connection-oriented logical link control (LLC) (IEEE 802.2 LLC Type 2) running on top of the MAC software. In addition, the operation over UTP (IBM Type 3) was announced. Interconnection of Token Ring to either STP or UTP was accomplished using a UTP filter, as shown in Figure 1.12. Token Ring had arrived in the marketplace.

1.2.3 Token Ring Early Growth (1985–1989)

With the announcement of Token Ring products in 1985, the next four years were a period of rapid expansion in the use of Token Ring technology and better understanding of compatibility. Besides having compatibility at the PHY and MAC levels, which enables products to interoperate on the ring, it became important to be able to utilize application software from one vendor over multiple hardware implementations of Token Ring. The networking industry concept of well-defined specific, inviolate software driver interfaces and *application programming interfaces* (APIs) was still at its infancy. IBM's implementation of a PC card utilized a shared RAM interface with LLC and MAC on card. TI's chipset implementation of a PC card utilized DMA or a memory-mapped I/O hardware interface with MAC on card and LLC off card. While at the

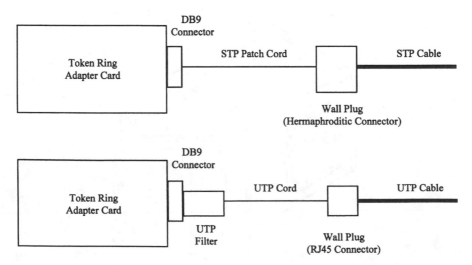

Figure 1.12 Token Ring adapter cabling attachment.

MAC level the implementations were compatible on the ring, the different hardware approaches and software partitioning prevented application software from being easily ported to different vendor's adapters. These software interoperability issues existed despite the fact that "on the ring" MAC code operation was interoperable according to the IEEE 802.5 standard. The Token Ring Forum was formed in 1986 in order to enable vendors to discuss how to better achieve multiple product availability from multiple vendors.

Early Token Ring products utilized connection-oriented LLC to establish a link between two stations. Connection-oriented LLC (also called Type 2 LLC) provides for frame error control by sequencing each data packet on the link, requiring positive frame acknowledgment of the sequence number. The LLC operation is specified by IEEE 802.2 and has similar characteristics to the family of high-level data link control (HDLC) protocols [6]. This Type 2 LLC was utilized in the protocol stacks for IBM's SNA stack and NetBIOS. With the advent of the Internet's Transmission Control Protocol/Internet Protocol (TCP/IP), most Token Ring traffic in use today has stopped using this connection-oriented protocol at the LLC layer in favor of a nonacknowledged LLC (Type 1 LLC) and depends on the upper level protocols of TCP/IP for frame sequence checking.

After the announcement of Token Ring, the U.S. Navy began a program to incorporate Token Ring products into another standard, called SAFENET. This standard aimed at providing automatic reconfiguration in the case of backbone cable failure by providing for a redundant backbone ring system.[3] The SAFENET I standard specified Token Ring with redundancy, which was eventually incorporated into the IEEE 802.5c Recommended Practice for dual-ring. The FDDI standard, for a Token Ring at 100 Mbit/s, was also developed at this time to enable both higher speed operation (100 Mbit/s) and fiber connection to stations (and later twisted pair) using redundant cables. FDDI, whose bit order is the same as Token Ring, became the backbone and high-speed upgrade path for Token Ring. Both FDDI and IEEE 802.5c enabled automatic ring reconfiguration and continued operation if the backbone cable was severed.

Source routing bridging was introduced with initial Token Ring products in 1985. An implementation from IBM used two normal PC cards with the addition of special bridge software to build a bridge product. Source routing depends upon the adapter stations to specify the route to take in a bridged network, with each bridge and ring assigned a ring specific number. In source routing, the route determination process is designed into each adapter and the

---

3. The intended use of SAFENET was a protocol for Navy battleships. The requirement was that the network had to remain operational even after a portion of the ship was struck by artillery. This requirement presented a whole new meaning to the term "fault tolerant."

bridges that forward frames need only develop tables of routing fields rather than tables of all adapter addresses as for transparent bridges. The source routing bridges were simple and provided a powerful method to increase network bandwidth by segregating stations into their user groups to limit networkwide traffic. However, because source routing was not utilized by Ethernet or FDDI, its use became difficult in heterogeneous networks, resulting in a decline in the use of source route bridging.

Token Ring has eight priority levels for prioritizing traffic. MAC traffic utilizes levels three and seven to ensure MAC processes are stable in the event of high ring utilization with data. Priority traffic is also used where bridges assign a priority of four to speed traffic requiring multiple hops through different rings. An algorithm for priority use was specified in the Token Ring Standard so that the same algorithm would be used by each adapter station in order to provide for fair use of the network by multiple vendors. Even though Token Ring had priority available since 1985, the applications use of quality of service mapping into time critical traffic was left unresolved so that this feature was not fully utilized in Token Ring networks. In 1995, ten years after Token Ring products were introduced, the use of priority in Ethernet networks is just beginning to develop.

With a deterministic Token Ring protocol, ring bandwidth could be demonstrated to be equal to the ring data rate so that a 4-Mbit/s Token Ring netted a true ring utilization of 4 Mbit/s. This is in contrast to Ethernet, where the CSMA/CD process results in about a 30 to 40% utilization. Token Ring at 4 Mbit/s thus has slightly more throughput capability than Ethernet at 10 Mbit/s. While this effective bandwidth similarity between Token Ring at 4 Mbit/s and Ethernet at 10 Mbit/s could be demonstrated, nevertheless, marketing statements from Ethernet vendors that Ethernet had a higher data rate caused Token Ring vendors to begin the development of a 16-Mbit/s standard in 1985. The concept for this higher speed version was to leave the MAC unchanged and increase the data rate by modifying only the PHY.

During the period between 1985 to 1989, Token Ring enjoyed rapid growth as the networking market grew in importance. Token Ring and Ethernet provided the dominant choices for networking, and many market predictors gave Token Ring a predicted growth curve that would exceed Ethernet by 1992, but 10BASE-T Ethernet was developed and Token Ring dominance did not occur.

1.2.4   Token Ring at 16 Mbit/s (1989–1992)

The first 16-Mbit/s Token Ring chipset was announced by IBM in 1988. The IEEE 802.5 standard for 16-Mbit/s Token Ring was completed in 1989. While this new standard was primarily based on an increased PHY data rate, the new MAC concept of *early Token release* (ETR) was also introduced. One additional change in the MAC specification of Token Ring was an increase in the elastic buffer size for 16-Mbit/s Token Ring to allow for increased accumulated jitter, thus increasing the node count. The elastic buffer at 4 Mbit/s had a minimum size of 6 bit times, while at 16 Mbit/s the minimum size was increased to 32 bits.

During 1989 and 1990, multiple vendors introduced 16-Mbit/s products and Token Ring was growing in importance. However, the operation of Token Ring products at 16 Mbit/s began to have inconsistent results. Large rings, especially, generally using UTP, started to exhibit high error rates that could not be understood. Vendors doing verification tests found that one implementation would not operate without errors when coexisting with another implementation for ring sizes greater than ten nodes. Various filters for UTP as shown in Figure 1.12 were being attached between the DB9 STP card connector and the UTP cable to try to improve interoperability.

The results were especially confusing because the same network would work perfectly one day and then errors would be observed the following day. The combination of the location of the AM station, the data pattern, and the specific implementations being tested was found to interact. Within six months, one company, Madge, recalled all its 16-Mbit/s products in the field [7]. The search for the compatibility demon had begun.

As compatibility issues arose, the IEEE 802.5 Standards Working Group took a major lead in coordinating an industrywide effort to resolve the problem. During 1991 and 1992, the IEEE 802.5 Working Group grew to over 100 members and over 100 papers and theories were put forth to explain the situation. A "deadly packet" algorithm was submitted in an IEEE 802.5 Working Group paper that showed how to identify which data patterns would cause the most errors. Almost simultaneously, four individuals (Dave Pearce, Kathy Otis, John Creigh, and Steve Hubbins) from four different companies realized that the jitter issue was due to accumulated jitter phase step at a change from data ones to data zeros, which could be both modeled and tested to demonstrate the error symptoms. The cooperation of individuals and companies during this period is one of the best examples of standards development benefits. Chapter 3 contains a summary of the analysis for the Token Ring jitter buildup.

Once the problem was identified, products were available in 1993 that provided a much wider operating margin for error-free behavior. A constant gain PLL phase detector was developed and the standard specifications on jitter buildup were tightened to ensure interoperability. However, the Token Ring marketplace momentum was damaged and the predictions made in 1988 for Token Ring to eclipse Ethernet were not fulfilled.

1.2.5   Token Ring Standard Solidified (1993–1995)

From 1993 to 1995, in order to enable multiple MAC implementations, the MAC standard was clarified and specificity added to more precisely describe the MAC processes. The original goal of this effort was to provide the bases for conformance tests for Token Ring; however, the conformance test methodology using internationally developed test cases and specifications was abandoned in place of real product interoperability testing. The resulting revision of this standard, nevertheless, was a much more specific Token Ring Standard that described the detailed state machine operations for the MAC including the improved PHY specifications. Table 1.2 illustrates the history for the base Token Ring Standard.

As illustrated by simple page count, the 8802-5:1995 specification contains greater details and specificity on Token Ring operation. The goal of a standard containing the details to enable multiple implementations was achieved.

1.2.6   Token Ring Switching and Dedicated Token Ring (1995–1997)

Dedicated Token Ring (DTR) defines signaling protocols and attachment topologies to enable a dedicated bandwidth for each station, which creates an evolutionary growth path for present Token Ring users. DTR is an enhancement to classical IEEE 802.5 Token Ring that greatly increases the attaching stations' available bandwidth. By connecting a DTR concentrator to a classic Token Ring concentrator, a small two-station ring is given the full ring bandwidth for each attaching station. By using the newly defined *transmit immediate* (TXI) protocol, full-duplex Token Ring attachment is provided. As an alternative to the token passing protocol (TKP), the TXI protocol effectively doubles the available bandwidth by allowing simultaneous transmission and reception of frames. Classic concentrators and stations can take advantage of the entire

Table 1.2
Token Ring Standards History

Date	Title	Pages
1984	IEEE Token Ring Draft Standard	64
1985	IEEE 802.5 Token Ring Standard	89
1989	IEEE 802.5 Token Ring Standard (16/4 Mbps)	95
1992	ISO/IEC 8802-5 Token Ring Standard (16/4 Mbps)	111
1995	ISO/IEC 8802-5 Token Ring Standard Revision 2	239
1998	ISO/IEC 8802-5 Token Ring Standard Revision 3	250

bandwidth improvements provided by DTR except for the final bandwidth doubling provided by the TXI (full-duplex) operating mode. In addition, many existing Classic Token Ring adapters were able to support the TXI operating mode with a microcode update.

Seven technical objectives for DTR were:

- To provide an evolutionary growth path to increased available bandwidth per station for Classic Token Ring hardware;
- To provide dedicated bandwidth for each attaching Classic Token Ring station;
- To provide dedicated bandwidth for each attaching DTR station to achieve simultaneous 16-Mbit/s transmit and receive;
- To address bandwidth requirements of emerging applications, that is, multimedia, thereby enabling real-time applications where bounded latency is required;
- To provide a migration path from shared-LAN to switched-LAN with standardized management;
- To preserve fundamental Token Ring principles, such as priority handling of designated traffic types, direct path routing via routing indicator (RI) field, and group addressing with broadcast/multicast capability;
- To provide an attaching station solution that can operate with present Token Ring hardware using installed transmission media and connectors and existing concentrators, bridges, and repeaters.

The basic DTR network consists of DTR concentrators, lobe cables, and attaching stations as shown in Figure 1.13. This figure shows both Classic Token Ring stations (A) and DTR stations (B) operating in the same network. For a DTR station, the TKP protocol for normal data transmission is identical to the classic Token Ring Operation while the TXI protocol does not require the capture of a token to permit information transfer. A DTR station in TXI mode may simultaneously transmit and receive different packets of information for full-duplex operation.

The DTR concentrator consists of C-Ports and a *data transfer unit* (DTU). The C-Port provides the basic connectivity between the DTR concentrator and the Token Ring stations or other DTR concentrators via lobe cables. Note that the C-Port can operate in either a port mode or in a station mode. The DTU provides connectivity between the C-Ports within a DTR concentrator by providing the functions of a switch fabric. In addition to direct connectivity between DTR C-Ports in either TKP or TXI mode, DTR concentrators may be connected by a nonstandardized *data transfer service* (DTS).

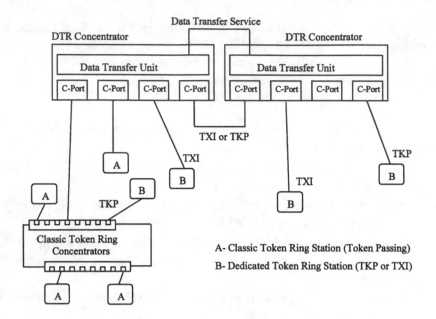

Figure 1.13 DTR network architecture.

DTR's basic compatibility with the IEEE 802.5 Token Ring installed base and applications will assure that a large number of users can migrate easily to this technology. Since DTR uses much of the same hardware that is used for present Token Ring networks, the costs for attaching stations are in line with this proven, accepted technology. DTR and its full-duplex option uses the 802.5 frame format and therefore is compatible with the LLC/MAC protocol boundary. Chapter 5 describes the detailed DTR protocols and provides insight to the DTR standards.

To better serve the Token Ring customer, a vendor organization was formed for Token Ring in 1994 called the *Alliance for Strategic Token Ring Advancement and Leadership* (ASTRAL). This organization aimed to better inform the Token Ring customer of developments in Token Ring and, in particular, the applications for Token Ring switching and DTR.

1.2.7 High-Speed Token Ring (1997–)

In mid-1997, due to significant continued demand for greater Token Ring performance, two initiatives aimed to increase the data rate of Token Ring and yet maintain the same Token Ring MAC processes. One of these projects sought to increase the data rate of Token Ring from 16 Mbit/s to 100 Mbit/s using the same physical layers as used in Ethernet for two-pair twisted pair cabling and fiber. The other project aimed to increase the data rate of Token Ring to 1,000 Mbit/s and is called Gigabit Token Ring. These projects are ongoing as this book is published and more details can be found in Chapter 6. Another vendor organization, the *High-Speed Token Ring Alliance* (HSTRA), was formed to promote higher data rate Token Ring technology.

1.2.7.1 Token Ring Standards Activity and Market Size

The IEEE 802.5 Working Group activities are based upon meeting three times a year at IEEE Plenary meetings and interim meetings held between each Plenary meeting. The attendance at these Plenary meetings since 1985 is shown in Table 1.3. Note that attendance grew substantially during 1991–1993 when the 8802-5:1995 standard was being developed with detailed MAC state tables and interoperable PHY. These attendance records foretell the tremendous growth in Token Ring adapters shipped during the 1993 to 1996 time frame. Data in the last column of Table 1.3 indicates the approximate shipped volume for adapter cards each year.

1.2.8 Ten Major Token Ring Standards and Papers

The major Token Ring standards and documents that provide information on Token Ring are listed for your reference. Because Token Ring has been a dominant network for the past ten years, a mixture of standards, collected papers, and organizational documents best describe Token Ring operation.

Table 1.3
IEEE 802.5 Attendance Records and Market Size

Year	Mar	Jul	Nov	Total Attendance	Shipped (Kunits)
1985	17	21	19	57	10
1986	22	28	28	78	50
1987	19	25	21	65	300
1988	30	39	38	117	600
1989	41	36	56	132	1,000
1990	57	84	94	235	1,500
1991	94	83	104	281	1,700
1992	110	104	109	323	2,000
1993	81	68	59	208	2,300
1994	49	51	48	148	2,700
1995	32	33	26	101	3,400
1996	32	22	16	70	3,900
1997	14	8	26	48	4,500

The ten documents listed here are required reading to acquire an understanding of the Token Ring specifications and the Token Ring lore. The reader is cautioned that many of these documents are updated and revised on a regular basis so the source for the document should be contacted to obtain the latest version of the document.

- ISO/IEC JTC1 11801 Building Wiring Guide provides commercial wiring guidelines for Token Ring using both STP (150Ω impedance) and UTP (100Ω impedance). This document, standardized in JTC1/SC25, is the basis for wiring Token Ring and Ethernet and is the source of the 100-m horizontal wiring guideline. The standard is available from ANSI at http://www.ansi.com.

- ISO/IEC JTC1 8802-5:1998 (3rd ed.) is the base Token Ring standard with detailed state tables and detailed PHY specifications. This standard is available from IEEE at http://standards.ieee.org.

- ISO/IEC JTC1 8802-5:1998 AM1 is the base DTR standard, which specifies DTR operation. This standard is available from the IEEE.

- ISO/IEC JTC1 8802-2:1998 is the base LLC standard, which specifies the connection and connectionless LLC used in all Token Ring networks. This standard also contains the route determination entity protocol, which is used for source routing route determination. This standard is available from the IEEE.

- ISO/IEC JTC1 15802-5:1998 specifies the source routing transparent bridging protocols. This standard was designated 10038:1993 and is available from the IEEE.

- IBM Token Ring Architecture Reference Manual, initially published in 1985, contains the detailed operation of MAC and LLC for IBM's Token Ring. This document provides considerable insight into Token Ring and is available from IBM at http://www.ibm.com.

- TI380 Token Ring User's Guide, initially published in 1985, contains excellent tutorial information on the operation of Token Ring and is available from TI at http://www.ti.com.

- 5.49 PHY Working Group Papers is a compendium of papers collected during 1990 and 1991 detailing the PHY operation and jitter control in Token Ring. These papers, presented by individual members of the IEEE 802.5 Working Group, contain details on jitter buildup that are not available in any other format. This document is available from Alpha Graphics at +1 (602) 863-0999.

- RFC 1743, 1748, and 1749 describe the management MIB using SMNPv2 for Token Ring. These document can be located using the WWW page at http://ietf.org.
- This book.

1.2.9 Token Ring Projects in the IEEE 802 Working Group

Activities in the IEEE 802.5 Working Groups leading to standards and supplements require the approval of a *project authorization request* (PAR) by the IEEE 802.5 Working Group, the IEEE 802 Executive Committee, the IEEE New Standards Committee (NesCom), and the IEEE Standards Board. A listing of past IEEE 802.5 PARs and their resultant standards are given here. The items with an asterisk (*) are the current standards or recommended practices in the IEEE, while items with a number (#) are current active PARs as of July 1998. Other PARs have been withdrawn and/or consolidated into other activities.

*802.5-1998, ISO/IEC 8802-5:1998. Base Standard.

802.5a, Station Management Supplement, included in 8802-5:1985

802.5b-1991, ISO/IEC TR 10738:1992. Withdrawn Recommended Practice for UTP.

*802.5c-1997, Recommended Practice for Dual Ring (IEEE only)

802.5d, Interconnected Token Ring, included in 8802-5:1995

802.5e, Management Entity Spec, included in 8802-5:1995

802.5f, 16-Mbps operation, included in 8802-5:1995

802.5g, Conformance and PICS, included in 8802-5:1995

802.5h, LLC III Operation, withdrawn from 8802-5:1995

802.5I, Early Token Release, included in 8802-5:1995

*802.5j-1994, ISO/IEC 8802-5 : 1998/AM1: 1998, Fiber Lobe Standard

802.5k, Media Specification, included in 8802-5:1995

802.5l, Maintenance, included in 8802-5:1995

802.5m-1993, ISO/IEC 15802-5:1998, included in Bridging Standard

802.5n, UTP at 16 Mbps, included in 8802-5:1995

802.5p, ISO/IEC 8802-2:1997, Route Determination Entity

802.5q, ISO/IEC 8802-5:1995, MAC Update to Base Standard

*802.5r, ISO/IEC 8802-5: 1998/AM1: 1998, Dedicated Token Ring

#802.5t, 100-Mbit/s Dedicated Token Ring

#802.5v, Gigabit Token Ring Operation

1.3 Bit Ordering in Token Ring

No introduction on Token Ring would be complete without a discussion of bit order in Token Ring and the differences between Ethernet bit order. Token

Ring utilizes a bit ordering where bits within a field are transmitted from left to right. The *most significant bit* (MSB) is on the left in figures and tables and for either hexadecimal format or binary format. An example of this Token Ring nomenclature, as standardized in 8802-5:1998, is shown in Figure 1.14.

In Figure 1.14, the first bit transmitted on the wire is B'0', the second bit is B'0', the third bit is B'1', and so forth. While Token Ring and FDDI utilize the bit order as shown, Ethernet uses the least significant bit (LSB) representation scheme. The Ethernet bit representation format uses the same standard format adopted by the IEEE when they administer addresses and is called the "Canonical" format or "IEEE Hex" format. Most Token Ring documentation, applications, the 8802-5:1998 standard, and this book use the "Token Ring" format. Occasionally, a document will use the Canonical format for Token Ring representations, so care must be exercised anytime Token Ring bit order is described.

If the bit representations were simply different and consistent, Token Ring and Ethernet would be able to coexist easily. However, the problem is exacerbated because the MAC address field in the frame must be transmitted with the "group bit first" on both Ethernet and Token Ring. This transmission of "group bit first" was done historically in order that hardware implementations could decode the address and determine whether the address is a group or multicast address in the first bit. Because the bit order on the wire is different, this leads to the problem of having to specify MAC addresses for Ethernet and Token Ring differently and to translate these addresses when sending frames across bridges from Ethernet to Token Ring.

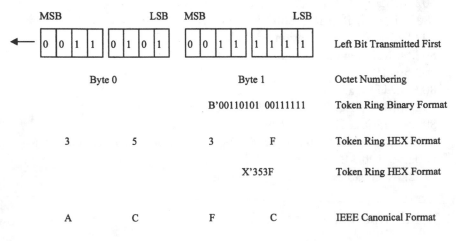

Figure 1.14 Token Ring bit order.

As an example for a specific address, the Token Ring AM address is written as X'C0 00 00 00 00 01' in the IEEE 802.5:1998 specification and this book. The first two bits transmitted on the wire are ones, followed by all zeros, with the last bit transmitted a one. However, in the Canonical format, this MAC address would be written as X'03 00 00 00 00 80.

The IEEE assigns an *organizational unique identifier* (OUI) as a three-octet address that can be used to generate unique MAC universal addresses and protocol identifiers per ANSI/IEEE Standard 802. A unique MAC universal address is formed by taking the first three octets assigned by the IEEE and adding three additional octets administered by the assignee.

For Token Ring applications, Figure 1.15 illustrates how an OUI address provided by the IEEE would be utilized to develop a universal address for Token Ring adapters. The OUI assigned by the IEEE is represented as three octets and is provided in the IEEE Hex or Canonical format. The bits in this address must be swapped to correspond to Token Ring format as shown in Figure 1.16. The assignee's three-byte address is then appended to the address to form the Token Ring universal address and transmitted on the ring. The first bit transmitted on the wire is B'0, the second bit is B'0', the third bit is B'1, and so forth.

The major impact of the different bit orders is that bridges that connect between an 802.3 LAN and an 802.5 LAN must correct for the address on each LAN by bit swapping within the address field. Bridging protocols, such as ISO/IEC 10038, are made much more complicated in order to assure correct representation of bit order in their respective media.

How did the standardization process of the 1980s end up with such a profound difference that would affect equipment for the next several decades? In fact, proponents for each format had good rationalization on their side. In the case of transmitting the least significant bit first, IEEE 802.3 standardized this format when there was only a single choice brought forward, so there was no alternative to discuss. When IEEE 802.5 developed the Token Ring standard it required bits within frames to be reset on the fly by receiving stations. Specifically, Token Ring protocols operate at the bit level (not the byte level) for token management and frame transmission. Being able to identify a header as a token or a frame as the frame was repeated by the station, and being able to change a token to a frame (see Figure 1.2) within the repeat path (yet minimizing the repeat path) forced the bit order transmission on Token Ring. A single bit within the token indicates whether a token is available for the station to transmit a frame inside the token. Also, the management of priority in the frame requires a bit significance and the bits must be transmitted MSB bits first. There were some decisions made in 1980 that will adversely

IEEE Octet Number	0	1	2	3	4	5
IEEE OUI Assignment	X'04	35	F0			
Assignee Address	X'			00	00	01
Token Ring Universal Address	X'20	AC	0F	00	00	01
Token Ring Binary Address	B'00100000	10101100	00001111	00000000	00000000	00000001

Figure 1.15 Token Ring universal address.

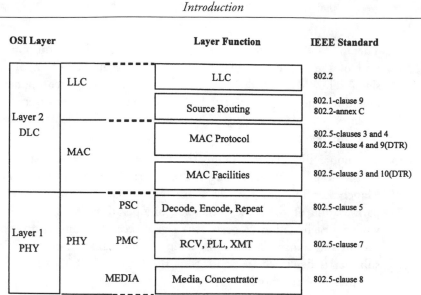

Figure 1.16 Token Ring relation to the OSI model.

affect the commercial LANs for the next 25 years—and bit order was one of them.

1.4 Open Systems Interconnection (OSI) Model and 8802-5:1995 Standard

Figure 1.16 illustrates the relationship between the formal ISO-defined OSI communications model and the IEEE 802 specifications for Token Ring. On the left in Figure 1.16 is shown the lower two levels of the OSI model, the PHY layer, and the *data link control* (DLC) layers. Each of these layers is described in the following two sections.

1.4.1 Layer 1: PHY Layer

The PHY layer is divided into three sublayers: media, *physical media components* (PMC), and *physical signaling components* (PSC).

1. The lower PHY layer contains the media, including the wire used to connect the station to the Token Ring. It also includes the concentrator (a single-speed connection device) used to connect stations to the Token Ring. This sublayer is described in clause 8 of ISO/IEC 8802-5:1998 and Chapter 2 of this book.

2. The middle PHY layer contains the PMC and provides for the application of the electrical signal to the wire and the reception of this signal. The PMC includes the analog transmit and receive components, including the clock recovery PLL and the receive equalizer. This sublayer is described in clause 7 of ISO/IEC 8802-5:1998 and Chapter 3 of this book.

3. The upper PHY layer contains the PSC sublayer and is responsible for the decoding and encoding of the electrical signal into Differential Manchester and provides the Token Ring repeat path. The repeat path is one of the basic principals of operation for Token Ring, whereby each station "repeats" the signal received from its upstream station and transmits this signal to its downstream station. The PSC sublayer is described in clause 5 of the ISO/IEC 8802-5:1998 and Chapter 3 of this book.

1.4.2   Layer 2: DLC Layer

The DLC layer is divided into the MAC and LLC sublayers.

1.4.2.1   MAC Sublayer

The MAC sublayer is the lower portion of the DLC layer. Token Ring, unlike Ethernet, contains a rich set of MAC frames that are exchanged between MAC peer entities within each station. These MAC frames provide for station management and control of the media and require less than 0.1% of the total available Token Ring bandwidth. MAC frames are specifically parsed by the MAC layer and are evaluated by the MAC layer to determine whether a MAC response frame is required or whether management needs to be informed of specific parameters in the MAC frame. The MAC frame protocol for Classical Token Ring is described in clause 4 of ISO/IEC 8802-5:1998 and Chapter 4 of this book. The MAC frame protocol for DTR is described in clause 9 of ISO/IEC 8802-5:1998/AM1 and Chapter 5 of this book.

The 802.5 Token Ring MAC sublayer is further divided into three functional areas.

1. The lower MAC sublayer is responsible for frame decoding, address recognition, FCS checking, FCS generation, and token transmit control. This functional area is described in clause 3 of ISO/IEC 8802-5:1998 and clause 10 of ISO/IEC 8802-5:1998/AM1.

2. The middle MAC sublayer is responsible for receiving, decoding, and transmitting frames used by the LLC and MAC sublayers. This

functional area is described in clause 4 of ISO/IEC 8802-5:1998 and clause 9 of ISO/IEC 8802-5:1998/AM1.

3. The upper MAC sublayer is optional and is present for Token Ring bridging using the source routing or transparent bridging protocols. This functional area is located at the boundary between the MAC protocol and the LLC protocol.

Source Routing

Clause 3 of IEEE802.5 defines the *routing information field* (RIF) used by the Token Ring MAC sublayer in support of the source routing protocol. Source routing using the RIF provides *station*-directed frame routing for Token Ring. This book discusses the RIF in Chapter 4. The location and size control of the RIF is specified by the ISO/IEC 8802-5:1998 Standard. Annex C of ISO/IEC 15802-5:1998 defines the MAC bridge that performs source routing when frames are received with a RIF.

Transparent Bridging

ISO/IEC 10038 (IEEE802.1D) defines the MAC bridge that performs transparent bridging when frames are received without a RIF. Transparent bridging provides MAC bridge-directed frame routing for the 802.5 Token Ring. The functional description of the MAC bridge is not covered in this book because this is not part of ISO/IEC 8802-5:1998. Clause 9 of ISO/IEC 8802-2 (IEEE802.2) defines how the *source routing transparent* (SRT) bridge connects multiple network segments into a single bridged network.

1.4.2.2 LLC Sublayer

The upper portion of the DLC layer is the LLC layer. This sublayer is responsible for establishing a connection between stations using either connectionless-oriented (LLC Type 1) or a connection-oriented (LLC Type 2 or 3) service. This book does not describe the operation of the LLC layer other than to describe the LLC frame's organization required by the 802.5 MAC sublayer. The reader is referred to the ISO/IEC 8802-2:1998 for a full operational description of LLC.

References

[1]   Bossen, D. C., and M. Y. Hsiao, "Model for Transient and Permanent Error-Detection and Fault Isolation Coverage," *IBM J. Res.,* Vol. 26, 1982, p. 67.

[2]   Bux, W., F. Closs, K. Kummerle, H. Keller, and H. R. Muller, "Architecture and Design Considerations for a Reliable Token-Ring Network," *J. Selected Areas in Comm.,* Vol. SAC-1, No. 5, Nov. 1983.

[3] Dixon, R. C., N. C. Strole, and J. D. Markov, "A Token-Ring Network for Local Data Communications," *IBM Systems J.,* Vol. 22, Nos. 1–2, 1983, p. 47.

[4] Carlo, J., and G. Samsen, "Chip Set Paves the Way to Token-Ring Access," *Mini-Micro Systems,* Sept. 1986, p. 121.

[5] "AT&T Wraps a Twisted-Pair Noose Around IBM's Neck," *Data Comm.,* June 1985, p. 45.

[6] ISO/IEC 4335, *High Level Data Link Control (HDLC) Procedures—Elements of Procedures,* ISO/IEC JTC1, 1993.

[7] Johnson, J. T., "Trouble in the Cards: Madge Points the Finger at the 802.5 Spec," *Data Comm.,* Aug. 1990, pp. 45–47.

2

Token Ring Cabling and Configurations

2.1 The Cabling Environment

Prior to the development of Token Ring, application-specific computer cabling was installed for each system and generally abandoned when that system was replaced. Typical office environments included IBM 3270 type "dumb terminals" connected to mainframe computers by coaxial cabling routed from the office areas directly to the computers in the "glass house." There were several problems with these installations including the following:

- The replacement of cabling for each computer system was wasteful and costly.
- Ceilings and wireways were beginning to fill with unlabeled coaxial cabling. Because it was unlabeled, it could not be reused for next generation applications.
- Much of the cabling was polyvinyl chloride (PVC) type, installed in building airways. Any new installations of this type of cabling would be in violation of recently passed fire safety codes.

To address these issues, the development of Token Ring was accompanied by the simultaneous development of a "universal cabling system" announced in 1984 as the IBM Cabling System. Patterned after the structured approach used for telephone cabling, the IBM Cabling System consisted primarily of permanently installed and labeled cabling from local wiring closets to the office areas. Unlike telephone cabling, there was no requirement to continue each of the cabling runs past the wiring closet to a building entry point, or to a

37

private branch exchange (PBX). Instead, a relatively small number of intercon-nections were required between wiring closets, with each of the interwiring closet runs carrying the data for dozens to hundreds of end stations.

Although the concept of structured wiring came from office telephone cabling, that cabling was deemed unsuitable for use in transmitting high-speed LAN signals. Originally designed to carry voice signals, the electrical transmission characteristics of existing telephone wiring were not even character-ized in the frequency range needed by Token Ring. Therefore, IBM, in conjunc-tion with various cabling manufacturers, embarked on a joint development to create cabling suitable for high-speed data transmission.

2.1.1 The IBM Cabling System

The cabling developed for Token Ring, as a universal telecommunications cabling system, pushed the state of the art in twisted-pair copper cable manufac-turing. The basic requirements for the cabling were:

- Two twisted pairs of wires: one for signal transmission, and one for signal reception;
- Very low signal attenuation;
- Excellent crosstalk immunity between the pairs;
- High immunity to external noise sources;
- Low radiation of high-frequency electrical energy from the network signals.

To achieve low signal attenuation, the cabling characteristic impedance was set at 150Ω rather than the 100Ω impedance of ordinary telephone wire. In addition, 22-gauge cabling was used rather than the narrower 24- or 26-gauge normally used for telephone wiring. Both these choices favored low signal attenuation over small cable diameter. The larger wire gauge obviously translates to a larger cable diameter. The larger characteristic impedance has the effect of increasing the diameter of the insulation around each wire, and therefore the center-to-center separation of the wires forming the twisted pairs. Note that signal attenuation is governed by the resistance of the wires per unit length divided by the cable characteristic impedance. The increased wire gauge reduces the resistance of the conductors, and the increased characteristic impedance further reduces the ratio of series resistance divided by characteristic impedance.

With cable that is well shielded from external sources of noise, the primary noise source at the receiver is pair-to-pair energy transfer coupled from the transmit pair to the receive pair. As shown in Figure 2.1, the transmitted signal continuously couples energy along the entire length of the cable.

For this configuration, the signal energy, dV_L, coupled into the receiver across the cabling interval dx at a length, L, from the transmitter, is proportional to both the interval length, dx, and the transmitted voltage level, V_0, which is attenuated from the original transmitted level based on that distance, L, that is,

$$dV_L = K \times dx \times V_0 \times e^{-\alpha L}$$

where K is an appropriate constant of proportionality for the particular cabling under consideration. The noise voltage reaching the receiver is then attenuated by the return distance, L, resulting in a noise contribution at the receiver

$$dN = K \times dx \times V_0 \times e^{-\alpha L} \times e^{-\alpha L} = K \times dx \times V_0 \times e^{-2\alpha L}$$

from the cabling interval dx. Because the noise contributions from each of the intervals, dx, are at different phase angles, the noise voltages do not add by simple addition. Instead, to a good approximation, the noise energies add. It is clear that the noise from elements close to the receiver will contribute more heavily to the overall noise. In fact, the crosstalk noise will be relatively independent of cable length as long as the cabling is long enough, that is, 30m or greater for Token Ring signals on the IBM Cabling System 150Ω cables. Because the noise source is near the receiver, this noise coupling mechanism is called *near end cross talk* or *NEXT*. Cables can be designed to minimize the noise-coupling factor K, thus minimizing this crosstalk noise. With such designs, the maximum attainable transmission distances can be greatly increased.[1] Therefore, the NEXT specification was set very aggressively for the IBM Cabling System, pushing industries' best practices at that time. In fact, early cable samples failed to meet the crosstalk specification because the different dyes

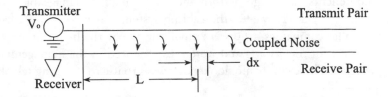

Figure 2.1 Noise coupling between transmission lines for Token Ring.

---

1. Transmission path analysis based both on cable parameters and on the effect of intersymbol interference was first presented at the APL '84 Symposium [1].

used to color the insulation of the four wires within the cable changed the insulation's capacitance differently for each wire and, therefore, the coupling of energy between the cable pairs. Early successful designs required an electrically conducting screen to separate the pairs. Later manufacturing techniques achieved the required crosstalk without requiring the pairs to be screened from each other.

In addition to developing a new cable for Token Ring and other high-speed telecommunications applications, a new connector had to be developed. The requirements on the connector were:

- Extremely high reliability, even after hundreds of connections and disconnections;
- Excellent transmission characteristics (low signal loss and low crosstalk between pairs);
- A single connector design that could mate with an identical connector (hermaphroditic[2]);
- Automatic wrapping of signals from transmit pair to receive pair when the connectors are not mated.

This last requirement to autowrap greatly improved overall system RAS and facilitated problem determination and resolution. The result of the specified performance requirements was an excellent performing, though somewhat bulky and expensive, data connector; the hermaphroditic connector is shown in Figure 1.11.

Because the IBM Cabling System was announced in 1984, a year before the announcement of Token Ring products, it was initially deployed supporting old technology, such as 3270 signal transmission. It received an enthusiastic welcome from many of the Fortune 500 companies, who were plagued with the problems caused by lack of a structured wiring approach. These companies understood the costs associated with cable management and the potential savings and benefits that a high-performance, high-quality, manageable cabling system would provide. However, the cabling system was never fully accepted by many smaller companies, who were unwilling to pay the high purchase and installation cost for the cabling and were inexperienced in the management of cabling systems. These companies looked for ways to use the existing telephone

---

2. Many electrical connectors are referred to as being either male or female depending on their mating geometry. Because an objective in the design of this connector was to have a single design that mates to itself, the terms *male* and *female* don't apply. Industry, however, must have clever terms to use to refer to their components, so *hermaphroditic* was adopted as a cute, descriptive cognomen for the connector.

wiring within their establishments and assumed the phone companies would manage the cabling for them.

2.1.2 Telephone Wiring: The Alternative Transmission Media

In the early 1980s, most telephone wiring was poorly suited for high-speed data transmission. The cabling was designed and installed to support telephone use. Little thought was given to alternate use of the cabling and to the new requirements that high-speed networking would generate. Still, telephone wiring was cheap and ubiquitous and, therefore, a very attractive alternative to the IBM Cabling System for data transmission. During this same time period, AT&T developed their *premises distribution system* (PDS) based on high-quality telephone wiring to support the 1-Mbit/s STARLAN. This capability gave further encouragement to the use of telephone wiring for LANs and increased the pressure to find a way to use existing cabling. In response to this pressure, IBM embarked on an extensive test program, and industry survey, to determine the characteristics of installed telephone wiring and the subset of that telephone wiring that could be used for Token Ring transmission. The results of this survey and development effort were encouraging but sobering. Indeed, it was possible to use much of the existing telephone wiring to support 4-Mbit/s Token Ring transmission. However, the electrical parameters of that cabling were not characterized in the frequency range of Token Ring signals and there was no guarantee or simple measurement method to determine that any particular cable could support the required signal transmissions. Ultimately, IBM developed a useable and verifiable minimal cabling specification that called for attenuation not to exceed 8 dB per 1,000 feet at 1 MHz, the highest frequency for which cable manufacturers had characterized their cables. In addition, IBM specified that the cable should have a characteristic impedance of 100Ω ±15% at 1 MHz. Because cable manufacturers had not characterized the NEXT for installed cabling, and because achievement of at least a minimum required level of NEXT attenuation is critical to proper signal reception, another way had to be found to determine if this noise parameter was within tolerable bounds. It was determined empirically that cabling with a twist rate of at least two twists per foot generally had adequate NEXT characteristics, so a mechanical requirement was added to the specification as a substitute for an appropriate NEXT attenuation requirement. The nominal twisting combined with a bounded attenuation would typically provide a low enough crosstalk noise and high enough signal strength for an acceptable signal-to-noise ratio at the Token Ring receivers. IBM codified the requirements for acceptable telephone wiring in their "Type 3 Media Specification" and published it within their manual describing use of telephone wiring for Token Ring transmission [2]. IBM, and later EIA/TIA, used the term *UTP* to easily distinguish the telephone wire from the IBM Cabling System's STP cabling. However, descriptions of UTP cabling were normally accompanied by a clause stating that shielded 100Ω cabling meeting the same electrical characteristics as their unshielded counterparts was covered within the specification.

Many caveats had to be placed on the use of existing UTP wiring. Even if the cabling had acceptable electrical characteristics, outside noise sources and installation practices could make the cable unusable for Token Ring signaling. Common telephone practices including running multiple extensions to a single phone number. This practice, known in the industry as utilizing bridge taps, rendered those runs unsuitable for high-speed data transmission. Use of interoffice buzzers, run over the phone lines and connected to the telephone equipment, to signal a secretary or manager produced enough high-frequency noise to totally swamp out the Token Ring data signals and was therefore prohibited for any cabling used to support Token Ring. External noise sources, which could be anything from an electric pencil sharpener near the phone wires to the ballasts in fluorescent light fixtures near the wiring runs, could create unacceptable noise levels and render the affected runs useless for Token Ring signaling.

Despite the limitations on the use of UTP for 4-Mbit/s Token Ring cabling, industry considered the fact that it could be used for this application as remarkable when IBM first announced that capability in October 1985. In fact, there was widespread skepticism expressed in trade press articles immediately following that announcement. Only later did the press, and the industry, come to accept the fact that ordinary telephone wiring could be used for high-speed data transmission (data rates in excess of 1 Mbit/s).

2.1.2.1 Advances in Telephone Wiring

Although the IBM Type 3 Media Specification for telephone wiring provided a usable and industry-accepted specification for reasonable-quality telephone wiring, that specification was clearly inadequate. Lack of available characterization required the upper frequency bound on attenuation to be limited to 1 MHz and precluded any specification of NEXT. This deficiency was soon to change. A competing LAN technology, Ethernet, then based on transmission over coaxial cable, was beginning to look at ways of utilizing telephone wiring in a standards development effort resulting in the highly successful IEEE 802.3 10BASE-T standard. Ethernet product developers, working closely with telephone cable manufacturers, developed a new telephone cabling specification that closely followed the capabilities of AT&T's new PDS cabling. The specification characterized all the critical electrical parameters up to 10 MHz, including characteristic impedance, attenuation, and NEXT. In addition, the 8-pin telephone connector was standardized rather than the more common 6-pin connector found on most private use telephones and telephone wall outlets. The new specification was adopted within the "Commercial Building Telecommunications Cabling Standard" EIA/TIA 568, along with IBM's 150Ω STP cabling. Cable meeting the telephone wire specification came to be known as "Category 3 UTP cable," which was derived from the original IBM label, "Type 3 Media Specification."

Political developments serendipitously worked with the growth of technology requirements to spur the development of superior telephone wiring. With the divestiture of AT&T came the transfer of telephone cabling ownership within a building from the phone company to the building's owner. Prior to divestiture, it had been in the interests of the phone companies to minimize the cost of telephone cabling and to use installed telephone cabling for as long as possible, fully exploiting their investment. Now that customers were beginning to own the telephone wiring within their buildings, it became economically attractive for the phone companies to develop higher performance, more expensive cabling. They marketed and sold that cabling as business-enhancing upgrades to customers that already had existing cabling plants, adequate for telephony, but only marginally acceptable for the transmission of high-speed LAN signals.

AT&T capitalized on this new business opportunity by developing a very high quality cable and associated components that it marketed as Systemax. With this product introduction, Category 5 cabling was born. By employing better insulating materials to separate the wire conductors, attenuation of the Systemax Category 5 cabling was reduced compared to the Category 3 PDS cabling that they had developed a few years earlier. By employing tight controlled twisting of the individual pairs and then laying those pairs in a relatively loose twist around each other, an exceptional NEXT performance was obtained. Through careful control of all the manufacturing processes, it was possible to characterize and guarantee performance of the Category 5 cabling up to 100 MHz rather than the 10-MHz upper bound initially specified in the EIA/TIA 568 standard. Systemax cabling, marketed heavily by AT&T, quickly achieved large market penetration. Its capabilities and performance parameters became de facto requirements in the industry and soon worked their way into the North American and international telecommunications cabling standards.

Because that cable was so difficult to manufacture, competing cable manufacturers that wanted to capitalize on the demand for better telephone wiring, but were initially unable to meet the Category 5 specifications, marketed their own data grade telephone cabling at a discount relative to Category 5 cable. It became known as Category 4 and also got incorporated into the cabling standards. EIA/TIA's TR 41.8.1 standards body initially specified Categories 4 and 5 cabling in *Technical Supplemental Bulletins* (TSBs) [3,4]. The TSBs were later incorporated in the second edition of the 568 cabling standard. The original edition of that standard included 150Ω STP, Category 3 UTP copper cabling (although the designation "Category 3" was not used in this first edition), 50Ω coaxial cable, and 62.5/125μ multimode optical fiber.

2.1.2.2   The Influence of European Cable Manufacturers

Telephone cabling is an important industry worldwide with different historical preferences and practices from country to country. Germany and France have a long history of using shielded telephone cabling. Germany has major cable manufacturers that make extensive use of quad construction rather than twisted pair. Some of the telephone cabling manufactured has a high-frequency characteristic impedance of 120Ω, rather than the 100Ω impedance specified in the North American Standard. Because of these historical differences and the strong national interests to protect native manufacturers, the stage was set for a heated battle in the development of an international cabling standard. That battle was further fueled by AT&T's aggressive international marketing of their highly successful Systemax Cabling System.

The international cable standardization battles took place from 1990 to 1995 and were pursued on three technical fronts: cable shielding, cable construction, and cable characteristic impedance.

Shielding

Shielding was popular in parts of Europe, especially France and Germany, where there were stringent requirements on electromagnetic radiation limits and a historical preference for shielded cable. Shielded telephone cabling had not won wide acceptance in the United States. Major arguments advanced against it were that it is more expensive than unshielded cable, much harder to terminate, has potentially higher signal attenuation, and when installed improperly can exhibit worse radiation characteristics than corresponding unshielded cabling. Although these arguments are all solidly backed by experiment and experience, they ignore the fact that the electrical performance issues are only potential problems that all can be overcome with good design. Properly designed and installed, shielded cabling can provide significantly higher immunity to external noise and greatly reduced electromagnetic radiation from the active devices attached to the cabling system. The real issues involved with choosing shielded versus unshielded cabling are the costs, complexity of installation, and increased bulk associated with the shielded systems. Part of the cost is because inclusion of the shield on the cabling increases manufacturing expense. Perhaps more significant is the increased cost of termination hardware and labor costs associated with the proper installation of the shielded cabling. The benefits of having a highly reliable shielded cabling system are that it is virtually immune to ordinary noise sources and simplifies equipment design by preventing signaling energy from radiating in violation of national electrical emission requirements. Without using shielded cabling, many LAN transceivers require special filters to limit radiated electrical energy. The potential savings in cable plant maintenance and LAN troubleshooting necessitated by cabling problems are the dividend for the additional component and installation costs of the shielded system. It must also be noted that the Unshielded Category 5 cabling requires far more skill to install properly than for telephone wiring. For example, a part of the improved crosstalk performance of the cable is a result of the pairs not being too close to each other. To maintain the pair-to-pair spacing, there is a requirement on Systemax cable that it be pulled with a force not exceeding 25 lbs during installation. In fact, most cable installers that are not specially trained or willing to take unusual care during cable installation will pull a cable with whatever force is necessary to pull it through raceways and around corners. Another component of the improved crosstalk is based on low crosstalk at the connectors. Category 5 installation practices demand that the cable is untwisted no more than a half inch (13 mm) at the connector when terminating. This level of care requires extra training for the installers and is more time consuming than the methods employed in the installation of cabling for voice applications.

Excluding the advantages of superior noise immunity that properly installed and terminated shielded cabling can offer, the electrical performance of shielded and unshielded cabling systems each meeting the same propagation and crosstalk characteristics is identical. Therefore, there is no technical basis to exclude shielded cable from the standard or to give it a second class status. However, the North American cabling standard, EIA/TIA 568A, does just that by referring to the telephone cabling as UTP, implying that telephone cabling is naturally unshielded. It goes on to state that shielded cabling meeting the same characteristics can be used. This language was totally unacceptable to the Europeans supporting shielded cabling. At their insistence, UTP was replaced in the international standard with the more neutral language "balanced transmission lines." By leaving out any reference to shielding, no bias is built into that terminology.

Cable Construction

By substituting "transmission lines" for "twisted pair," a second point of contention was satisfied. There are two primary ways to provide electrical isolation from cable pairs within the same sheath. The first is twisting pairs together, as implied by the term *UTP*. The second technique is to employ quad construction. Again, the term *transmission lines* remains neutral to the physical distinctions in these two cable construction techniques. The benefits associated with both twisted-pair and quad construction are based on the use of differential mode rather than common mode signaling to transmit the desired information. Both signaling modes are shown in Figure 2.2. For common mode excitation, all conductors are excited to the same voltage level relative to ground. With differential excitation, the two wires of the pair are excited with equal and opposite voltages: +V/2 on one of the transmit wires, T1, and −V/2 on the other transmit wire, T2. To detect differential mode signals, the receive circuitry is designed to filter out common mode energy so that noise signals of the same voltage and polarity on both R1 and R2 will not be detected at the receiver. Only the differential component of the coupled noise will interfere with signal reception. Because the primary noise-inducing mechanism is capacitive coupling from the two transmit wires, having opposite voltages on T1 and T2 will induce opposite polarity noise voltages on R1 and R2 that will tend to cancel each other out. The cancellation mechanism can be better understood by examining the twisted-pair and quad cable topologies in Figures 2.3 and 2.4.

There are two results of twisting cable pairs, each of which minimizes differential mode noise at the receiver as seen in Figure 2.3. First, where there is a positive voltage on one of the transmit wires, for example, on T1, there will be an equal negative voltage on the other wire, T2. Therefore, the noise coupling from T1 and T2 will be of opposite polarity as indicated by the noise coupling arrows shown in the diagram. These noise signals will partially cancel each other, greatly reducing the induced noise voltages on the adjacent receive

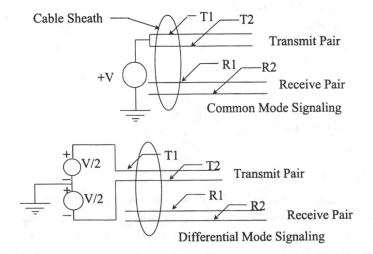

Figure 2.2 Common and differential mode signaling.

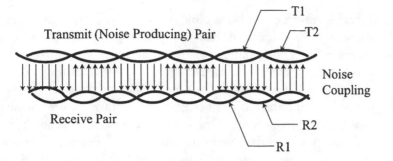

Figure 2.3 Detail of noise coupling between twisted-pair transmission lines.

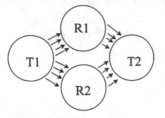

Figure 2.4 Detail of noise coupling between quad transmission lines.

wires. In addition, the noise coupling from T1 to R1 will be of the same polarity as the noise coupling from T1 to R2, thereby impressing a noise on the receive pair, which is primarily common mode. That noise component will be filtered at the receiver and will not interfere with signal reception. Only the differential component of the resultant noise will degrade the detected signal. The effectiveness of the noise cancellation produced by twisting pairs is governed by the carefully controlled geometry of the cable. The twists of the transmit pair must be at a different pitch from the twists of the receive pair. Further, the lengths of the two pitches should not be simple ratios. Finally, both pitch lengths should be very small compared to the highest frequency noise that is of concern. For multipair cables weaving the pairs around each other further reduces crosstalk so that no two pairs stay in close proximity over the cable run. Advances in cable manufacturing have resulted in a phenomenal decrease in cable crosstalk that can be achieved using twisted pair construction.

Quad cable construction, as shown in Figure 2.4, uses a geometry that theoretically provides perfect balance. The coupling capacitance from T1 to R1 is equal to the coupling capacitance from T2 to R1, and similarly for the capacitances from T1 and T2 to R2. Because the signals on T1 and T2 are equal in magnitude and opposite polarity, the net noise voltage induced onto R1 and R2 will each theoretically be zero. This result is for every cross section of the quad cable and applies to all frequencies. It is, however, highly dependent on careful cable construction. If the cable construction is not uniform and one of the wires is slightly displaced, then the coupling capacitances will not be equal and the induced noise voltages will not cancel. In this case, large resultant noise levels can result. Early quad cable designs had poor uniformity in construction that resulted in large crosstalk noise levels, making these cables unreliable for high-speed signal transmission. Today, there are quad cables available that provide excellent crosstalk levels. One note of caution should be observed in using quad cables to populate 8-pin telephone jacks. Because there are only four wires in a quad cable, if two cables are used to fully populate the 8-position outlet, the lengths of those cables may not be identical. Some transmission protocols utilize all four pair and require propagation delay to be the same within close tolerance for all of them. The quad cable construction could be problematic for these protocols. Fortunately, Token Ring has no such restrictions.

Characteristic Impedance

France Telecom responded to the challenge of AT&T's Systemax cabling introduction by developing a 120Ω shielded cabling system for both telephone and high-speed data transmission. The higher characteristic impedance provided them with a lower attenuation than 100Ω cabling for the same conductor diameters. The shielding provided them with an advantage in parts of Europe, especially France and Germany, where there were stringent requirements on electromagnetic radiation limits, and where there was a historical preference for shielded cable. The small difference in characteristic impedance from 100Ω could generally be tolerated by transmitter and receiver designs that already had to accommodate a ±15Ω variation in the cables used.

International cabling standardization took place under the auspices of the Joint Technical Committee 1 of the International Standards Organization and International Electrotechnical Committee, Sub Committee 25, Working Group 3 (ISO/IEC JTC 1/ SC 25/ WG 3). Within that working group there was great reluctance to add yet another cable type and further complicate the decision of choosing a cable for data communications within an establishment. There was also concern about the additional requirement that the new cable type would potentially impose on equipment manufacturers to now test their equipment over an extended range of interfaces. However, 120Ω cable was already the only cabling choice approved for ISDN primary rate 2-Mbit/s transmission. Based on that existing requirement, France Telecom was successful in getting approval for inclusion of their 120Ω cabling in the developing standard. France Telecom's cabling offering added further fuel to the topic of appropriate cable choices in the early 1990s. This interest was reflected in requests for papers on cabling at symposiums [5] and in trade publications [6].

In the International Standard, 100Ω and 120Ω cabling, shielded and unshielded cabling, twisted pair and quad construction, all came under the universally recognized Categories 3, 4, and 5 initially specified in the North American Cabling Standard. The basis for meeting the category requirements for the international standard was based primarily on meeting the attenuation and crosstalk specifications already set by North American Standard and additionally requiring the high-frequency characteristic impedance to be either 100 ± 15Ω or 120 ± 15Ω. In fact, the 120Ω cabling marketed by France Telecom was not able to fully meet the Category 5 requirements for NEXT at high frequencies. Fortunately, this higher impedance cable was less lossy than the corresponding 100Ω cabling, so the international specification for Category 5 cable has an alternate specification to allow a lower loss, lower (poorer) NEXT attenuation, 120Ω cable to meet the standard. In fact, this trade-off is a reasonable one, because most signal transmissions will either be limited by total signal attenuation or the ratio of signal to crosstalk noise, and the alternate specification will provide the performance requirements to meet both these criteria.

As a result of the proliferation of available cabling types, the Token Ring Standards body, IEEE 802.5, addressed support of Token Ring transmission on all popular copper-balanced transmission media. The present standard specifies Token Ring transmission at both 4 and 16 Mbit/s over all of the following copper cabling types: IBM Cabling System's 150Ω STP cabling; Categories 3, 4, and 5 100Ω cabling; and Category 5 120Ω cabling. Within the IEEE 802.5 Token Ring Working Group, presentations demonstrated analyses that showed the signal degradation that would result from the impedance mismatches at the transmitters and receivers when running over 120Ω cabling systems was at least offset by the lower attenuation of the signals. Neither the presentations nor the analysis provided exhaustive proof that the 120Ω cabling was equivalent to or better than 100Ω cabling. However, the preponderance of facts was such as to move the IEEE 802.5 Working Group to add an informative annex to the standard stating that the use of 120Ω cabling for Token Ring transmission should work satisfactorily. Large Token Ring vendors also added claims to their product announcements that their products would operate properly when attached to 120Ω cabling systems.

2.1.3 Optical Fiber Cabling

Along with the IBM Cabling System in 1984 came the introduction of a new 100/140μ multimode glass optical fiber cable. The 100/140μ refers to the fiber's inner and outer diameters as shown in Figure 2.5. The inner diameter has a higher dielectric constant than the surrounding glass. Because of this difference, light rays launched into the fiber's inner core at a sufficiently shallow launch angle bounce off the dielectric interface and remain trapped within the inner fiber core, which thereby becomes a waveguide for the light energy. The newly introduced $100/140\mu$ fiber size was significantly larger than the $62.5/125\mu$ and $50/125\mu$ glass multimode fibers common at that time. IBM chose the larger diameter for its greater light gathering powers from large LED transmitters and for the greater ease in building and mating connectors with relatively low transmitted energy loss. The state of optical fiber technology at that time was such that for a 2-km transmission with four mated optical fiber pairs in the link, the majority of the attenuation in a worst case link design would be the result of connector losses rather than attenuation from the fiber. Optical fiber cabling was envisioned to be primarily of use for long transmission runs that could not be supported by copper links. Typical uses for fiber links would be between buildings on an academic or industrial campus, for long runs within very large buildings, and to interconnect wiring closets that had to be electrically isolated because they did not have a common ground potential. Initially, no lobe attachment requirement was seen for optical fiber, and no early attempts were made to standardize that attachment capability.

Because of the lack of sustained marketing thrust for the $100/140\mu$ optical fiber, and because of very significant improvements in fiber termination technology, $100/140\mu$ fiber never achieved significant market penetration. By the early 1990s, IBM had abandoned its efforts to advance the $100/140\mu$ fiber, although it continues to support its use for LAN applications. Today $62.5/125\mu$ and $50/125\mu$ are the two accepted standard multimode glass fiber sizes. However, optical fiber transmission for Token Ring did not follow the

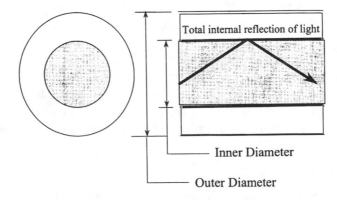

Figure 2.5 Optical fiber cable geometry.

path of the $100/140\mu$ fiber that was designed for it. Rather, it has become increasingly more important, especially for 16-Mbit/s transmission, and has had its application range extended to include lobe cabling. The IEEE first issued a trial use standard for Token Ring Optical Fiber Lobe Attachment in 1993, which has been updated as a full standard (see the list of standards in Section 1.2.9).

2.1.4 Optical Fiber Attachment

A major concern in the development of the optical fiber standard for lobe attachment was the emulation of the requirement for an attaching device to signal its readiness to join the ring and follow the required protocols. For a copper cable attachment, data is transmitted using a differentially balanced signal with no dc component, and the request to insert into the ring is done with a separate dc current called "phantom." The attaching device goes through a self-test followed by a test of the lobe by transmitting test frames to ensure that it can detect its own transmitted signal. Only after successful completion of this test does it request ring attachment by turning on its phantom current to signal its concentrator port that it is ready to join the ring. The station maintains its phantom current as long as it wants to continue to participate in ring activity. It removes the phantom current to signal its intention to imminently detach from the ring. Because the data signal has no long-term dc component and the phantom current has no significant high-frequency ac component, both data and phantom signaling can be carried simultaneously with no mutual interference. Unfortunately, there is no corresponding inexpensive and relatively available "out of band" signaling technique available to fiber optic stations to provide a one-to-one replacement for the phantom current used by stations transmitting on copper wires. Instead, the fiber station must transmit a ring insertion key to signal its intention to join the ring.

The insertion process for a fiber station is:

- The station carries out an internal self-test of its circuitry.
- Lobe testing is carried out by transmitting a test signal and determining that the station can accurately detect the receipt of that signal, wrapped or regenerated at the concentrator port.
- A ring insertion key (shown in Figure 2.6(a)) is transmitted.
- The station detects the ring insertion key, which has been echoed by the concentrator.
- The station enters repeat mode to begin ring operation.
- When the station is ready to remove itself from the ring, a ring bypass key, as shown in Figure 2.6(b), is transmitted.

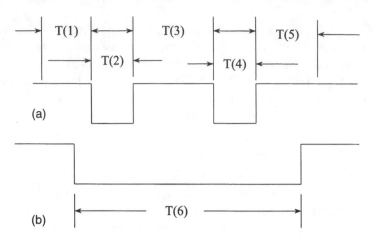

Figure 2.6 Optical fiber (a) insertion and (b) bypass keys.

Ring insertion and bypass keys are independent of ring speed. The times for key elements (1) through (6) as specified by [7] are shown in Table 2.1, all times are in microseconds.

T(1) > 5000 μs is a stipulation that the insertion key shall be preceded by an interval of at least 5 ms with optical transmit power in the "on" state. This interval provides adequate time for the *fiber optic trunk coupling unit* (FOTCU) to become aware that the attached device is transmitting and to be ready to detect a transmitted insertion key. The insertion key is then transmitted by turning off optical power for two intervals T(2) and T(4), each 0.833 ms nominal, separated by a powered interval T(3) of 1.667 ms nominal. Finally, power is again asserted for a period, T(5), of at least 1.616 ms. When the insert key is recognized by the FOTCU, it initiates the appropriate ring insertion procedure. That procedure includes acknowledging receipt of the insertion key by echoing it back to the station and inserting the station onto the ring within one second of the end of insertion key receipt by the FOTCU.

Fiber optic stations indicate their intention to detach from the ring by sending out the signal T(6), which is turning off transmitted power for at least

Table 2.1
Key Element Timing Parameters for Ring Insertion and Bypass

Time Interval (μs)	State	Min	Typical	Max
T(1)	Po>Po_OFF	5000		
T(2) and T(4)	Po<Po_OFF	808	833	858
T(3)	Po>Po_OFF	1616	1667	1717
T(5)	Po>Po_OFF	1616	1667	
T(6)	Po<Po_OFF	4850	5000	

4.85 ms, as shown in Figure 2.6(b). Upon detection of the bypass key, the FOTCU must quickly remove the station from the ring. Until it has done so, the ring is "broken" with all transmissions terminating at the station signaling request for removal. Therefore, bypass must be completed, with the ring operational, within 10 ms of the onset of the received power loss caused by the bypass key transmission.

Although it is typically the station that signals its insertion or removal from the ring, there are instances when the station may stop receiving a transmitted optical signal from the FOTCU. One such instance is when there is a break in the fiber carrying the FOTCU's transmit signal to the station but no break in the fiber carrying the station's transmitted signal to the FOTCU. For this occurrence, the ring is broken because all signals fail to reach the station. However, the station will send out idles as required by the standard, causing the ring to go into beaconing. A corresponding condition does not exist for copper transmission because any break in the cabling from the concentrator to the station would be accompanied by a break in the phantom current path, causing the station to detect wire fault and remove itself from the ring. Station removal for this fiber optic case is accomplished by requiring the station to transmit a bypass key echo to the FOTCU when it detects power loss. Based on receipt of this echo signal, the FOTCU will remove the station and the remainder of the ring will be operational, all within 10 ms of the fiber break.

The phantom emulation protocol employed for fiber optic stations is also applicable to fiber repeaters in the trunk, or main ring path. Consider a four wiring closet ring connected by fiber cabling. Further assume that a break occurs in one of the fiber links in the main ring path, as shown in Figure 2.7. For this configuration, the down stream station at C will detect a signal loss. To prevent the ring from becoming disabled, the fiber repeater detecting that signal loss can wrap. However, the ring will not recover until the upstream fiber repeater at B also wraps. Therefore, the repeater at C will "echo" the signal break by transmitting a bypass key on the back-up path to the repeater at B. Upon receipt of this key, the repeater at B will wrap and the ring will be healed. Note that proper repeater design should include the ability of the repeaters to notify network management that they are in a wrapped state. That way, ring maintenance personnel will be able to repair the ring on a nonemergency basis with no significant loss in ring availability.

2.1.5 Specifications for Optical Fiber Operation

Table 2.2 presents the 850-nm wavelength optical fiber specifications for Token Ring operation based on transmission over $62.5/125\mu$ fiber as specified in both the U.S. Cabling Standard [8] and the International Standard [9].

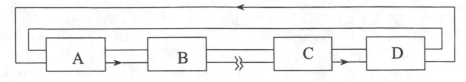

Break Detected by Fiber Repeater in C

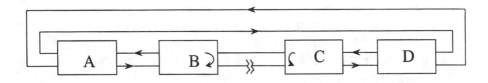

Repeater in C "echoes" break and both B and C wrap

Figure 2.7 Automatic wrapping upon detection of fiber break in main ring path.

Table 2.2
Fiber Optic Specifications at 850-nm Wavelength to Support Token Ring Transmission

Fiber Type	Numerical Aperture Max	Numerical Aperture Nominal	Modal Bandwidth MHz · km (Min)	Attenuation dB/km (Max)
62.5/125	0.31	0.275 ± 0.015	160	3.75
50/125		0.2 ± 0.02		
100/140		0.29 ± 0.02		

In addition, numerical aperture specifications are set for both 50/125- and 100/140μ fiber as:

- Numerical aperture for 50/125μ fiber = 0.2 ± 0.02;
- Numerical aperture for 100/140μ fiber = 0.29 ± 0.02.

The transmit and receive requirements for the optical fiber signal are based on the use of 850-nm LED transmitters as well as the transmitter and receiver requirements. Specifications are based on the requirement of being able to transmit signals at least 2 km on the three multimode fiber types: 62.5/125-, 50/125-, and 100/140μ fiber. Actual distances supported are not specified within the standard and are left to the equipment designer and/or system designer to compute based on the available power budget.

For ease in establishing specifications, all requirements are based on using 62.5/125 multimode optical fiber as the transmission medium. It is up to the individual manufacturers to specify the performance of their equipment using alternate fiber sizes. However, some care was taken to limit the maximum power that could be received so that no receivers would be overdriven when receiving a signal from any conformant transmitter attached to any of the allowable fibers.

Summary specifications call for the transmitter to have a center wavelength between 800 and 910 nm, an average power in its on state of between −19 and −12 dBm, and to have its power off state limit any leakage power to no more than −38 dBm. The receivers must detect −32 dBm as a valid "on state" and must not overload with −12 dBm incident power level. They must detect as an off state any incident power below −38 dBm. These specifications guarantee that conformant stations can operate on links with a loss of up to 13 dB including fiber trunk, fiber patch cables, and all connectors in the channel. The specification also provides manufacturers the freedom to develop fiber optic transceivers with significantly more capability. For example, if a transceiver is designed within the requirements and with transmitted power restricted to the range −12 to −15 dBm rather than −12 to −19 dBm, then links with up to 17-dB attenuation could be accommodated. To understand the implication of this improvement consider the following sample link budget assumptions:

- Maximum *fiber loss* (FL), 3.75 dB/km;
- Maximum *connector loss* (CL), 1 dB/connector;
- Maximum *connectors/link* (Cn), 4;
- Minimum design margin (Margin), 1.5 dB;
- Minimum transmitted Power Case A (Ptxa), −19 dBm;
- Required receive signal level (Prx), −32 dBm.

Under these assumptions, a 2-km link with four fiber connections can be supported as shown:

$$Ptxa - Prx = 2 \text{ km} \times FL \text{ dB/km} + Cn \times 1 \text{ dB/Connector} + \text{Margin}$$

$$-19 - (-32) = 2 \text{ km} \times 3.75 \text{ dB/km} +$$
$$4 \text{ Connectors} \times 1 \text{ dB/Connector} + 1.5 \text{ dB}$$

$$13 \text{ dB} = 7.5 + 4 + 1.5 = 13 \text{ dB}$$

For the next example, assume a link has transceivers each with a minimum transmitted power of −15 dBm. Further assume that the link is carefully designed with fiber having a loss of 3.5 dB/km and using only two connector pairs in the link and that those connector pairs have a maximum attenuation of only 0.3 dB/pair. Let D equal the maximum supportable fiber length and we get

$$-15 - (-32) = D \text{ km} \times 3.5 \text{ dB/km} + 2 \text{ Connectors}$$
$$\times 0.3 \text{ dB/Connector} + 1.5 \text{ dB}$$

$$17 \text{ dB} = (3.5D + 0.6 + 1.5) \text{ dB}$$

or

$$3.5D = 14.9$$

$$D = 4.25 \text{ km}$$

Even without using special transceivers, with careful link design as described previously we have

$$-19 - (-32) = D \text{ km} \times 3.5 \text{ dB/km} + 2 \text{ Connectors}$$
$$\times 0.3 \text{ dB/Connector} + 1.5 \text{ dB}$$

$$13 \text{ dB} = (3.5D + 0.6 + 1.5) \text{ dB}$$

$$3.5D = 10.9 \text{ dB}$$

$$D = 3.1 \text{ km}$$

These examples assume that the links are loss limited. In fact, there are two potential limits: channel attenuation and channel bandwidth. The latter restriction will derive from the fiber's distance bandwidth limitation, which for conformant fiber links is a minimum of 160 MHz·km. This limit imposes an upper bound of at least 10 km for 16-Mbit/s Token Ring operation, well beyond the range of channel attenuation limits. Note that the requirements specified previously are summary requirements for fiber operation. Detailed specifications are contained in the Supplement to the IEEE 802.5 Standard [7].

2.2 Physical Configuration of a Token Ring

For any given cabling plant, there is normally a wide range of options in configuring a Token Ring network. To determine the best network configuration, one must understand both the physical attachment restrictions forced by considerations of signal propagation and the range of logical connectivity options that meet one's requirements. The physical restrictions are most easily understood by beginning with the logical network and understanding how each physical limitation manifests itself for that network. Additionally, network behavior under fault and problem determination must be factored in to understand how the transmission path may change for these conditions. Logical connectivity options have been developed in response to the wide range of network requirements found in practice.[3]

2.2.1 Physical Attachment Considerations and Requirements

Figure 2.8 shows the logical connection of stations on a Token Ring. Each station must receive every incoming message and transmit that same message to the next station. When a message arrives at a station, it is converted from an analog to a digital signal for interpretation. Whether or not the received signal is intended for that station, the digital signal is reconverted back to an analog data stream at full transmit voltage levels for retransmission to the next downstream station on the ring. In this way, each station receives a data stream generated by its upstream neighbor. Because of the dependence of each station on its upstream neighbor, it is imperative that all stations on the ring continuously repeat the incoming signal. Whenever a station is unable to meet this requirement, such as when it is powered down, it must remove itself from the ring. The station is then bypassed and the ring is automatically reconfigured without that station. When the station is once again ready to fulfill its obligation as a good citizen of the ring, it signals its intention and is re-inserted. Control of station insertion and removal of the ring is handled by the Token Ring concentrators, as depicted in Figure 2.9. A discussion of the details of concentrator operation will be presented later in this section.

---

3. An excellent reference for the physical configurations for a Token Ring network is the International Technical Report [10]. This report was developed by ISO/IEC JTC 1/SC 25/WG 3, the same working group that later created IS11801, the international cabling standard. The technical report was based on the 1992 Token Ring standard that provided support for operation over 100Ω UTP only at 4-Mbit/s data rate. It has not been updated since support of 16-Mbit/s Token Ring using 100Ω and 120Ω cabling is, by design, in accordance with the cabling guidelines laid out in both IS11801 [9] and the U.S. standard EIA/TIA 568A [8].

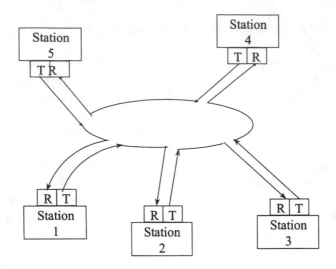

Figure 2.8 Logical connection of stations on a Token Ring.

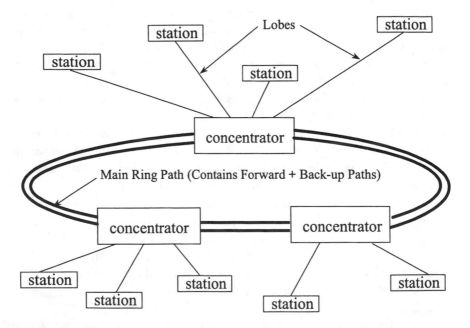

Figure 2.9 Concentrators connect stations to the Token Ring.

When configuring a few Token Ring stations in an office, or in a "bull pen" work area, the concentrators that interconnect the stations can often be collocated with the stations. However, when interconnecting Token Ring stations in a large office environment with multiple rooms, the structured wiring approach described previously is typically used. For these configurations, the Token Ring concentrators are generally rack mounted in wiring closets. Typical run lengths, from the wiring closet to the office areas supported by those closets, are in the range of 30m to 70m. Cabling distances between wiring closets vary from about 5m for wiring closets vertically stacked in a multistory building to 150m or more in sprawling buildings and factories. The total length of a large Token Ring LAN, including all the lobe cabling to the stations and the cabling between wiring closets, may be tens of kilometers.

As an example of the size of a large Token Ring LAN, consider that shown in Figure 2.10, a 200 station Token Ring that uses 25 concentrators placed in four wiring closets. Further assume that the average wiring distance between the stations and their attaching concentrators is 50m. Neglecting the wiring between concentrators, what is the size of the ring? Because each signal must go from the concentrator to each attaching station and back again, the total amount of lobe wiring to be traversed is 200 stations × 50 m/station × 2 paths/lobe = 20 km. Fortunately, the Token Ring signal does not have to travel the entire 20 km before it is detected, as each attaching station will receive the signal and regenerate that signal at full signal strength to continue the journey around the ring. Therefore, the basic transmission requirement is that the signal strength be adequate from one station to the next station. One would expect that this simple requirement would lead to a simple wiring rule for Token Ring. However, the Token Ring wiring rules are more complicated.

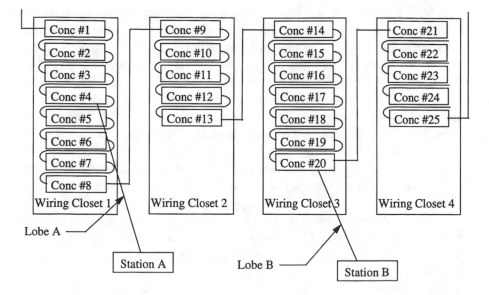

Figure 2.10 Large multiwiring closet Token Ring.

Again, referring to Figure 2.10, consider the case when there are no active stations between stations A and B. Then, the signal originating at A must traverse the lobe wiring A, connecting station A to concentrator #4. It then must pass through concentrators 4 through 20, including all the cabling connecting concentrators 4 to 5, 5 to 6, . . . and finally 19 to 20, before proceeding down lobe wiring B from concentrator #20 to station B. Transmitted signal strength must be sufficient to overcome the attenuation of each of the concentrators; each of the interconcentrator attaching cables; and the attenuation of the two lobes, those connecting station A to concentrator #4 and B to concentrator #20. In addition to designing the network to allow station A to connect to station B, we must ensure that a single station can exist on the ring, thus requiring that station A must be able to transmit a signal back to itself.

This detailed look at the set of paths requiring support has brought us from the expectation of having a relatively simple wiring rule for Token Ring to having to verify that each of the 40,000 possible signal paths will be valid. In fact, another situation must be considered. Token Ring was developed as a high-reliability, high-availability network. Therefore, network physical design is based on minimizing the probability of a fault bringing down the network and minimizing the time the network is down even when there are singular catastrophic faults. Consider the following catastrophic fault: Building maintenance personnel are renovating the building and accidentally cut the cable between wiring closets 1 and 2, which connects concentrators 13 and 14. There are no spare cables between those wiring closets, and it will take three days to pull another cable. How do we get the ring back up and operating in minutes, or worst case, in hours, rather than in days? The Token Ring wiring rules were designed to allow the ring to continue to operate properly if any one of the cables between any pair of concentrators is removed. Figure 2.11 shows the ring with that cable removed and the path traversed by the signal going from station A to station B. Note that appropriate problem determination and resolution procedures would require that the location of the break be identified. The broken cable is removed from concentrators 15 and 16, thereby activating the back-up path of the ring. The ring-in port on concentrator 16 wraps the signals from the back-up path to the forward path, and the ring-out on concentrator 15 wraps the signals from the forward path to the back-up path.

Under normal operation, the Token Ring signaling propagates on only one pair of wires in the cabling between the concentrators. When the backup path is used, however, as described previously, both pair in the main ring path cabling are used. The transmission path from station A to station B now includes traversing every leg of the main ring path twice except for that one leg that has been removed. The length of the transmission path shown in Figure 2.11 is

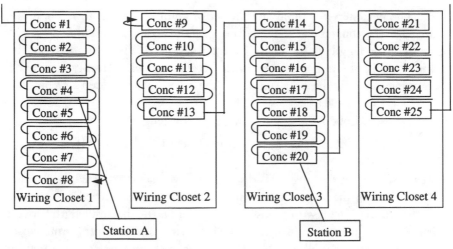

With broken lobe cable removed, signal wraps at
concentrator #8 ring-out port and concentrator #9 ring-in port

Figure 2.11 Large ring with cable removed.

$$(L\_A) + (L\_B) + 2 \times [(\text{Main ring path length}) - \tag{2.1}$$
$$(\text{Segment between wiring closets 1 and 2})] = \text{TotalL}$$

where L_A and L_B are the cabling lengths of lobes A and B, respectively,
and TotalL is the total transmission path length.

Consideration of each possible removal of a faulty cable in the main ring
path requires a computation similar to (2.1) for each of the 20 main ring path
segments that could be removed. This new complexity has just added to
our required 40,000 transmission length calculations another 20 × 40,000.
Therefore, to guarantee the proper received signal level for all possible transmission paths in the Token Ring depicted in Figure 2.10, under both normal and
problem determination and resolution conditions, we must be able to verify
that all 840,000 potential signal paths are within the maximum transmission
capability. Obviously, considerable simplification is required to make this
problem tractable.

The first step in simplifying the problem comes by observing that the
main ring path cables between concentrators in the same wiring closet are
patch cables that are easy to replace. Therefore, if one of those patch cables
was to become faulty, substituting a new patch cable for the bad one could
repair the ring quickly and simply. This observation eliminates consideration
of removing the shortest connections between concentrators, and therefore,
potentially the longest ring lengths.

The next step in simplifying the set of computations is to methodically search for the longest path length and guarantee that it is within allowable bounds. This length is found by considering the requirement that the first station to join the ring must be able to receive its own transmitted signal to determine that it is the only station on the ring. For this case (2.1) becomes

$$(2 \times \text{Lobe A}) + 2 \times [(\text{Main ring path length}) - \newline (\text{Segment between wiring closets 1 and 2})] \leq \text{MDL} \qquad (2.2)$$

for each station on the ring. This distance is obviously maximized by choosing the station with the longest lobe length. It is further maximized by assuming that the removed main wiring closet ring segment is the shortest length between wiring closets. Therefore, for a Token Ring that spans multiple wiring closets we can guarantee that all possible transmission paths will be contained within allowable bounds by verifying that

$$[2 \times (\text{MLL} + \text{MRP} - \text{SS})] \leq \text{MDL} \qquad (2.3)$$

where

- MLL is the maximum lobe length;
- MRP is the main ring path length;
- SS is the shortest segment between wiring closets;
- MDL is the allowable maximum drive length.

For a Token Ring whose main ring path wiring is confined to a single wiring closet, the ring recovery procedure calls for replacing any suspected or known faulty patch cables interconnecting concentrators. Following this procedure, the main ring backup path is never used; hence,

$$\text{The maximum transmission path length} = (2 \times \text{Lobe A}) + \newline (\text{Main ring path length}) \qquad (2.4)$$

One final factor to be accounted for is the attenuation associated with each of the concentrators. The signal traversing the concentrator's back-up path is attenuated based on the resistance of the circuit board wiring within the concentrator. The signal in the forward direction additionally encounters the resistance of each set of relays to each of the concentrator's ports. To a first-order approximation, the concentrator's attenuation characteristics can be modeled as two different cabling lengths, one for the forward and one for the back-up path. Define FLConc as the equivalent cable length of the concentrator forward path attenuation and TLConc as the equivalent forward path + backup path cable length of the concentrator. Then, for a particular choice in cable, that is, 150Ω STP cabling, (2.3) and (2.4) modified to include this additional attenuation become for a multiwiring closet Token Ring

$$\text{TLConc} + [2 \times (\text{MLL} + \text{MRP} - \text{SS})] \leq \text{MDL} \qquad (2.5)$$

Defining MRP − SS as *adjusted ring length* (ARL), (2.5) becomes

$$\text{TLConc} + [2 \times (\text{MLL} + \text{ARL})] \leq \text{MDL} \qquad (2.6)$$

and for a single wiring closet Token Ring

$$\text{FLConc} + (2 \times \text{MLL}) + (\text{MRP}) \leq \text{MDL} \qquad (2.7)$$

Finally, because the patch cabling used to attach both concentrators and end stations to the permanently installed cabling is constructed using narrow-gauge stranded conductors for better flexibility, it has higher attenuation per unit length than the in-wall cable. Consequently, the equivalent length of permanent cable with the same attenuation must be used to represent the length of those links.

IBM simplified the task of determining the allowable cabling limits by characterizing the loss of both its 8228 multistation access unit concentrators and standard-sized (8-foot) patch cables. It published tables [11] of (MLL + ARL) for various multicloset ring configurations accounting for all patch cabling so that only the permanently installed cabling need to be accounted for by the network planner. Those tables, a subset of which are reproduced here as Tables 2.3 to 2.8, assume a passive concentrator with transmission characteristics like the IBM 8228 MSAU. Also assumed are the use of 150Ω 22-gauge STP cabling; 8-foot, 26-gauge patch cables connecting concentrator ring-in to ring-out; and connecting lobe cabling both to the concentrator in the wiring closet and to the attaching equipment in the office. When there are multiple equipment racks within a single wiring closet, 30-foot patch cables are assumed for interconnection.

Tables 2.3 and 2.6 present the allowable lobe lengths. They were derived based on the solution of (2.7) with FLConc, ARP, and MDL known. The tables provide length as a function of the number of racks used to hold the concentrators, based on the assumption that additional 30-foot jumper cables

Table 2.3
Single Wiring Closet Lobe Lengths in Meters (Type 1 or Type 2 Cable) for 4-Mbit/s Rings as a Function of the Number of Concentrators [11]

	Number of Racks									
	1	2	3	4	5	6	7	8	9	10
1	385									
2	380	370								
3	370	365	360							
4	370	360	355	350						
5	365	355	350	345	340					
6	360	350	345	340	335	330				
7	355	345	340	335	330	325	320			
8	350	340	335	330	325	320	315	310		
9	345	335	330	325	320	315	310	305	300	
10	340	330	325	320	315	310	305	300	295	290
11	335	325	320	315	310	305	300	295	290	285
12	330	320	315	310	305	300	295	290	285	280
13		315	310	305	300	295	290	285	280	275
14		310	305	300	295	290	285	280	275	270
15		305	300	295	290	285	280	275	270	265
16		300	295	290	285	280	275	270	265	260
17		295	290	285	280	275	270	265	260	255
18		290	285	280	275	270	265	260	255	250
19		285	280	275	270	265	260	255	250	245
20		280	275	270	265	260	255	250	245	240
21		275	270	265	260	255	250	245	240	235
22		270	265	260	255	250	245	240	235	230
23		265	260	255	250	245	240	235	230	225
24		260	255	250	245	240	235	230	225	220
25			250	245	240	235	230	225	220	215
26			245	240	235	230	225	220	215	210
27			240	235	230	225	220	215	210	205
28			235	230	225	220	215	210	205	200
29			230	225	220	215	210	205	200	195
30			225	220	215	210	205	200	195	190
31			220	215	210	205	200	195	190	185
32			215	210	205	200	195	190	185	180
33			210	205	200	195	190	185	180	175

will be needed to interconnect the wiring racks with the cabling properly dressed.

Tables 2.4 and 2.7 provide the maximum value for the sum of the Lobe Length + ARL as a function of the number of wiring closets with TLConc known. The dependency on number of wiring closets may seem unusual

Table 2.4

4-Mbit/s Allowable Wiring Distances in Meters (Type 1 or Type 2 Cable) Without Repeaters or Converters as a Function of the Number of Concentrators [11]

	Number of Wiring Closets										
	2	3	4	5	6	7	8	9	10	11	12
2	363										
3	354	350									
4	346	341	336								
5	337	332	328	323							
6	328	324	319	314	310						
7	319	315	310	306	301	296					
8	311	306	302	297	292	288	283				
9	302	297	293	288	284	279	274	270			
10	293	289	284	280	275	270	268	261	257		
11	285	280	275	271	266	262	257	252	248	243	
12	276	271	267	262	258	253	248	244	239	235	230
13	267	263	258	253	248	244	240	235	230	226	221
14	259	254	249	245	240	236	231	226	222	217	213
15	250	245	241	236	231	227	222	218	213	208	204
16	241	237	232	227	223	218	214	209	204	200	195
17	232	228	223	219	214	209	205	200	196	191	186
18	224	219	215	210	205	201	196	192	187	282	178
19	215	210	206	201	197	192	187	183	178	174	169
20	206	202	197	193	188	184	179	174	170	165	160
21	198	193	188	184	179	175	170	165	161	156	152
22	189	184	180	175	171	166	161	157	152	148	143
23	180	176	171	166	162	157	153	148	143	139	134
24	172	167	162	158	153	149	144	139	135	130	126
25	153	158	154	149	144	140	135	131	126	121	117
26	144	150	145	140	136	131	127	122	117	113	108
27	135	141	136	132	127	122	118	113	109	104	99

considering that (2.6) does not explicitly show it. In fact, each wiring closet requires that an additional patch cable be used. Therefore, the associated loss of the cable plus the loss of an additional pair of connectors to be used in cabling between wiring closets must be factored in. Because the drive distances listed are for the permanently installed cabling only, the table must account for each of these extra factors that are topologically based.

In addition to the factors presented previously, numbers presented in the table accounted for a nonlinearity in the response of the Token Ring receivers. Specifically, receivers were designed to compensate for some of the cabling loss when it existed. To do this, the receivers had a different response to weak

Table 2.5

4-Mbit/s Allowable Wiring Distances in Meters (Type 1 or Type 2 Cable) as a Function of the Number of Concentrators With Repeaters or Converters [11]

	Number of Wiring Closets											
	1	2	3	4	5	6	7	8	9	10	11	12
1	376											
2	368	363										
3	359	354	350									
4	350	346	341	336								
5	341	337	332	328	323							
6	333	328	324	319	314	310						
7	324	319	315	310	306	301	296					
8	315	311	306	302	297	292	288	283				
9	307	302	297	293	288	284	279	274	270			
10	298	293	289	284	280	275	270	268	261	257		
11	289	285	280	275	271	266	262	257	252	248	243	
12	281	276	271	267	262	258	253	248	244	239	235	230
13	262	267	263	258	253	248	244	240	235	230	226	221
14	253	259	254	249	245	240	236	231	226	222	217	213
15	244	250	245	241	236	231	227	222	218	213	208	204
16	236	241	237	232	227	223	218	214	209	204	200	195
17	227	232	228	223	219	214	209	205	200	196	191	186
18	218	224	219	215	210	205	201	196	192	187	282	178
19	210	215	210	206	201	197	192	187	183	178	174	169
20	201	206	202	197	193	188	184	179	174	170	165	160
21	192	198	193	188	184	179	175	170	165	161	156	152
22	184	189	184	180	175	171	166	161	157	152	148	143
23	175	180	176	171	166	162	157	153	148	143	139	134
24	166	172	167	162	158	153	149	144	139	135	130	126
25	147	153	158	154	149	144	140	135	131	126	121	117
26	139	144	150	145	140	136	131	127	122	117	113	108
27	130	135	141	136	132	127	122	118	113	109	104	99

signals than to strong ones. However, the loss from concentrators did not have the same result on the signals as did lengths of line. To account for this difference, adjustments were made in the supported drive lengths. These adjustments appear as anomalies in the supported wiring distance.[4]

---

4. For example, in Table 2.5, columns 1 and 2, line 13, allowable drive distance increases for two wiring closets compared to one closet because the single wiring closet configuration assumes two racks connected with two 30-foot patch cables that are not required for the two-closet configuration.

Table 2.6
Single Wiring Closet Lobe Lengths in Meters (Type 1 or Type 2 Cable) for 16-Mbit/s Rings as a Function of the Number of Concentrators [11]

	Number of Racks									
	1	**2**	**3**	**4**	**5**	**6**	**7**	**8**	**9**	**10**
1	173									
2	170	160								
3	166	156	151							
4	162	152	147	142						
5	158	148	143	138	133					
6	154	144	139	134	129	124				
7	150	140	135	130	125	120	115			
8	146	136	131	126	121	116	111	106		
9	142	132	127	122	117	112	107	102	97	
10	138	128	123	118	113	108	103	98	93	88
11	134	124	119	114	109	104	99	94	89	84
12	131	121	116	111	106	101	96	91	86	81
13		117	112	107	102	97	92	87	82	77
14		113	108	103	98	93	88	83	78	73
15		109	104	99	94	89	84	79	74	69
16		105	100	95	90	85	80	75	70	65
17		101	96	91	86	81	76	71	66	61
18		97	93	87	83	77	72	67	62	57
19		93	88	83	78	73	68	63	58	53
20		85	80	75	70	65	60	55	50	45
21		77	72	67	62	57	52	47	42	37
22		69	65	59	54	49	44	39	34	29
23		61	56	51	46	41	36	31	26	21
24		53	48	43	38	33	28	23	18	13
25			40	35	30	25	20	15	10	5
26			32	27	22	17	12	7	—	—
27			23	18	13	8	3	—	—	—
28			15	10	5	—	—	—	—	—
29			7	—	—	—	—	—	—	—

2.2.1.1 Example[5]

Consider the three wiring closet ring shown in Figure 2.12. Assume all cabling is Type 1 150Ω STP. For 4-Mbit/s Token Ring operation, can the ring be

---

5. Additional detailed examples for design of Token Rings as 4 Mbit/s and 16 Mbit/s appear in two articles in *Data Communications* [12,13].

Table 2.7

16-Mbit/s Allowable Wiring Distances in Meters (Type 1 or Type 2 Cable) Without Using Repeaters or Converters as a Function of the Number of Concentrators [11]

	Number of Wiring Closets								
	2	**3**	**4**	**5**	**6**	**7**	**8**	**9**	**10**
2	162								
3	155	150							
4	149	144	139						
5	142	137	132	127					
6	135	130	125	120	115				
7	129	124	119	114	109	104			
8	122	117	112	107	102	97	92		
9	115	110	105	100	95	90	85	85	
10	109	104	99	94	89	84	79	74	69
11	102	97	92	87	82	77	72	67	62
12	95	90	85	80	75	70	65	60	55
13	82	77	72	67	62	57	52	47	42
14	69	64	59	54	49	44	39	34	29
15	56	51	46	41	36	31	26	21	16
16	43	38	33	28	23	18	13	8	3
17	30	25	20	15	10	5	—	—	—
18	17	12	7	—	—	—	—	—	—

configured without using repeaters? If so, what is the longest lobe length that will be allowed? How about for 16-Mbit/s operation?

From Figure 2.12 we see the interwiring closet cabling distances are 25m, 30m, and 50m. The ARL for this layout will be $(50 + 30 + 25) - 25 = 80$m. For 4-Mbit/s operation, Table 2.4 specifies the maximum length of ARL + the longest lobe. For the configuration shown with a total of nine concentrators distributed in three wiring closets we read the maximum value of 297m. Therefore, for this example, at 4-Mbit/s operation,

$$ARL + MLL = 297; \qquad 80 + MLL = 297; \qquad MLL = 217m$$

Because 297m is significantly longer than installed maximum cabling lengths, no repeaters will be necessary for this ring. Furthermore, if more stations and more concentrators are added over time, there will be no need to worry about exceeding the wiring rules.

Now look at the same configuration for 16-Mbit/s operation. We still have ARL = 80m. Now, however, we go to Table 2.7 and see the limit for ARL + MLL = 110m. Therefore, for 16-Mbit/s operation,

Table 2.8

16-Mbit/s Allowable Wiring Distances in Meters (Type 1 or Type 2 Cable) With Repeaters or Converters as a Function of the Number of Concentrators [11]

	1	2	3	4	5	6	7	8	9	10
				Number of Wiring Closets						
	173									
2	167	162								
3	160	155	150							
4	154	149	144	139						
5	147	142	137	132	127					
6	140	135	130	125	120	115				
7	134	129	124	119	114	109	104			
8	127	122	117	112	107	102	97	92		
9	120	115	110	105	100	95	90	85	85	
10	114	109	104	99	94	89	84	79	74	69
11	107	102	97	92	87	82	77	72	67	62
12	100	95	90	85	80	75	70	65	60	55
13	77	82	77	72	67	62	57	52	47	42
14	64	69	64	59	54	49	44	39	34	29
15	51	56	51	46	41	36	31	26	21	16
16	38	43	38	33	28	23	18	13	8	3
17	25	30	25	20	15	10	5	—	—	—
18	12	17	12	7	—	—	—	—	—	—

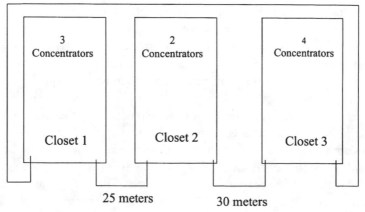

Figure 2.12 Example: three wiring closet ring.

$$ARL + MLL = 110; \qquad 80 + MLL = 110; \qquad MLL = 30m$$

Unless there are all unusually short lobe lengths, repeaters would be required for this configuration.

2.2.2   Use of Repeaters for Large Rings

If the maximum distance requirement determined from Tables 2.3, 2.4, 2.6, and 2.7 is not sufficient, then "repeaters" or "converters" can be added in the ring backbone to increase cabling distances. The repeater is a device that recovers the incoming data and retransmits this data to increase the signaling strength and quality on a long cable run. Because both the forward and backup paths are used to recover the ring under fault conditions, repeaters are required on both paths and are often installed or built into active equipment in pairs. Converters are like repeaters but are used for a change of signal media, for example, when going from copper-to-fiber optic transmission. When very long main ring path distances between wiring closets are required, fiber connection is the practical solution, allowing separations of up to 2 km. Note that the use of copper-to-optical fiber converters is the only solution for extending lobe segments because the copper repeaters do not pass phantom current. Many copper-to-fiber converters support phantom transmission by generating phantom at the converter closest to the concentrator whenever there is an incoming signal.[6]

Repeaters must actively participate in ring activity at all times and thus remain powered-up in the wiring closet. When they are used, the ring is considered to be divided into segments as shown in Figure 2.13. Each segment is treated independently for purposes of determining maximum drive distance and for certain problem determination procedures. However, each of the segments is still part of the same ring, and signals generated in any segment propagate to all of them. Because of the increased complexity of the ring with repeaters, problem determination procedures established to quickly locate the failing ring element require each segment to operate as a separate ring when all the repeaters connecting that segment are removed so that the signal wraps at these points. Under this condition, all devices within a segment will be able to communicate over the LAN with each other but will not be able to communicate with devices in a LAN segment from which they are isolated. Note that isolation of a segment will only occur if both the input and output repeaters are removed. With any single set of repeaters removed, the ring will

---

6. The use of the term   *converters*   to describe Token Ring devices that convert between media types, such as from copper to fiber, first appeared in [10].

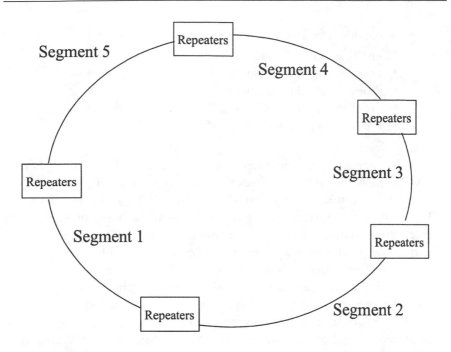

Figure 2.13 Ring with repeaters.

wrap at both sides of the repeaters and all devices on the ring will be able to communicate with each other.

For Token Ring segments, Tables 2.5 and 2.8 present the maximum length for total main ring path cabling in the segment plus the longest lobe length. The concept of ARL is not used for the segments because the segment must propagate Token Ring signals from any station back to the same station with repeaters removed and with no main lobe cabling removed. For this case, therefore, instead of (2.5)

$$\text{TLConc} + [2 \times (\text{MLL} + \text{MRP} - \text{SS})] \leq \text{MDL} \qquad (2.5)$$

there is no SS to eliminate and we get for ring segments bounded by repeaters or converters

$$\text{TLConc} + [2 \times (\text{MLL} + \text{MRP})] \leq \text{MDL} \qquad (2.8)$$

With this understanding of the cabling rules required for rings with segments, let's revisit the ring shown in Figure 2.10 to determine where repeaters would be needed for 4-Mbit/s operation over 22-gauge 150Ω STP cabling given:

- Maximum lobe length for all stations on the ring is 85m.
- Interwiring closet distances are:
  - From wiring closet 1 to 2, 30m;
  - From wiring closet 2 to 3, 55m;
  - From wiring closet 3 to 4, 60m;
  - From wiring closet 4 to 1, 145m.

First we must determine if repeaters are needed at all. Assuming they are not, we figure MLL + ARL and check that distance with the maximum allowed in Table 2.4. For this configuration MLL + ARL = 85 + (30 + 55 + 60 + 145 − 30) = 345m. However, according to Table 2.4, for a four wiring closet ring with 25 concentrators, the maximum allowable value is 154m. Therefore, this ring must be divided into segments using repeaters. The placement of those repeaters remains a variable with considerable flexibility; they may be placed based on minimization of the number of repeaters required, ease of access to a particular wiring closets by cabling and LAN specialists, and availability of space and power in the various wiring closets. Further, for simplification of adds, moves, and changes, and in keeping with minimizing the number of repeaters required, it is recommended that they be placed at the input or outputs of wiring closets rather than separating the concentrators within a single closet into separate LAN segments.

Two sets of repeaters, appropriately placed, will enable this configuration to meet the wiring restrictions as shown in Figure 2.14. Specifically, place one set of repeaters at the input of wiring closet 1 and the second set at the output

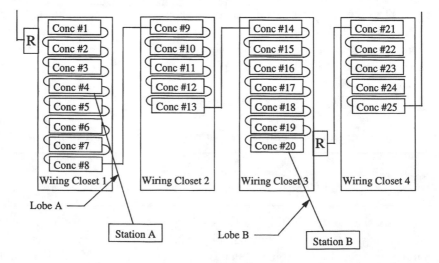

Figure 2.14 Repeaters (R) added to satisfy distance limitations for 4-Mbit/s operation.

of wiring closet 3. The ring is now divided into two segments. The first, a three wiring closet segment, has 20 concentrators, a total main ring path length of 30 + 55 = 85m, and a MLL of 85m. To check this configuration we must now turn to Table 2.5. It shows that for a ring segment with repeaters or concentrators, the drive capability for a three wiring closet configuration with 20 concentrators is 202m, providing about a 30-m margin over the 170m of drive that must be supported. The second Token Ring segment includes a single wiring closet and its attaching cabling at the wiring closets input and output. For this case, the total main ring path length is 60 + 145 = 205m. The longest lobe length remains at 85m for a total drive requirement of 290m. From Table 2.5, we see that a single wiring closet configuration with five concentrators can support a maximum drive of 341m, so each of the two Token Ring segments satisfies the drive requirements. The 4-Mbit/s Token Ring physical design is complete. Because each of the repeaters adds jitter to the transmitted signal, they must be included in the total device count when determining that the maximum device attachment limit has not been exceeded.

2.2.3 Considerations When Using Telephone Wiring

A wide performance range exists with installed telephone wiring. Rules for wiring the original 4-Mbit/s Token Ring were dependent on the wire meeting the IBM Type 3 Media Specifications. These specifications predate the standardization of cabling categories and were created by IBM to reflect commonly used telephone wiring that could accommodate 4-Mbit/s Token Ring signaling. The maximum station attachment for 4-Mbit/s Token Ring operation on UTP cable was 72 stations. This limitation was set because of UTP's high crosstalk and external noise and its lack of transmission uniformity for installed cabling compared to today's better performing wire. That limit was also required because 4-Mbit/s Token Ring was designed with just enough elastic buffer (See Section 3.3.2) to accommodate the jitter buildup using 150Ω STP cabling with its very low crosstalk and external noise and highly uniform transmission characteristics.[7] The tradeoff sacrificing high station count when using poorer transmission media had to be made to get a usable design point for Token Ring over UTP. With these limits, maximum lobe cabling could be set to 100m (the "magic" design number to cover over 95% of known lobe cabling runs) and the maximum ring size was limited to a single wiring closet.

---

7. It should be noted that the specification for 16-Mbit/s Token Ring included a significantly larger elastic buffer requirement to easily accommodate the latency associated with a 250-station ring. Adapters built after that specification all used the larger elastic buffer. Because 4-Mbit/s rings might include legacy adapters built to the original specifications, the 72-station limit for 4-Mbit/s rings on UTP remains today.

With these restrictions, IBM published simple, straightforward wiring rules [2] with no requirement for calculations or tables. Token Rings using telephone twisted pair meeting the IBM Type 3 Media Specification must be contained in a single wiring closet, are limited to nine 8-port passive concentrators, and may have a maximum lobe length (MLL) not exceeding 100m. 16-Mbit/s Token Ring could not reasonably be supported by Type 3 Telephone Twisted Pair. Reliable transmission would require that lobe lengths be limited to 20m to 30m for rings with no more than a couple of dozen devices with all concentrators contained in a single wiring closet. Early claims of UTP support for 16-Mbit/s Token Ring were followed by many field calls and problems. The rings would work well for most configurations, most of the time, but the cable had too much variability. Too many 16-Mbit/s rings using telephone wire were plagued by unreliable operation. 16-Mbit/s support was not practical for enterprises that had to depend on reliable operation of their network. Fortunately, better grades of telephone wire were becoming available along with the growing popularity of 16-Mbit/s Token Ring.

Upon completion of the recommended practice for 4-Mbit/s Token Ring operation using telephone wire [14],[8] the IEEE 802.5 Standards committee embarked on a project to determine specifically what support could be given for 4- and 16-Mbit/s Token Ring operation over the various grades of telephone wire that had just become standardized by the EIA/TIA.[9] Active concentrators were introduced with repeaters built into their ring-in and ring-out ports as shown in Figure 2.15. Using these devices, ring segments were restricted to a single concentrator and wiring rules were replaced with a simple MLL specification for each of the cable types supported. These maximum distances were about the same length allowed by the wiring rules for a ring that uses a single concentrator. The next generation of active concentrators employed active redrive at each of the lobe ports. For these devices, the signal is regenerated each time it reaches the concentrator or a station. The MLL using these active concentrators is equal to the total signal drive capability on the cable type within that lobe. This distance is more than twice that for active concentrators with only ring-in and ring-out redrive because the path losses that must be considered for this configuration no longer include the relay contact losses within the concentrators. Because the redrive circuitry adds jitter to the transmit-

8. This recommended practice was in support of Type 3 media as well as Category 3 cabling.
9. Under the auspices of ANSI (American National Standards Institute), the Telecommunications Industry Association (TIA), and the Electronic Industries Association (EIA) developed cabling standards in a joint technical committee, TR41.8.1. The committee first published the Commercial Building Telecommunications Cabling Standard, TIA/EIA 568, followed by Technical Supplemental Bulletins defining additional cabling types and their connectors, TSB-36 and TSB-40.

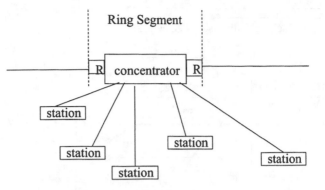

Figure 2.15 Active concentrators with built-in repeaters at ring-in and ring-out ports.

ted signal in the same way that a station does, active redrive concentrator ports are included in the total station count. Therefore, the total number of supported end stations is effectively reduced in half. With this understanding, maximum station attachment for rings employing active concentrators is limited to 132 devices.

2.3 Token Ring Concentrator Operation

One of the main physical layer functions unique to Token Ring consisted of the architecture for inserting and bypassing stations using phantom drive signaling to the Token Ring concentrators shown earlier in Figure 2.9. Figure 2.16 shows the signal path within the concentrator to effect ring insertion and removal of the stations.

The first Token Ring concentrators available were the IBM 8228 *multiple station access units* (MSAUs). They are 8-port passive devices with relays controlled by the attaching stations and ring-in and ring-out ports for extending the ring. Each station signals its readiness to participate in the Token Ring protocol by raising a dc phantom current. With these concentrators the phantom current charges a capacitor, which in turn provides the voltage required to toggle the relays effecting ring insertion. When the phantom current is removed, the capacitor discharges, the voltage drops, and the relays toggle back to the home position, removing the station from the ring. Because the relays are bistable devices, it is possible for some of them to be in the wrong state when they are initially installed due to vibration sustained during shipping. Therefore, the concentrators normally come with a reset tool to force the relays to their home state before initial usage. Relays (typically 4FORMC, a package consisting of four double-throw relays in a single module) were used rather

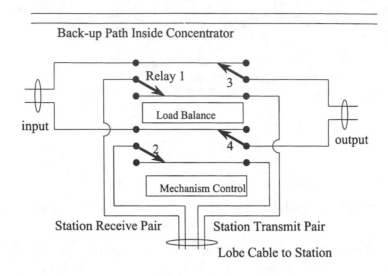

Figure 2.16 Concentrator relays provide signal connectivity to ring.

than active circuits to provide for highest possible ring availability. In the early days of Token Ring, many of the wiring closets were unpowered and most of them were lacking supplemental power that would continue, even during power outages. The unpowered box was ideally suited to provide high reliability and availability in this environment. Additionally, almost all powered devices had higher failure rates than the components used in the all-passive IBM 8228 MSAU. It is a tribute to the original architects and designers of the early concentrators of Token Ring that the passive concentrator design turned out to be the most reliable system component (in the authors' experience, an electrically failed concentrator was never verified).

It is imperative to isolate the phantom current originating at the stations from the main ring path to prevent the various stations' phantom drives from interfering with each other. Figure 2.17 is a schematic of the circuitry used to maintain isolation of the phantom drive, while permitting transmission of the high data rate signal. Transformer coupling of the data signal is possible because there is no net dc component in the Token Ring signaling.

The use of phantom signaling for ring insertion and removal, and as a way to monitor lobe cabling, greatly enhanced the reliability of Token Ring operation. Unfortunately, even the best engineering designs are hit with problems from totally unexpected sources. For phantom signaling, an early problem resulted from the choice of connector at the adapter card. A 9-pin D-shell connector was chosen because of its reliability, availability, and low cost. Unfortunately, the same DB9 connector was used to attach the old-style CGA monitors; more unfortunately, these monitor ports impressed a dc voltage on

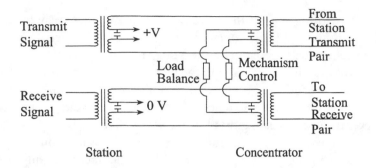

Figure 2.17 Phantom control circuitry.

the Token Ring pins that would energize the phantom drive relay in the passive concentrator. Therefore, when an end user mistakenly attached his Token Ring cabling to the monitor port rather than to the Token Ring card, it resulted in the monitor port inserting into the ring, taking the ring down until the offending station was located. The perpetrator of this LAN outage was completely unaware of the problems he or she caused until the perpetrator heard from an irate LAN administrator. The solution to this problem was the use of a "green dot" on each Token Ring adapter PC card that matched the "green dot" on the patch cord to make it easier to always connect the patch cord to the right connector. The problem has long since gone away both with the advent of next generation monitors that use a 15-port connector and with the increased use of modular telephone connector for Token Ring ports.

2.3.1　Phantom Signaling and Wire Fault Detection

In addition to using the phantom current to signal readiness of the Token Ring adapter to insert and remove itself from the ring, it is also used to continuously monitor the state of the lobe cabling attaching the station to its concentrator port. Two separate but equal dc signal voltages are transmitted on the two wires of the stations transmit pair, with the two wires of the receive pair serving as returns. Because both signal paths have the same length wire, the resistance of the two paths caused by the lobe wires should be substantially the same. In addition, at the concentrator, one of the signals drives the relay circuitry that controls ring insertion. The other drives a dummy load nominally equal to the load caused by the control circuitry. If there are no fault conditions, the two signal paths should have approximately the same total load and therefore should carry approximately the same current. By monitoring the current on these two signal paths, a determination of the condition of the lobe cabling path can be made. Specifically, if the currents are too high, then the wires may be shorted together. If the currents are too low, there may be an open-circuit condition or a faulty contact causing a high-impedance connection. If the currents are not substantially the same, there may be a fault condition along one of the wires or contacts in the lobe. Any of these conditions signifies a fault condition and would activate the removal of the station from the ring.

The specific values and ranges for the phantom voltages and currents were specified in excruciating detail in the 1995 standard. This detailed write-up is in marked contrast with the dozen lines of specification in the 1989 standard and highlights a trend in standards development showing the additional complexity of specification required to help guarantee interoperability under all conditions. Among the cases that were considered in developing the 1995 standard description was the use of splitters, devices plugged into the office data outlet to attach two stations to a single lobe cable. The splitters have their own phantom control circuitry and balance loads. The allowable phantom current ranges had to allow for the possible presence of a lobe splitter.

2.3.2 Concentrators With Active Redrive

The next generation of concentrators used active devices employing redrive at ring-in and ring-out. They required no change to the lobe attachment circuitry from the passive concentrators but required power for their repeaters. The third generation concentrators employed active redrive at each lobe port. This configuration was required to provide reliable support for 16-Mbit/s Token Ring transmission on Category 3 UTP cabling. Figure 2.18 is a schematic diagram of the data path for the lobe signaling with stations attached to ports 1 and 3 inserted into the ring and stations attached to ports 2 and 4 in bypass. Actual circuitry for these devices often includes active electronics in place of the mechanical relays shown in the figure.

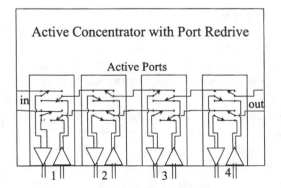

Figure 2.18 Active concentrators with redrive at each port.

2.3.3 Concentration Devices for Dedicated Token Ring Operation

DTR operation and concentration device requirements for that operation are discussed in detail in Chapter 5. The DTR concentration devices are active devices with four modes of operation for their ports depending on the port's attachment. Therefore, it is critical that their ports be able to determine if their attachment configuration is to:

1. A Classic Token Ring station;

2. A Classic Token Ring concentrator;

3. A DTR station;

4. Another DTR concentration device.

The first attachment mode, configuration 1, is in support of Classic Token Ring. In this first mode, the port operation from the station's point of view is as described previously for active concentrators. The other configurations are in support of DTR operation. Configuration 2, attachment directly to a port of a Classic Token Ring concentrator, requires the port to be in Classic Token Ring station emulation mode. This mode is critical for using the DTR concentrator to microsegment the network into small rings. (The implications and use of station emulation mode is described in [15], which is included as Appendix C.) In configuration 3, the port is attached to a single DTR station in full-duplex mode. In configuration 4, the two DTR concentration devices go through a "handshaking" procedure that will result in one going into DTR station emulation and the other in support of the DTR station.

DTR concentration devices must be able to configure each of their ports independently based solely on the device attachment to that port.

2.4 Logical Design of a Token Ring Network

Physical placement of stations and concentrators is determined primarily by topology considerations and building constraints. While it is often convenient to design the logical connectivity of a Token Ring network based on the physical topology, much wider latitude is possible; careful consideration of all the requirements for the network is strongly recommended. For small LANs with few attached stations, planning is straightforward. Choose a convenient wiring closet area for the concentrators, install cabling from each potential work area to the wiring closet, attach the stations and concentrators to the cabling, and interconnect the concentrators. All stations will communicate over a single Token Ring network. Because there are not many stations on the ring, the capacity of the network should be more than adequate to support the most common applications.

As soon as the network grows in the number of stations being attached or in the geographic area required to encompass all stations, network planning becomes essential. Primary considerations in designing a logical network include:

- Maximum number of stations to be attached;
- Expected LAN traffic patterns;
- Equipment costs;
- Physical constraints, including type of LAN cabling available;
- Plans for moves, adds, and changes;
- Administration costs;
- Availability requirements.

Next, consider a large number of stations all wired to a single wiring closet. The maximum number of stations supported on a single ring according to the IEEE 802.5 standard is 250. If concentrators with active retiming elements at each port are used, that number decreases to 132. If there are more stations than can be attached to a single ring, then the network must be subdivided into multiple rings. In addition to the physical attachment limit, the network administrator must account for peak bandwidth demands and network management considerations in determining a suitable upper bound to the number of stations attached to a single ring. A single Token Ring maximum data transfer bandwidth is equal to the ring speed.

As we have seen earlier, a repeater can be used to improve maximum cable lengths of a single ring. Another method to achieve greater total cable length and simultaneously increase the total data transfer bandwidth is to divide the large ring into two rings using a bridge. This division can be done in the wiring closet. As an example, a single bridge as shown in Figure 2.19 could replace the two repeaters used in Figure 2.14.

For the configuration in Figure 2.19, if traffic is segregated between the two rings, only that traffic destined for stations between rings needs to be sent through the bridge. Because work groups, where most traffic occurs, can be connected to the same wiring closet, this bridge combination enhances network bandwidth. Specifically, the total network bandwidth for the configuration shown in Figure 2.19, assuming both rings operate at 16 Mbit/s, will be 32 Mbit/s minus the traffic through the bridge. As the network grows, by the

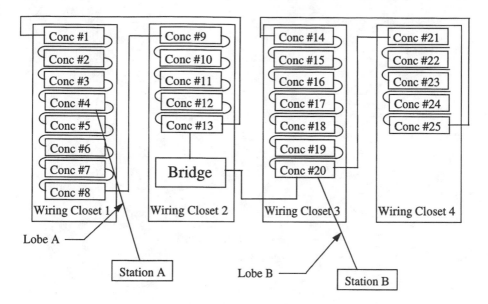

Figure 2.19 Bridge added to divide LAN into two rings.

addition of a third ring or by splitting one of the rings into two, network topology choices arise that will fundamentally affect the amount of traffic on each of the rings and the total traffic potentially supported by the network. For example, consider two ways of interconnecting three rings shown in Figure 2.20. In the first instance, ring 2 must carry all of the traffic between rings. In the second configuration, the placement of a third bridge allows communication between rings 1 and 3 to bypass ring 2 completely, thereby decreasing the total traffic on that ring.

An entirely different topology for ring interconnection employs the concept of a backbone ring as shown in Figure 2.21. This topology is suitable for very large and dispersed networks.

This topology has the advantages of being highly expandable, providing independence in configuration of bridges for the local rings, and being straightforward to manage and troubleshoot. Servers can be placed on the rings they serve or on a separate server ring located conveniently for network administration. Bridges can all be placed in the wiring closet(s) of the backbone ring, and the backbone ring can often be a collapsed ring located entirely within a single wiring closet. For example, consider the design of a Token Ring network to service an entire 10 story building with 50 stations on each floor. An obvious configuration would be to have a single ring service each floor and to have a backbone ring that goes to each floor. However, a better configuration would be to have the backbone ring on a single floor near the network administrators.

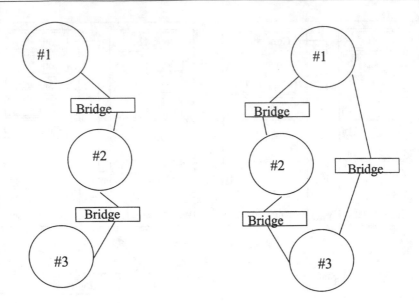

Figure 2.20 Bridged rings, two configurations.

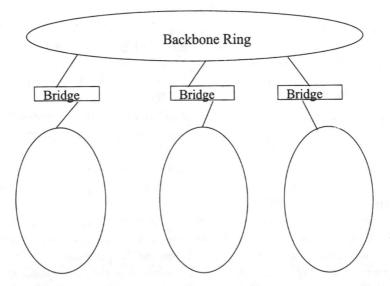

Figure 2.21 Local rings bridged to a backbone ring.

All of the bridges would be placed within the wiring closet of this collapsed backbone, with lobe cable connections attaching each of the bridges to a different local ring. Servers that are accessed by stations on all the rings could be placed directly on the backbone ring or on a separate ring in the same wiring closet. Local servers would be best placed on the rings of the stations they serve to minimize the traffic that must go onto the backbone ring. Planning for this configuration would require extra cabling from the backbone ring location to each of the wiring closets on every floor. This cabling is for lobe attachment of the bridges rather than for main ring path wiring.

As new rings are added to the configuration each one requires a single bridge to the backbone ring, without regard to the placement or existence of the other rings and bridges. The only planning consideration in adding the new ring is a determination of the traffic it will add to the backbone ring and that impact on the ability of that backbone ring to carry all the inter-ring traffic without exceeding its capacity.

2.4.1   Network Topology Design for Fault Tolerance and Redundancy

The preceding configurations all provide for the connectivity necessary for communication between any pair of stations on the ring. However, any hardware problems in the backbone ring or in any of the bridges will likely cause at least one ring to be isolated from the others until the fault can be corrected. Because any replacement of faulty equipment requires (1) notification of the problem, (2) availability of the repair personnel, (3) location of the fault, (4) availability of the replacement hardware, and (5) replacement of the faulty equipment, bringing the ring back up to full operational capability would take hours or days. Even eliminating steps (4) and (5), only bringing the ring up to partial capability by working around the fault, can take hours. When network integrity is vital to the smooth operation of the business, a better solution must be implemented.

Two principle guidelines for planning fault tolerant networks are:

1. Minimize the likelihood that any outage producing fault will occur on the network;
2. Design the network so that no single fault will take down more than a single station for more than a very limited time (to be measured in seconds or minutes rather than hours).

The design of Token Ring's architecture and its hardware included fault tolerance as a requirement. For example, the first generation concentrators were passive devices that relied on phantom current from the attached station to actuate the ring insertion relays. This design eliminated the requirement for power supplies with their associated failure rate and reliance on electrical power availability in the wiring closet. Wrap plugs provided some measure of fault tolerance, allowing the ring to enter a wrap state and keep operating even if one of the connections along the main ring path was disconnected. However, any break in the main ring path cable other than at the wrap plug would take down the entire ring until the broken cable was found and removed from the ring. Any break in the backbone ring would stop all inter-ring traffic. A break in the cable connecting a bridge to either ring, a failure in the bridge hardware, or a power outage at a bridge would isolate the associated ring from the rest of the network. The likelihood of an actual cable break is remote but still possible. For rings whose main ring path is contained within a single wiring closet, the probability of such a break within the main ring path becomes insignificant. However, having a cable disconnected by mistake or having hardware failures are events that are to be expected and planned for in the operation of any large network.

Token Ring architecture will force failing end stations and end stations with faulty cables to detach from the network in order to maintain ring integrity for the rest of the network.[10] However, the architecture is unable to directly protect from the less likely event of a severed cable or bad connection within the main ring path or from a fault occurring in a vital networking component like a bridge. What the architecture has done is allow for flexible design to work around these problems. To adhere to the second guideline of fault-tolerant design, we must create a configuration that is not dependent on any single bridge or on a single backbone ring. This guideline can be met by using a dual-backbone configuration as shown in Figure 2.22. The local rings can be designed as single wiring closet rings, with no main ring path wiring other than patch cables. If there is a problem with any of those cables, they can be quickly, easily, and inexpensively replaced. If there is a problem with either of the backbone rings, the bridges that are attached to the rings will send a beacon message that can be interpreted by appropriate network management software to pinpoint the location of the problem for quick repair action. In addition, before that repair action occurs, the traffic on the now faulty ring is shifted to the other backbone ring, thereby maintaining the orderly flow of a business dependent on its LAN traffic.

If there is a failure in any bridge or in the cabling connecting that bridge to either the backbone or its local ring, a new data session can be established using the other bridge connected to the same local ring. The effect of this

---

10. The Token Ring architecture is unable to automatically protect against stations that violate this architecture. One important example is the insertion of a station at the wrong speed. It will not detect the ring's proper condition and will remain inserted, taking down the entire ring, until removed, unless the concentrator to which it attaches or the station's Token Ring adapter card have additional features to automatically detect speed mismatch.

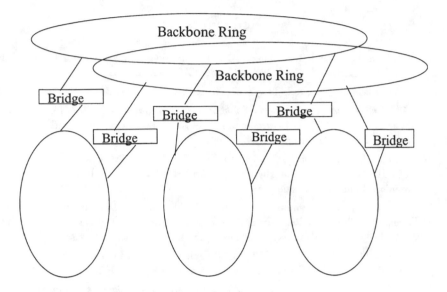

Figure 2.22 Network design employing dual-backbone rings.

network session is the same as if the backbone to which the failing bridge is attached to had itself become faulty.

Another network failure mode is a power outage in one section of the building. If that power outage affects the wiring closet that holds the two backbone rings, all of the bridges could lose power, effectively severing network communications. There are two divergent solutions to this scenario. One is to have uninterrupted power in the crucial wiring closet. Taking this approach, it is effective to bring all critical network resources including servers and bridges into this one area with a protected source of power. The other tact is to locate the two backbones in separate and possibly widely spaced wiring closets to protect the network from a power outage that affects only one part of the building.

2.4.2 Network Design With Dedicated Token Ring

Even with dual-backbone network designs, Token Ring traffic congestion is likely to be observed first in the backbones. Using DTR concentrators, described in Chapter 5, a migration path is established for increasing overall network bandwidth as described in Appendix C. The first step uses the high-speed backplane within the DTR concentrators as the network's backbone. This critical next step in Token Ring design can satisfy network bandwidth requirements for the next several years.

References

[1]   Friedrich, A. A., and R. D. Love, "APL Transmission Path Analysis for Design of a Local Area Network," *Conf. Proc. APL '84 ACM,* Finland, June 1984.

[2]   IBM Token-Ring Network Telephone Twisted Pair Media Guide GA27-3714, 1985.

[3]   *Additional Cable Specifications for Unshielded Twisted Pair Cables,* EIA/TIA Technical Systems Bulletin TSB 36, Electronic Industries Association/Telecommunications Industries Association, Arlington, VA, Nov. 1991.

[4]   *Additional Transmission Specifications for Unshielded Twisted-Pair Connecting Hardware,* TIA/EIA Telecommunications Systems Bulletin TSB 40, 1992, Published by Telecommunications Industry Association.

[5]   Love, R. D., "Telecommunications Building Cabling for the '90s," presented at 3ème Symposium Précâblage Réseaux "Immeubles Intelligents," New Orleans, Feb. 1992, and published in the conference proceedings, Actes des Conférences.

[6]   Love, R. D., T. Toher, and L. C. Haas, "Planning Guidelines for Token-Ring Cabling," *IBM Personal Systems Technical Solutions,* Roanoke, TX, July 1992.

[7]   ISO/IEC 8802-5: 1998 Amd. 1: 1998 Information technology—telecommunications and information exchange between systems—Local and Metropolitan Area Networks—specific requirements—Part 5: Token Ring access method and physical layer specifications Amendment 1: Dedicated Token Ring operation and fiber optic media.

[8]   *Commercial Building Telecommunications Cabling Standard,* TIA/EIA Standard TIA/EIA 568A, Telecommunications Industry Association, Arlington, VA, 1995.

[9]   ISO/IEC 11801:1995(E) Information technology—generic cabling for customer premises, 7/15/95.

[10] "Information Technology—Customer Premises Cabling—Planning and Installation Guide to Support ISO/IEC 8802-5 Token Ring Stations," ISO/IEC TR 12075, reference no. 12075(E), 1994.

[11] IBM Token Ring Network Introduction and Planning Guide GA27-3677, 1992.

[12] Love, R. D., and T. Toher, "How To Design and Build a 4-Mbit/s Token-Ring LAN," *Data Comm.,* May 1987.

[13] Love, R. D., and T. Toher, "How To Design and Build a 16-Mbit/s Token-Ring LAN," *Data Comm.,* July 1989.

[14] IEEE Std. 802.5b-1991 IEEE Recommended Practice for Use of Unshielded Twisted Pair Cable (UTP) for Token Ring Transmission at 4 Mb/s, 1991.

[15] Beyer-Ebbesen, B., M. Cowtan, S. Hakimi, and R. D. Love, "Migration Strategies for Token Ring Networks," *Int. J. Network Management,* Spring 1997.

3

Physical Layer

This chapter describes the PHY layer components of the Token Ring adapter as shown in Figure 1.16 with an emphasis on the key interoperability PHY principles for Token Ring. First, the architectural structure of a station will be described to show the relationship between the MAC and the PHY layer. Second, the *physical signaling components* (PSC) responsible for the Differential Manchester symbol encoding and decoding will be described as well as the elastic buffer, which is responsible for accommodating the total accumulated jitter by all the stations on a ring. Third, the *physical media components* (PMC) responsible for transmitting the electrical signals and receiving the data will be described. The PMC contains the critical component, PLL, in the PHY. Finally, this chapter will summarize the interoperability principles for Token Ring.

3.1 Station Structure

Figure 3.1 illustrates the station structure, showing the PHY and MAC layers. The PHY layer consists of the receiver, repeater, and transmitter. The receiver is responsible for monitoring the ring and deriving receive clock from the ring's Differential Manchester code; decoding the Differential Manchester codes as a token, frame, or abort sequence; and providing the station and/or the repeater with the data or signals derived from the Differential Manchester code. The repeater is responsible for taking the signals from receiver and providing these signals to the transmitter. The transmitter is responsible for deriving its transmit clock from either the station's internal crystal clock (an *Active Monitor,* or AM) or the receiver's receive clock (a *Standby Monitor,* or SM), and for transmitting

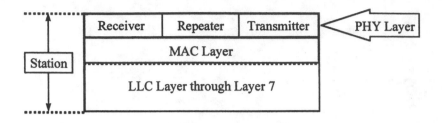

Figure 3.1 Station structure.

the signal from either the repeater (repeat mode) or the station's internal buffers (transmit mode).

Also shown in Figure 3.1, the MAC layer controls the Access Protocol and uses the PHY layer to receive and transmit frames. The LLC and higher layer functions control and use the MAC and PHY layers to receive and transmit frames. Now that the functions of the station have been examined, the operational modes of the station will be described in more detail.

3.1.1   Station PHY Operational Modes

This section will provide a basic understanding of how the station PHY operates. The station PHY supports one of three modes: *repeat, repeat and copy,* or *transmit.*

Repeat mode, illustrated by Figure 3.2, allows the station to repeat the ring signal without change other than setting the end delimiter's (ED) *error bit* (E-bit) (see Section 4.4.5.1) if an error is detected as required by frame or

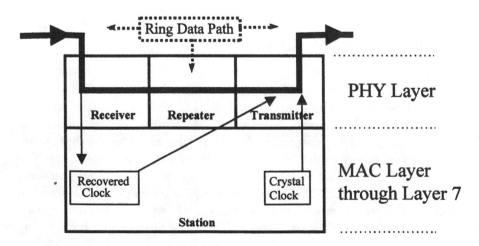

Figure 3.2 Repeat mode overview.

token checking. The repeat mode is the normal mode of the station and is used to repeat abort sequences, frames, or tokens. Repeat mode generates the output signal of the transmitter, which may use either the station's crystal clock (an AM station) or its recovered clock (an SM station).

The repeat and copy mode, illustrated by Figure 3.3, allows the station to repeat and copy the ring signal without change other than setting the frame control field's *address recognized indicators* (ARI) and *frame copied indicators* (FCI) and setting the ED's E-bit if required by frame or token checking. The repeat and copy mode is activated when the station recognizes the frame as having a destination address equal to one of its internal recognizable receive addresses. The repeat and copy mode allows the station to repeat a frame and at the same time put the frame into the station's receive buffers (these receive buffers can be in either the station or the station's attached product). The repeat and copy mode generates the output signal of the transmitter, which may use either the station's crystal clock (an AM station) or its recovered clock (an SM station).

The transmit mode is illustrated in Figure 3.4. The transmit mode is activated when the station has captured a token or when the station enters the ring recovery mode. The transmit mode has two modes of operation: transmit normal and transmit immediate. The transmit normal mode operates when transmitting a frame as the result of token capture. During transmit normal mode, the station's transmitter clocks its output using either its receiver's recovered clock (an SM station) or its crystal clock (an AM station), and the station's receiver derives it receive clock from the signal on the ring. The transmit immediate mode operates when transmitting a frame in Dedicated

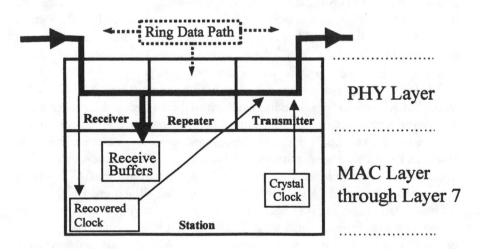

Figure 3.3 Repeat and copy mode overview.

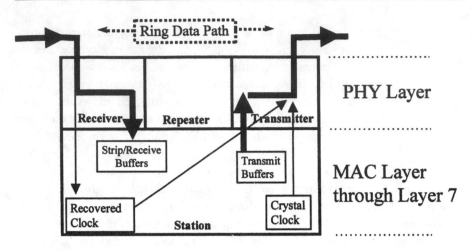

Figure 3.4 Transmit mode overview.

Token Ring transmit immediate mode or as the result of the ring recovery process. During transmit immediate mode, the station's transmitter clocks its output using its crystal clock and the station's receiver derives it receive clock from the signal on the ring.

3.2 Physical Signaling Components

The PSC of the Token Ring PHY includes those components responsible for the signal encoding/decoding, clocking, data repeating and transmission on a token, and latency control. These blocks are illustrated in Figure 3.5, and the detailed specifications appear in the Token Ring Standard, IEEE 8802-5:1998 clause 5. The Differential Manchester symbol decoder at the receiver converts the encoded symbols into a data stream and recognizes special, nondata symbols. These special symbols are the *start delimiter* (SD) and *end delimiter* (ED), which form the start and end of a frame or a token. The XMT/REPEAT MUX block provides for selection of transmitted data between either the data received in repeat mode or the transmit data generated during transmit mode. The station has two operating clock modes: SM mode (transmit on recovered clock) and AM mode (transmit on crystal clock). Only one station on the ring operates in AM mode during normal Token Ring operation. The AM's internal crystal oscillator is the basis for the clock timing of all stations on the ring. All other stations on the ring operate in SM mode, deriving their transmitted signals clocking from the incoming bit stream. While in this repeat mode, each station is also monitoring the ring to ensure all AM functions are occurring. If not, then each station stands ready to take over the task of becoming the

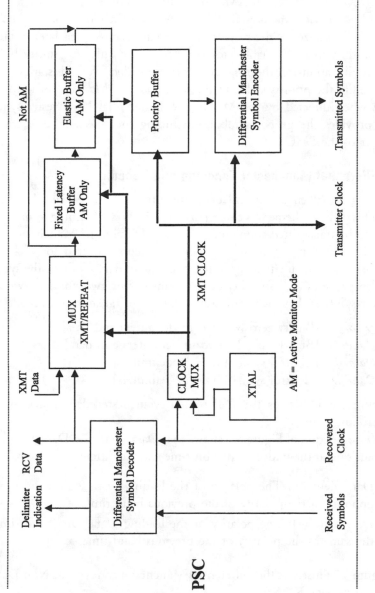

Figure 3.5 Physical signaling components.

AM following an orderly determination procedure to select a successor AM. The AM mode operation is controlled by the MAC protocols, which physically set the CLOCK MUX block in Figure 3.5. This block provides the transmitter clocking (XMT CLOCK). In AM mode, the transmitted clock provided by the AM MUX block is the crystal (XTAL) oscillator, while for SM mode, the XMT CLOCK utilized is the received recovered clock. A fixed latency buffer and an elastic buffer are enabled for the AM mode only. A priority buffer, enabled in all stations on the ring, adds sufficient delay for the station to be able to modify the priority fields in the *access control field* (AC) of the frame or token when directed by the MAC. The Differential Manchester symbol encoder provides the proper symbols, including the SDs and EDs, to the transmitter of the PMC.

3.2.1   Differential Manchester Encoding and Decoding

Token Ring signal encoding and decoding utilizes Differential Manchester coding. This coding scheme was chosen to enable both clock timing information and data to be transmitted simultaneously. The Differential Manchester code is relatively inefficient, with two signaling elements required for each data symbol. Because of its inefficiency in transmission of data symbols, the symbol stream also provides the capability of also transmitting two nondata symbols. Thus, four symbols that each station is required to transmit and decode are

> *Data_Zero*   [Binary zero (0) used in delimiters and data]
> *Data_One*   [Binary one (1) used in delimiters and data]
> *Non-Data_J*   [Control symbol used in delimiters]
> *Non-Data_K*   [Control symbol used in delimiters]

Figure 3.6 illustrates the Differential Manchester data codes for the Data_Zero and Data_One.

An examination of Figure 3.6 shows that Data_Zero and Data_One both have a transition at the midpoint of a bit time and are differentiated as follows:

- Data_Zero (0): The polarity of the leading Data_Zero baud time is opposite of the polarity of the previous baud time.
- Data_One (1): The polarity of the leading Data_One baud time is the same as the polarity of the previous baud time.

Figure 3.7 illustrates the Differential Manchester code for the Non-Data_J ('J') and Non-Data_K ('K').

An examination of Figure 3.7 shows that Non-Data_J ('J') and the Non-Data_K ('K') both lack transition at the midpoint of a bit time and are differentiated as follows:

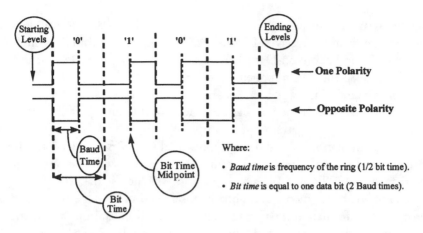

Figure 3.6 Differential Manchester coding of '0' and '1'.

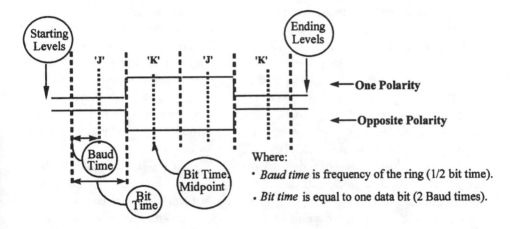

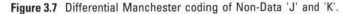

Figure 3.7 Differential Manchester coding of Non-Data 'J' and 'K'.

- Non-Data_J: The polarity of the leading Non-Data_J baud time is the same as the polarity of the previous baud time.

- Non-Data_K: The polarity of the leading Non-Data_K baud time is opposite of the polarity of the previous baud time.

Unique delimiters described in Section 4.4.2 using the characteristics of the Data_Zero, Data_One, Non-Data_J, and Non-Data_K are defined: one to start a structured message, called an SD, and one to end a structured message, called an ED.

In Differential Manchester coding, the data symbols are not dependent on polarity and have no net dc steady-state component. For data symbols, there is always a transition at the middle of the symbol period to enable clock regeneration from the received PLL in the PMC as described in Section 3.3.4. The Non-Data_J and Non-Data_K used in a starting or an ED occur in pairs of 'J-K' to keep the dc component zero.

In Token Ring, a period of nonclock transitions is designated as "BURSTx," where x is the number of periods of the same polarity without a transition in the data pattern. An example of a BURST4 is show in Figure 3.7. The combinations of J and K symbols used to designate token and frame start and end limit the maximum number of periods with the same voltage polarity to BURST3 and always dc balance the net signal level. A higher number of burst patterns are occasionally observed during ring error conditions and provide information for ring recovery. Token Ring stations automatically limit the length of transmitted ring burst errors to BURST5 to enable the receiver PLL to maintain phase lock.

In Differential Manchester symbol encoding and decoding, because the maximum voltage transition rate is two voltage transitions per symbol, the highest fundamental frequency associated with Differential Manchester is equal to the data rate. For a long pattern of Data_Ones there will be transitions only at the midsymbol time, resulting in a data transition frequency that is half the data signaling rate, that is, 8 MHz for 16-Mbit/s Token Ring operation. For a long pattern of Data_Zeros there will be transitions at the midsymbol and start symbol times resulting in a data transition frequency equal to the signaling rate, in this case 16 MHz.

This frequency spectrum dependence on data pattern is highly significant in Token Ring data clock and data recovery design. The natural occurrence of large strings of Data_Ones and large strings of Data_Zeros in data patterns results in a data pattern-dependent frequency spectrum. For a Data_Ones signal, the main frequency component of the resulting signal energy spectrum is at 8 MHz. For a Data_Zeros signal, the main frequency component of the resulting signal energy spectrum is at 16 MHz. This spectrum dependency, in turn, causes accumulated jitter dependent on data pattern as the data propagates around all stations in the ring as described in Section 3.3. Unlike some LANs, such as 100BASE-TX, the data stream is not scrambled between the raw data symbols and the encoding and decoding circuits. The absence of a scrambler, not recognized during early developments of 4- and 16-Mbit/s Token Ring, severely complicated the design of the receiver PLL and added to *electromagnetic interference* (EMC) emission problems for 16-Mbit/s Token Ring.

3.2.2 Elastic Buffer and Fixed Latency Buffer

In order to achieve maximum performance on the Token Ring (as discussed in Appendix A by Bux et al.), the repeat path of each Token Ring station is minimized to reduce ring latency (the time it takes for a frame or token to circulate to all the stations around the ring from the transmitter and back to the transmitter's receiver). As described in Section 3.3, the effect on accumulated jitter is also very dependent on this repeat path delay. There are three occasional cases where an additional delay may be introduced into the station that are too infrequent to affect jitter accumulation but important to understand in the Token Ring protocol. A block diagram of the AM station repeat path summarizing the various components of latency delay is shown in Figure 3.8 and subsequently described.

At the AM station, a fixed latency buffer is inserted into the repeat path. The latency buffer guarantees that the minimum length of a ring, measured in the number of bits circulating, is as large or larger than a token bit length. This minimum length is a requirement for proper Token Ring operation to allow a token to circulate on a single station ring.

At the AM station, an elastic buffer is added to compensate for jitter accumulation around the ring. As shown in Section 3.3, there may be a large accumulation of jitter as the token circulates around the ring. Because the AM transmits data on its crystal clock but receives data on its recovered clock, during long periods of clock jitter, the mismatch in frequency between the crystal clock and the recovered clock causes the synchronization of the transmit and received signals to get further and further out of phase. Because the received signal must be retransmitted based on the timing of the crystal, this synchronization error must be accommodated by the elastic buffer. This elastic buffer must accommodate a jitter accumulation up to a minimum of B symbols (in each direction as specified in clause 5.8.3.1 of 8802-5:1998). For a 16-Mbit/s Token Ring, the minimum B for jitter accommodation of the elastic buffer is 15 symbols. For a 4-Mbit/s Token Ring, the minimum B for jitter accommodation of the elastic buffer is 3 symbols. This elastic buffer is generally reset to zero just after the transmission of a token to avoid long-term drift in elastic buffer centering.

An additional delay is added at any station that has priority stacking because of the transmit token reservation protocol. A station that is participating in priority stacking may need to introduce added delay in its repeat path in order to properly modify the priority reservation bits in the AC. Typically, up to 8 symbols of delay are added to enable proper operation of the priority MAC protocol.

3.3 Physical Media Components

The function of the PMC in the PHY is to enable a station to transmit signals with controlled characteristics to its downstream neighbor and to receive the

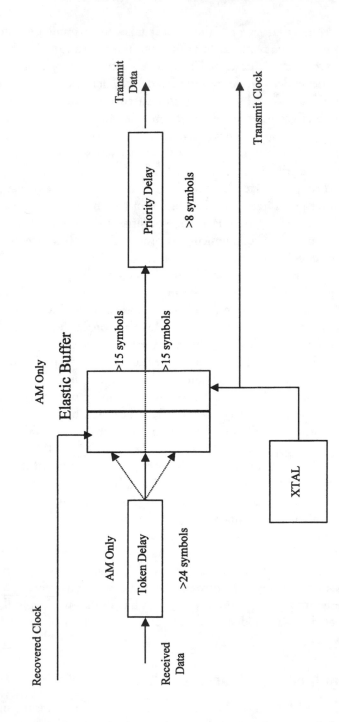

Figure 3.8 Active monitor repeat path delay in a Token Ring station.

signals transmitted by its upstream neighbor. The main components of the transmit portion of the PMC are the transmitter and transmit filter. The main components of the receive portion of the PMC are the receiver equalizer, the receiver PLL, and the data reclock. In addition, the PMC contains the ring access control for enabling phantom drive station insertion as illustrated in Figure 3.9.

Table 3.1 lists the PMC functions and summarizes their role in a Token Ring station.

The PMC specifications ensure that receiving stations from one vendor may receive transmitted signals from another vendor after transmission on a

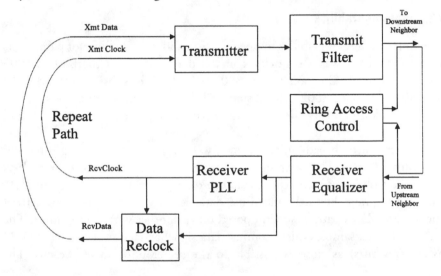

Figure 3.9 Physical media block components.

Table 3.1
PMC Functions

Name	Function
Transmitter	Transmits data synchronized to either the recovered clock (for SM stations) or crystal (for the AM station)
Transmit filter	Reduces EMC radiation from transmit signal
Receiver equalizer	Compensates for received signal frequency dependent cable losses
Receiver PLL	Derives a recovered clock from the received data
Data reclock	Recovers data using the recovered clock
Ring access control	Provides insert current (phantom drive) for station insertion and wire fault diagnosis

specified cabling channel. This chapter describes the parameters of the receiver and the transmitter that are specified in 8802-5:1998 to achieve interoperability. The PLL receiver bandwidth is a key parameter in the design for Token Ring (and was the basis for over 100 papers and presentations in the IEEE 802.5 meetings between 1990 and 1993) [1]. In Token Ring, because all stations are SMs except for the single AM station, all stations except one transmit using the received recovered clock. The stations on the ring form a chain of interacting clock repeater stations that must be modeled together in order to guarantee interoperability.

3.3.1 Interoperability

Guaranteeing interoperability by specification and measurement of parameters at the PHY layer was a major task in the development of the 8802-5:1995 standard. The PMC parameters in 8802-5:1995 differ substantially from the specifications in the 8802-5:1992 standard due to an increased understanding of jitter buildup and its effects on interoperability gained between 1990 and 1993.

Jitter can be broadly divided into two components: sampling jitter and accumulated jitter. The sampling jitter is due to noise in the recovered clock that is best observed using the classic eye diagram, as shown in Figure 3.10. This eye diagram shows the incoming data signal, with a Data_Ones pattern and a Data_Zeros pattern superimposed on each other with timing jitter. The recovered clock samples this incoming data signal at each unit internal (31.25 ns at 16 Mbit/s) as close as possible to the center position of the eye. The

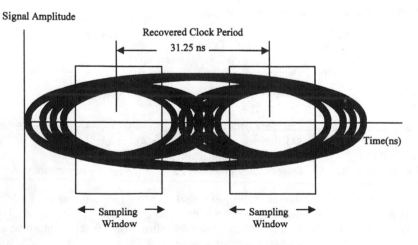

Figure 3.10 Incoming data signal eye diagram.

sampling window designates the range in sampling timing error that would still result in correctly sampling the data, as shown in the eye diagram. Details of the sampling window depend on the minimum signal level, which allows for correct data recovery and the capability to center the sampling clock within the received signal eye.

The second component of jitter, accumulated jitter, is the jitter that accumulates from station to station in the ring. This jitter produces an accumulated frequency error (or data phase error) that shifts both the incoming data signal and the recovered clock. This jitter is described in the *accumulated phase slope* (APS) section (Section 3.3.2). Both jitter types must be analyzed and limited in order to achieve interoperability.

The following statement is the summary requirement for limitation of jitter buildup.

> *Interoperability depends upon the margin between receiver jitter toler-ance (the ability of a receiver to recover data in the presence of noise) and the transmitted APS (how much jitter is transmitted by each station in the chain of PLLs).*

The remaining sections of this chapter will aim to describe the significance of the preceding statement in Token Ring.

3.3.2 Accumulated Phase Slope

APS is a measure of the magnitude of the frequency difference between the crystal clock and the recovered clock; it can be understood by analyzing the waveforms shown in Figure 3.11. Typically, jitter is measured in nanoseconds and the jitter versus time (in units of UI) is plotted in the lower graph in Figure 3.11. The *unit interval* (UI) is equivalent to the recovered clock period, or UI = 31.25 ns for a 16-Mbit/s ring with 32-MHz recovered clock.

The top waveform in Figure 3.11 illustrates the reference clock, which is controlled by the AM station and is usually produced by a crystal oscillator with frequency F0. Waveforms A1 and A2 in Figure 3.11 represent possible timing examples for the PLL recovered clock, as might be observed at a ring station. In these waveforms, it is assumed that the rising edge is used to recover the data. In waveform A1, for a few transitions, the rising edge timing phase has a momentary phase error (defined as timing jitter) for three clock transitions. This could be caused by the received PLL recovered clock noise due to incoming received data noise or data pattern change. In waveform A1, there is no long-term timing jitter since the recovered clock returns to the crystal clock frequency after the short-term disturbances.

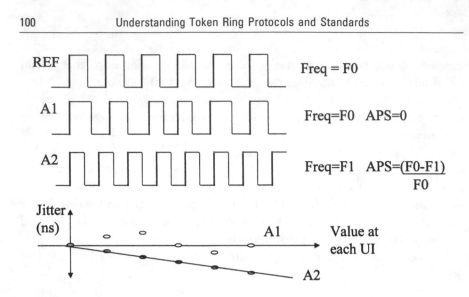

Figure 3.11 Accumulated jitter waveform example.

In waveform A2, the recovered clock phase error is accumulating at a fixed rate at each data transition. This constant difference between transmitted and received clock frequencies results in a fixed timing error per clock transition. This jitter per timing UI can be called the phase slope and represented as ns/UI. From the viewpoint of the frequency domain, the recovered clock waveform A2 has a frequency (f1) while the crystal reference clock has a frequency (f0). In the frequency domain, the normalized frequency difference is (f1 − f0)/f0. Note that the term *APS* is utilized to describe this accumulated jitter increase per timing period (a better term may have been *average jitter slope*).

Consider a specific example for a 16-Mbit/s Token Ring. The AM crystal clock has a frequency, F0, that is 32.00 MHz and a period of 31.25 ns with a 0.01% tolerance. The reference UI or clock period is thus equal to 31.25 ns. Now assume that the timing error (jitter) at each timing edge in waveform A2 is decreasing at a rate of 0.5 ns per pulse. The period of waveform A2 is 30.75 ns with a frequency, f1, equal to 32.52 MHz. The APS of waveform A2 is therefore −0.5 ns/UI, or −0.016 ns/ns. This computation could also have been made using the formula shown in Figure 3.11 as

$$APS = (F0 − F1)/F0 = (32 − 32.52)/32 = −0.016 \text{ ns/ns}$$

The APS is filtered to focus on the long-term frequency error (waveform A2) rather than short-term jitter effects (waveform A1). Thus, the *filtered accumulated phase slope* (FAPS) is specified after filtering by a low-pass filter with frequency F6. Referring to Table 13 in 8802-5:1998, for 16-Mbit/s Token

Ring, F6 is 360 KHz and the absolute value of the maximum value of FAPS is 0.0082 ns/ns which is equivalent to a maximum timing jitter of 0.25 ns per UI (or clock transition).

3.3.3 Receiver Jitter Tolerance

Receiver *jitter tolerance* (JTOL) is the ability of a receiving station to receive error free data in the presence of an APS and other jitter noise sources. This jitter, as shown in waveform A1 in Figure 3.11, can be either for a few cycles or have a much longer time period as in the case of the APS shown in waveform A2. The receiver JTOL can be represented in the frequency domain. The approximate JTOL specification (derived from the IEEE 802.5 standard JTOL values) is plotted, measured as a function of frequency as shown in Figure 3.12.

In Figure 3.12 the *x*-axis is the frequency of the sine wave jitter pattern applied to the receiver. The *y*-axis is the amplitude (zero to peak) of the JTOL, measured in UI. The maximum effective frequency of the jitter pattern is 16 MHz for a 32-MHz recovered clock. For high-frequency jitter, as shown in region C in Figure 3.12, the timing jitter is caused by a combination of noise with little correlation between clock cycles (referred to as uncorrelated jitter) and data pattern-dependent jitter correlated to the data pattern (correlated jitter). This edge jitter is above the PLL bandwidth (BW). The JTOL depends on how well the recovered clock timing edges are centered to the data signal and on the PLL receiver clock noise.

In region B, where the frequency of the jitter is below the PLL BW, the recovered clock will track the timing edge change. This region of the JTOL curve is important for tracking the FAPS described in waveform A2 of Figure 3.11. The value is directly related to the maximum FAPS buildup that can be

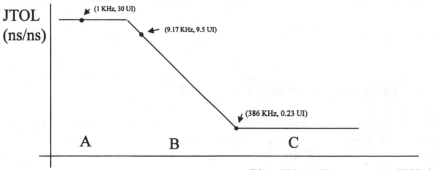

Figure 3.12 JTOL frequency specification.

received without error. The amplitude of the sine-wave jitter times the frequency is a constant in region B. For a jitter amplitude of $B\cos(\omega t)$, the effective maximum jitter slope tolerance is given by $B\omega$, which is equivalent to a square-wave JTOL in ns/ns. For a JTOL of 0.0172 ns/ns for a 16-Mbit/s ring for $f = 9.17$ KHz, this corresponds to $B = 298$ ns or 9.5 UI. Assuming the amplitude times frequency remains constant up to the PLL BW (which is the approximate demarcation point between regions A and B), for a PLL $BW_{3dB} = 382$ KHz, $B = 7.2$ ns or 0.23 UI. Note that region B is directly related to the PLL BW_{3dB} and increasing BW_{3dB} increases the height of the curve in region B. This analysis was focused on the receiver being able to track the incoming jitter. However, one must also look at the effect this station has on the next down-stream station. An ability to be able to receive an incoming signal with large FAPS translates into an ability to transmit that signal with large FAPS, making it harder for the next downstream station to receive the data stream error free.

In an attempt to break this tight coupling between incoming FAPS tolerance and transmitted FAPS, some designs, especially those deployed in active concentrators, will include a two-stage PLL receiver. The first stage has a relatively wide bandwidth to receive whatever jitter the upstream station can transmit within the limits of the standard. The next stage has a very low PLL bandwidth and includes a short elastic buffer. This stage transmits signals with very low FAPS, easily received by the downstream station, and uses its elastic buffer to smooth out the long-term jitter swings.

Finally, in region C, where the jitter frequency is very low, the JTOL is set by the elastic buffer at each station that must accommodate long-term data transition frequency errors. As seen in the prior section, the elastic buffer minimum jitter accommodation is 15 symbols, or 30 clock periods (positive and negative), so the height of region C is 30 UI.

In 8802-5:1992, the JTOL was specified by tolerance to sine-wave jitter as shown previously. In the 8802-5:1995, the JTOL is specified using a square-wave frequency disturbance to better correlate the FAPS buildup that corresponds to region B.

3.3.4 Token Ring Phase Lock Loop Model

The goal of the receiver PLL is to produce a recovered clock where the clock transitions are centered within each data period to allow maximum capability to recover the data. Figure 3.13 illustrates the PLL receiver block diagram showing both the recovered clock and the data pattern. It is assumed that the data is sampled at the rising edge of the recovered clock to determine whether the incoming data is high or low.

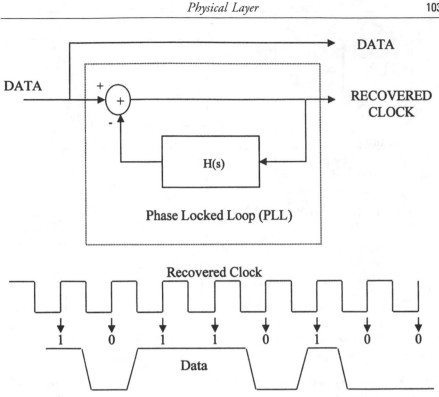

Figure 3.13 PLL block diagram.

Figure 3.14 shows the recovered clock centering in more detail. For an ideal situation, noise in the incoming data timing edges are accommodated by the PLL as is the case in B and C in Figure 3.14. However, a waveform D reference to the recovered clock would result in a data error.

For a 16-Mbit/s data stream the ideal PLL recovered clock is 32 MHz and the effective timing window to recover the data reliably is 31.25 ns or ± 15.625 ns. The lack of centering of the clock to the data is referred to as the recovered clock phase error, and there are generally three components of this timing noise: a fixed static component related to the PLL design tolerances, a noise component (uncorrelated jitter), and a dynamic component related to the incoming data pattern (correlated jitter).

A key feature of clock recovery in Token Ring PLL design is the *dependency of the recovered clock phase upon the incoming data pattern called correlated jitter.* Because the incoming data frequency components of Data_Zeros (16-MHz signal) and Data_Ones (8-MHz signal) are different, the recovered clock phase becomes data-dependent. The receiver equalizer aims to reduce this data dependency phase shift by compensating for the cable loss (which is frequency-dependent), but the effect cannot be reduced to zero across all cable lengths.

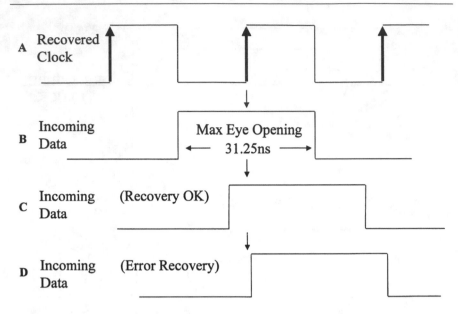

Figure 3.14 PLL recovered clock centering.

This phase change in a station due to a long series of Data_Ones and Data_Zeros is called the *mean filtered correlated jitter* (mFCJ) and is specified by item 4) in Annex C of 8802-5:1995. Figure 3.15 illustrates this phase shift reference to the recovered clock.

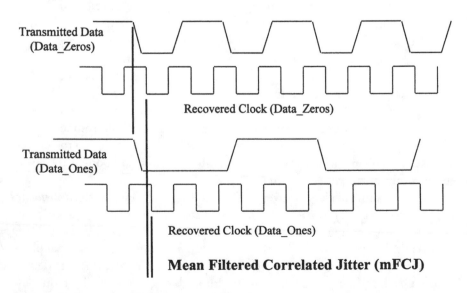

Figure 3.15 Data-dependent correlated jitter.

The Data_Ones and the Data_Zeros are assumed to be perfectly in-phase as transmitted by the AM station. However, the resulting recovered clock exhibits a phase change, which can be on the order of 1 ns for a single station; but because each station transmitter's clock is based on the recovered clock, this correlated pattern jitter buildup will accumulate on successive stations in the ring as discussed in the next section.

A simplified PLL model is shown in Figure 3.16. The top portion of Figure 3.16 illustrates the basic building blocks for the Token Ring PLL, and the lower portion illustrates the component blocks for the PLL. The PLL is referred to as a charge-pump PLL [2]; although the presence of two capacitors result in a third-order loop, the basic analysis is similar to a first-order loop. In Figure 3.16, the INPUT DATA timing phase edge and the RECOVERED CLOCK timing phase edge are subtracted by a phase detector (PHASE DET), which produces an error signal proportional to the phase timing difference of the input edges. A charge pump is utilized that produces either a negative current or a positive current based on the phase timing difference. Capacitor C2 is a small value (less than 500 pf) and is present to reduce the voltage ripple caused by the charge pump, which outputs a specific current proportional to the phase deviation input at each phase detector sampling period. This charge pump has a current gain, *Kd*, measured in Amps/radian. A PLL filter formed by resistor R1 and capacitor C1 sets the bandwidth of the PLL using

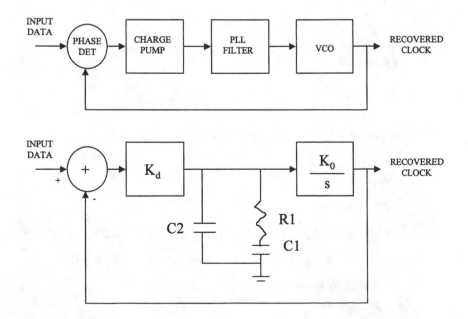

Figure 3.16 Analytical PLL model.

an R/C network. Finally, a voltage controlled oscillator (VCO), or simple integrator, is shown with gain *Ko*, where *Ko* is measured in radian/(Volt-seconds).

Of key importance is the rate of the timing edge comparisons made by the phase detector because the data transitions for a long string of Data_Zeros (16 MHz) is twice the data transitions for a long string of Data_Ones (8 MHz) as seen in Figure 3.15. If the phase detector comparisons are based solely on the number of data edges, then the resultant gain of the PLL (and thus bandwidth) is data-dependent and is twice the value for Data_Zeros as for Data_Ones. While PLLs used in Token Ring before 1992 had this variable dependency with data pattern, later PLLs utilize a "constant gain" approach that enables the phase detector to maintain constant timing edge comparisons independent of the data edge transitions.

Resistor *R*1 sets the closed-loop PLL bandwidth while capacitor *C*1 sets the damping factor, which if less than 15 can create jitter peaking after a long series of cascaded PLLs. The closed-loop PLL transfer function for this PLL is

$$H(s) = \frac{(2 \times D \times W \times s)}{(s^2 + 2 \times D \times W \times s)}$$

where

$K = Ko \times Kd \times R1$ (PLL effective gain)

$W = \sqrt{\dfrac{K}{R1 \times C1}}$ (PLL natural frequency)

$D = \sqrt{\dfrac{K \times R1 \times C1}{4}}$ (PLL effective damping factor)

As an example, at a 16-Mbit/s ring data rate:

$R1 = 800\Omega$

$C1 = 100$ uF

$C2 = 100$ pf

$Kd = 3.0 \times 10^{-5}$ Amp/rad   (for the TI TMS38054)

$Ko = 9.4 \times 10^7$ rad/Volt-sec   (for the TI TMS38054)

$K = 2.26 \times 10^6$ rad/s

$D = 70$

$W = 17$ KHz

As a first approximation, for an effective damping factor greater than 50 ($D > 50$) and frequencies greater than the PLL natural frequency ($s > W$), the PLL equations simplify to those for a first-order filter

$$H(s) = \frac{K}{s + K} = \frac{1}{(1 + s/K)}$$

where the 3-dB BW (BW_{3dB}) is given by

$$BW_{3dB} = \frac{K}{2\pi}$$

Using $K = 2.26$ Mrad/s at 16 Mbit/s (see example in 8802-5:1998 Annex C, equation 15), gives

$$BW_{3dB} = 382 \text{ KHz}$$

Conceptually, the PLL provides a stable recovered clock by filtering data phase deviations above BW_{3dB} and tracking data phase deviations below this frequency. For the 16-Mbit/s Token Ring, the BW_{3dB} is approximately a factor of a hundred less than the recovered clock data rate. This enables the PLL to be insensitive to high-frequency jitter due to the cable noise, data pattern changes, and internal PLL noise. On the other hand, the BW_{3dB} is high enough that phase tracking to long-term data phase variation caused by the FAPS would be tracked. The JTOL measure of capability to follow timing errors due to FAPS is directly proportional to the BW_{3dB}; thus, high BW_{3dB} results in high phase slope JTOL (region B in Figure 3.12).

As an example, Figure 3.17 shows the recovered clock phase versus input data phase. For rapid changes in data transition frequency or period (A), the recovered clock frequency remains the same and does not track the data frequency. For longer time period changes in data transition frequency, the recovered clock frequency will tend to the data transition frequency as seen by (B). The PLL BW_{3dB} is the demarcation point between these two regions.

3.3.5 Receiver Jitter Tolerance Measurement

The receiver specifications are found in 8802-5:1998 (clause 7.2.3.1, Table 26). Two parameters are specified: the measurement of the JTOL over a range of channel attenuation with no additional noise and the measurement of the JTOLX, which includes the presence of additional noise. Note that while this is referred to as JTOL, in reality the specifications apply to the tolerance to a

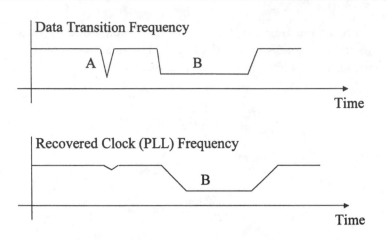

Figure 3.17 Recovered clock phase versus input data.

phase slope and, therefore, more properly should be termed jitter phase slope tolerance. However, because the IEEE 802.5:1998 uses the term *JTOL*, this is the term that will also be used in this book.

The measurement of JTOL can be made using the setup in Figure 3.18 consisting of a two-station Token Ring. The test reference Token Ring adapter is established as the AM station and the *unit under test* (UUT) adapter is the SM station. To establish the test reference adapter as the AM, simply insert the test reference station first into the ring and, after the test reference station has completed the insertion process, insert the UUT adapter onto the ring.

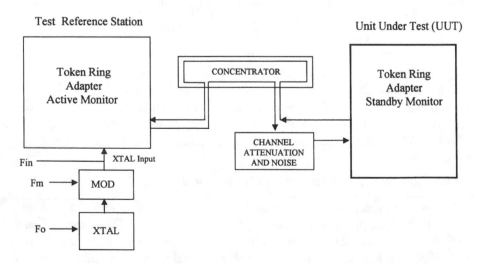

Figure 3.18 Jitter tolerance frequency measurement.

The AM station transmits data from its crystal oscillator, which in the test adapter is actually a modified clock generator that is set at a specific frequency (f0) modulated by the frequency Fm. The AM receiver PLL is modified to achieve a very high BW to guarantee that the test reference station has a higher JTOL than the UUT. Data frames (typically with long sequences of Data_Zeros and Data_Ones) are transmitted by the test adapter, and normal Token Ring MAC protocols (such as FCS checking on circulating transmitted frames) are used to determine whether the transmitted data is recovered without error by the UUT, retransmitted, and then received by the test reference station. Because the test reference adapter has a high-BW PLL with high JTOL, any errors in the frames are assumed to be caused by the UUT adapter.

Rather than utilize a square-wave frequency modulation technique to vary the frequency of the transmitted data clock, most Token Ring testing strategies use a frequency-modulated pulse waveshape so that the positive and negative JTOL can be determined separately. This technique is especially enlightening for phase detectors that have different gains dependent on data pattern. Figure 3.19 shows an example frequency modulation that is input into the test adapter for both square-wave modulation (A) and pulse modulation (B).

In case A, a square wave, centered at 32 MHz, is used to clock the transmitted data in the test adapter. The test adapter transmits a large number

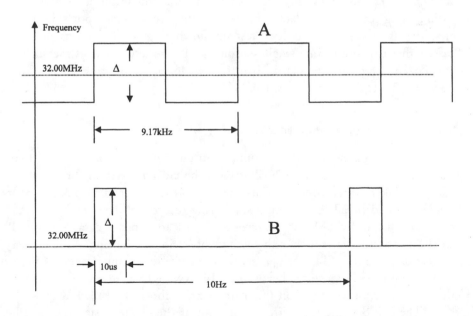

Figure 3.19 Jitter tolerance frequency modulation.

of frames at amplitude Δ at a rate of a 9.17 KHz as specified in Table 26 of 8802-5:1998. The amplitude Δ is increased to the point where data errors occur. This zero-to-peak amplitude divided by the reference frequency, $fo = 32$ MHz, is the JTOL. Actual 150Ω STP or 100Ω UTP cable and attenuators (to simulate transmit voltage level variation) are placed in the lobe cable of the UUT adapter to test for operation at various cable attenuation. Noise can also be injected at the receiver of the unit under test to determine the value for JTOLX according to clause 7.2.4.4 of 8802-5:1998.

In case B, a pulse frequency modulation is used to modify the clock transmitted data. This method allows the measurement of both positive and negative JTOL, which enables a more detailed investigation of the UUT adapter behavior. This measurement must be done for both positive and negative amplitude, and the minimum absolute magnitude of Δ that results in data errors is the measured JTOL value.

The test reference station can be a commercial Token Ring adapter PC card with the crystal oscillator output bypassed to the frequency-modulated input and the receiver PLL BW_{3dB} increased. A second PC card in the test reference station containing an IEEE 488 controller can be connected to a programmable frequency generator for modulating the frequency amplitude input in order to automate the jitter tolerance measurement process. Having associated adapter software that transmits frames and verifies transmit errors while the IEEE 488 controller increases the modulation amplitude is essential to obtaining reproducible data. This test system can be made portable, thus enabling its use as a field unit to test unknown ring configurations. In this case, rather than measuring single-station JTOL, the test system is measuring the Token Ring network JTOL. In this case, for a reliable ring system, the network JTOL should be at least 0.004 ns/ns.

3.3.6   Jitter Accumulation Model

This section explains the jitter accumulation buildup for a chain of multiple stations in the ring. Figure 3.20 illustrates the clocking system for a Token Ring. In Token Ring, a single station, the AM, provides a crystal clock (XTAL) that is retimed by all stations on a single ring (up to 250-receiver PLLs). All other stations, called SM, use the recovered clock to transmit data. Recovered clock jitter accumulates with each subsequent station in the ring. As discussed in Section 3.3.3, the PLL BW_{3dB} serves to separate those components of jitter that accumulate (frequencies below the PLL BW_{3dB}) and those components that are filtered at each station (high frequencies above the PLL BW_{3dB}).

The FAPS is the total jitter at the end of the chain of stations (which could be from different implementations with different lobe cable lengths and

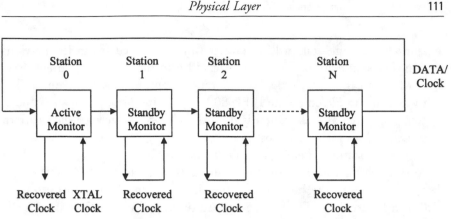

Figure 3.20 Token Ring clocking architecture.

different PLLs at each station). During the development of 8802-5:1998 it was found necessary to develop a model for APS based on specifications and measurements found on a single station to ensure interoperability. In addition, it was found that each Token Ring implementation must be tested at different node count configurations with different cable to experimentally confirm the FAPS calculation buildup and ensure compliance to the specification.

The total jitter transfer function for Token Ring can be developed using the jitter accumulation model shown in Figure 3.21. A similar analysis was performed in the 1960s for a chain of repeaters used in early digital telephone transmission [3,4] with the initial model for Token Ring accumulated jitter buildup published in 1983 [5]. The PLL transfer functions at each node, $H(s)$,

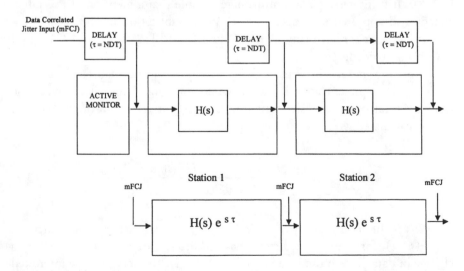

Figure 3.21 Jitter accumulation model.

are simply the single-station PLL transfer functions discussed in the previous section. For simplicity in the model, each PLL in the chain is assumed to be identical (the case where PLLs are different is important for interoperability and was also considered by the IEEE 802.5 Working Group). In this model, a jitter phase step at each station is due to a change in received data pattern from Data_Ones to Data_Zeros.

At each station, there is a small *net data delay time,* NDT = τ, which represents the data delay within the station (usually in the MAC) after the data recovery until the point where the data is retimed for transmission as shown in Figure 3.22. The phase response timing to a phase step is critically related to the data change that created the phase step. Analytically, this can be modeled by replacing the normal jitter transfer function, $H(s)$, by the transfer function $H(s)e^{s\tau}$. If the data repeat path (in the Token Ring MAC) consists of the equivalent of four shift register delays and the recovered clock has period UI, then NDT would equal 4 UI, as shown in Figure 3.22. This delay time is *not* equivalent to the cable delay because cable delay affects both the data and the phase buildup synchronously.

The inclusion of this delay time in the jitter accumulation model for jitter buildup in a chain of repeaters surprisingly results in a potentially large increase in jitter accumulation around the ring. Most Token Ring stations have a nominally small NDT, typically 1 to 2.5 symbols. NDT is kept small to keep the total ring latency time small and to maximize performance as described in Appendix A. The effect of NDT, an increase in accumulated jitter, was first reported by a number of IEEE 802.5 Token Ring developers in 1990. It is interesting and unfortunate to note that a mention of this effect already had appeared unnoticed in a Japanese publication in 1985 [6]. Were this article seen earlier, many of the Token Ring PHY issues in 1991 could have been avoided.

For the model in Figure 3.21, assuming a normalized jitter step (mFCJ = 1) is present at each station when the data change is received, the resultant total system transfer function $H_{ACC}(s)$ for the chain of N PLLs [7] can be calculated as

$$H_{ACC}(s, N) = \frac{H(s) \times e^{sN\tau} \times (H^N(s) \times e^{sN\tau} - 1)}{(H(s)e^{sN\tau} - 1)} \qquad (3.1)$$

where N is the total number of stations. The resulting jitter transfer function, $H_{ACC}(s, N)$, for a normalized step phase in jitter at each station caused by a change in data pattern can be analyzed using (3.1). Readers who are interested in more details of the results of the analysis of this equation are referred to

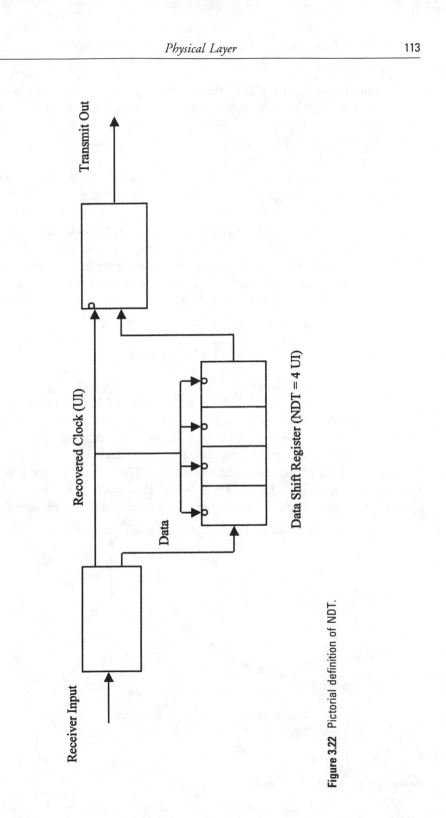

Figure 3.22 Pictorial definition of NDT.

the Compendium of IEEE 802.5 Working Group papers [1]. Experimentally and by analysis, it has been shown that for most implementations FAPS buildup with station count reaches an asymptotic value after about 15 stations. While the total accumulated phase change increases with station count (because each station contributes mFCJ jitter), the FAPS stays constant.

Looking at the effect of NDT from a time reference point, after a station receives a change in data pattern at its receiver PLL, there is a delay before the recovered clock step change of mFCJ begins to occur. The NDT term delays the data pattern, causing the phase step so that the delay in phase step of the clock becomes closer in time to the delay in reception of the data. Therefore, the two terms may act synchronously together to increase the APS. As an approximation, for small NDT and the number of ring stations, $N > 15$, the resultant FAPS is approximately

$$FAPS = \frac{mFCJ \times 2\pi BW_{3dB}}{(1 - NDT \times 2\pi BW_{3dB}} \tag{3.2}$$

Equation (3.2) is plotted in Figure 3.23.

At the point where $NDT = K = 2\pi BW_{3dB}$, the FAPS increases rapidly. For NDT greater than K (to the right of the values plotted in Figure 3.23), FAPS decreases. In this case, where the station MAC delay is much greater than the delay in phase change, the APS from each station does not build up synchronously. To ensure interoperability, the 8802-5:1998 standard specifies that all implementations should operate with NDT small (the left side of the peak point, where $NDT < K$). Otherwise, the combination of alternating

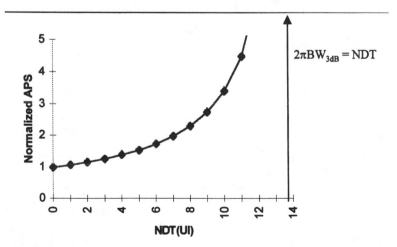

Figure 3.23 Normalized APS versus NDT.

stations, one on each side of the NDT = *K* point, can result in high APS buildup.

Summarizing this analysis, data-dependent phase jitter (where long strings of Data_Ones and Data_Zeros are transmitted in the data pattern) will result in a phase step being produced at each station as the data changes. This phase step and the data change will circulate around the ring. *The relative timing between the phase step change and the data change as they circulate around the ring has a major effect on the build-up of the APS.*

Figure 3.24 summarizes how jitter builds up around the ring for the case with the AM station transmitting data consisting of a large sequence of Data_Ones followed by Data_Zeros and monitoring the recovered clock phase at stations N30 and N100.

In Figure 3.24, note three particular items in the recovered clock phase that have already been discussed. First, the change in phase occurs earlier for station N30 than for station N100 due to the fact that the data reaches station N30 before reaching station N100. Secondly, the resulting total phase difference is 30 mFCJ for station N30 and 100 mFCJ for station N100. Finally, the slope, the change in recovered clock phase versus time (which is the APS), is the same for both the N30 and N100 stations assuming all PLLs are identical.

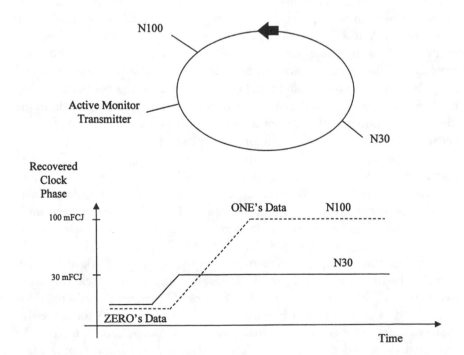

Figure 3.24 Correlated jitter build-up around the ring.

3.3.7 Accumulated Phase Slope Measurement

The measurement of APS for Token Ring requires the establishment of a ring with multiple nodes with each node having a well-controlled and repeatable lobe cable. This measurement should be made on at least one through 15 stations (to observe the APS buildup with station count), with occasional verification at the full 250-station count in a ring. The measurement is based on measuring the recovered clock phase for a well-controlled, repetitive, specific data pattern referenced to the AM's crystal clock phase. Figure 3.25 shows the experimental setup for measurement of the APS.

The measurement procedure is as follows:

The N stations (UUT) are inserted into a Token Ring concentrator bank with minimum cable between concentrators (out-port to in-port on the concentrator), with the selected lobe cable length at each station. Cable lobe lengths and cable type (UTP or STP) are modified to measure the APS as a function of lobe channel attenuation. Typically, all stations are measured with the same cable length. Once results have been measured for one cable length and different station counts, the cable length is modified for all stations.

The pattern generator–driven transmitter is inserted onto the ring. This station should have a compliant transmitter waveform pulse shape and lobe insertion phantom drive signal. The pattern generator should be set to generate a Differential Manchester–coded data pattern including an embedded beacon frame. The beacon frame, when received by the other stations on the ring, force each station to enter beacon repeat mode, where incoming data is repeated from the receiver to the transmitter. This repeat mode is the normal operating mode for most stations in the ring. The data pattern outside the beacon frame can be programmed with different data patterns to determine the phase slope versus correlated data. Code violation effects caused by the SD and ED can also be measured. A typical data pattern is shown in Figure 3.26. This data pattern shows a long series of Data_Ones followed by a long series of Data_Zeros to provide a phase step that accumulates around the ring. The Manchester-encoded beacon frame data pattern is listed in Table 3.2. This pattern must be programmed exactly since the FCS must be correct for the frame to be received by all ring stations correctly.

A time interval analyzer is connected to the recovered clock of the Nth station. The time interval analyzer measures the phase of this clock in relation to the phase of the timing reference XTAL clock. The trigger signal is provided from the data pattern to synchronize the recovered clock data for successive measurements. Alternatively, a digital timing oscilloscope can be used at the recovered clock and the successive timing edges measured and compared to the reference XTAL clock.

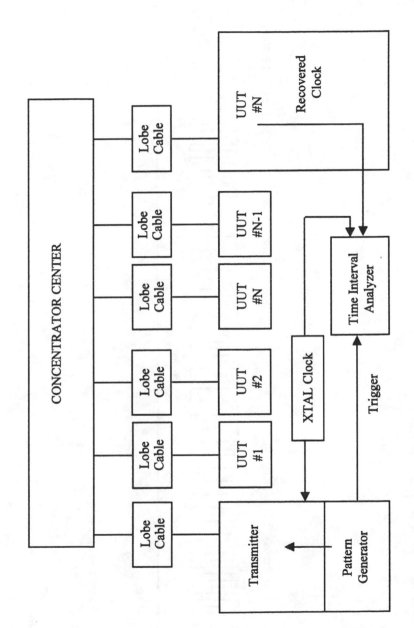

Figure 3.25 APS measurement test setup.

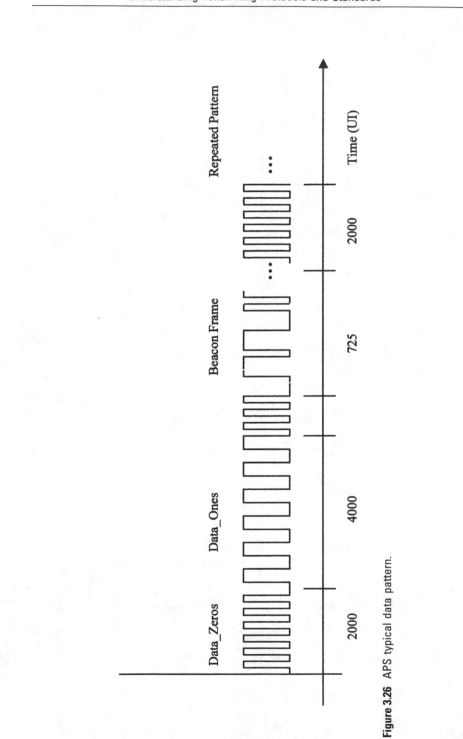

Figure 3.26 APS typical data pattern.

Table 3.2
Beacon Frame Pattern

Hex	Manchester Encoded	Comment
00SD	0101010101010101 1100100011010101	Start Delimiter
1002	0101011010101010 1010101010100101	AC = >10 (Frame) FC =>02
C000	1001010101010101 0101010101010101	DA = >C000 FFFF FFFF Broadcast
FFFF	1001100110011001 1001100110011001	
FFFF	1001100110011001 1001100110011001	
0400	0101010101101010 1010101010101010	SA = >4000 0000 0000
0000	1010101010101010 1010101010101010	
0000	1010101010101010 1010101010101010	
0016	1010101010101010 1010100101100101	MAC Length = >0016
0002	0101010101010101 0101010101011010	Major Vector = >0002 Beacon
0401	1010101010010101 0101010101010110	
0002	1010101010101010 1010101010100101	
0802	0101010110101010 1010101010100101	
0000	0101010101010101 0101010101010101	
0000	0101010101010101 0101010101010101	
0000	0101010101010101 0101010101010101	
060B	0101010101100101 0101010110100110	
C000	0110101010101010 1010101010101010	
0000	1010101010101010 1010101010101010	
50B2	1001011010101010 0101100101011010	FCS Value = >50B2 A37C
A37C	0101101010100110 1001100110010101	
EDFS	1100011100010101 0101010101010101	End Delimiter, FS = 0
0000	0101010101010101 0101010101010101	

The slope of this phase change is the APS, measured in ns/UI, which is converted to ns/ns units in the specification by dividing by 31.25 ns/UI. This APS value can be digitally filtered by the required F6 single pole filter specified in 8802-5:1998 (Table 13) to yield the FAPS. The filtering removes high-frequency phase changes, above the PLL BW_{3dB}, to show only the accumulating jitter component.

Figure 3.27 shows an example of the processed digital oscilloscope phase of the recovered clock versus time (measured in UIs for a 16-Mbit/s Token Ring) for a 15-station ring using 50-m STP lobe cable at each lobe. The cables must be carefully controlled and set up between each run to keep a sense of integrity in the measurement. This control burden and effort cannot be over-stated because the long times required to obtain the data can be wasted if one station has an unknown and unrecorded cable length.

The measurement is typically repeated at least ten times for each data point because of the variability in jitter that can be caused by the polarity of

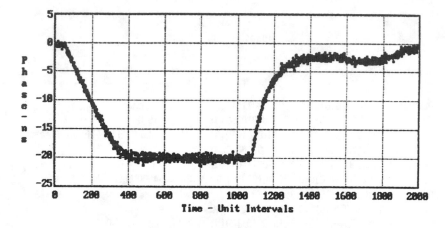

Figure 3.27 APS measurement example for a 15-station ring.

the transmitted Manchester-coded pattern. The polarity of the transmitted Manchester data is randomly established by each receiving station after its insertion, and while the signal polarity does not affect the data decoding, it can affect the PLL recovered clock jitter.

In Figure 3.27, the filtered reference phase error (filtered accumulated pattern jitter—FAPS) of the recovered clock is the vertical axis and the time (measured in UI) is the horizontal axis. A long pattern of Data_Zeros was being received by the fifteenth station until the time 20 UI when the pattern changed to Data_Ones. Then at approximately 1100 UI, the pattern changed back to Data_Zeros. The relative phase error of the recovered clock is measured at each rising edge and subsequently digitally processed with a low-pass filter of 360-KHz bandwidth.

Several parameters can be determined from this single measurement. The total accumulated jitter is approximately 20 ns. The accumulated jitter divided by the number of stations (all identical with identical lobe cables) is $20/15 = 1.3$ ns/station, which is the value of mFCJ under these conditions. The maximum APS is approximately −0.17 ns/UI (−0.0054 ns/ns) for the negative slope and 0.08 ns/UI (0.0026 ns/ns) for the positive slope. The phase change is asymmetric with direction due to the PLL design. This data was taken on the TMS38054 from Texas Instruments, which has a "constant gain" phase detector but still exhibits some asymmetry in the phase slope. The APS data in Figure 3.27 is actually taken four separate times, and thus several separate curves are discernible. Note that the phase, while reasonably consistent past 1200 UI for different sample points, has some variation and does not actually return to zero, the reference phase for 0-UI Data_Zeros pattern. This anomalous behavior is due to hysteresis in the PLL receivers.

The APS example shown in Figure 3.27 meets all accumulated jitter specifications established within the 8802-5:1998 Standard, and these stations give error free operation in a ring environment.

3.4 Interoperability Summary

This chapter has explained the main PHY issues for interoperability in Token Ring. The FAPS specifies the maximum phase slope that can be transmitted by a station. The receiver JTOL specifies how large a phase slope must be received by a station without the station producing data errors. Both FAPS and JTOL increase with PLL BW_{3dB} (in fact, FAPS may increase faster because of NDT effects). To allow for noise and for both correlated and uncorrelated data effects that also reduce the effective jitter tolerance value, a safety margin of a factor of two between receiver JTOL and FAPS is needed.

In summary, *FAPS must be less than two times the JTOL values over all possible lobe cable lengths for best interoperability.*

Table 3.3 summarizes the JTOL and FAPS specifications from 8802-8:1995. The JTOL-APS interoperability equation is not much concern at 4 Mbit/s, where the JTOL/FAPS ratio is large. However, in the case of 4-Mbit/s rings, because the minimum size of the elastic buffer is only ± 3 symbols or ± 6 UI, the total accumulated jitter is the more stringent limiting parameter and the maximum station count is limited to 72 nodes for UTP cable for early Token Ring designs.

As a final test of Token Ring PHY interoperability, specific worst case data patterns in an operational Token Ring can be generated to verify robustness. These data patterns stress the receivers in the ring by providing high phase slope "waves" in the presence of worst case data jitter sequences, such as the start delimiter. The data patterns consist of frames with a long pattern of Data_Ones and a long pattern of Data_Zeros, such that the timing of the Data_Ones signal fills the ring completely from the transmitting station to the AM station. In an example test, 1,000 frames are transmitted with a specific long pattern of Data_Ones and Data_Zeros. Then the number of Data_Ones

Table 3.3
JTOL and FAPS Specifications

Ring Speed	JTOL	FAPS	JTOL/FAPS
16 Mbit/s	0.0172	0.0082	2.1
4 Mbit/s	0.007	0.0012	5.8

and Data_Zeros in the pattern is increased and another 1,000 frames are transmitted, and so forth. At a specific pattern length, dependent on ring size and AM location, frame errors in a marginally operational Token Ring system will occur. These frames have been called the "symptomatic frames" in Token Ring because, while normal ring operation appears acceptable, the presence of these symptomatic frames can rapidly increase the ring frame error rate.[1] Experience has shown that provided the JTOL and FAPS values are met per the 8802-5:1998 specifications, these "symptomatic frames" do not cause ring errors. A ring that cannot circulate these "symptomatic frames" could operate relatively error free in transmitting data until a data sequence resembling the "symptomatic frame" is transmitted. Diagnosis of this problem in the field, with changing data patterns and AMs, is very difficult or impossible.

In retrospect, the reader might ask why Token Ring clocking architecture used the recovered clock for data transmission, thus allowing propagation of accumulated jitter around the ring. Besides not being aware of the significance of the station delay (NDT) impact, there were two principle reasons. The first is that the most obvious alternative, requiring each station to transmit on its own crystal, requires each station to have a buffer to accommodate the difference in crystal frequency between the transmit and receive stations. In addition, a significant buffer would be required between frames, also to accommodate the tolerance of the crystals. Both the required buffer and the additional latency delay built into each station would decrease the maximum performance capability of Token Ring (see Appendix A). A second reason was to provide for increased fault isolation at each node. Where the AM station actually receives its own crystal generated clock in its recovered clock at the receiver, the AM can verify that the frequency is equal to its crystal frequency. In addition, each SM station verifies that the AM station crystal is within the limits defined by the frequency error determination hardware described in clause 5.7.2 of 8802-5:1998 (1.76% at 16 Mbit/s). These checks enable diagnosis of a faulty AM station. In other token systems, such as fiber distributed data interface (FDDI), data is transmitted at each station using that station's crystal clock and, thus, accumulated clock jitter is eliminated. FDDI does not suffer from the accumulated clock jitter problem in Token Ring. The DTR architecture using a point-to-point system for full-duplex operation, with only a single node transmitting on its recovered clock, also does not suffer from an accumulated clock jitter problem but, in fact, enjoys an improved margin at 16 Mbit/s.

Summarizing the key points of this section:

---

1. Use of symptomatic worst case frames is not unique to Token Ring. In 100BASE-TX certain worst case data patterns (based on the inverse scrambling algorithm) can create worst case baseline wander and threshold shifts in the receiver.

- The receiver PLL BW_{3dB} affects both the FAPS and JTOL parameters of Token Ring. Increasing PLL BW_{3dB} increases JTOL and also increases FAPS (slightly greater due to NDT effects).

- The FAPS measurements must be taken on multiple stations to verify accumulation.

- The NDT can have a major increase on FAPS buildup when it is near $2\pi BW_{3dB}$. Stations should aim to keep NDT small in order to minimize FAPS buildup and reduce ring latency to achieve better performance.

- The standard has set JTOL to be more than a factor of 2 greater than FAPS for guaranteed interoperability.

- The elastic buffer size limits the total number of stations on the ring due to accumulated jitter buildup, which limits station count for 4-Mbit/s Token Rings.

References

[1] Carlo, J., "Key Presentations on 802.5 Revised PHY Standard," Nov. 9, 1992, IEEE 802.5 Working Papers available from Alpha Graphics, 1-602-863-0999.

[2] Gardner, F., "Charge-Pump Phase-Lock Loops," *IEEE Trans. Communications,* Vol. COM-28, Nov. 1980, p 1849.

[3] Byrne, C. J., B. J. Karafin, and D. B. Robinson, "Systematic Jitter in a Chain of Digital Regenerators," *Bell System Tech. J.,* Nov. 1963, pp. 2679–2714.

[4] Manley, J. M., "The Generation and Accumulation of Timing Noise in PCM Systems— An Experimental and Theoretical Study," *Bell System Tech. J.,* Mar. 1969, pp. 541–612.

[5] Keller, H. J., H. Meyr, and H. R. Mueller, "Transmission Design Criteria for a Synchronous Token Ring," *J. on Sel. Areas of Communications,* Vol. SAC-1, Nov. 1983, pp. 721–732.

[6] Takasaki, Y., and A. Hori, "Fundamental Study of Dependence of Jitter Accumulation on Signal Path Delay," 1985 Natl. Conv. Rec., IECE Japan Paper 2553, Mar. 1985.

[7] Otis, K., "Jitter Accumulation in Token Ring UTP Networks," key paper in [1].

4

Classic Token Ring

This chapter describes the Classic Token Ring's token passing protocol defined in the ISO/IEC 8802-5:1998 Standard clauses 3, 4, 5, 7, and 8. This standard is normally referred to as "the Standard," except when it is important to identify the exact standard.

The token passing protocol is referred to in this chapter as the "TKP Access protocol" to differentiate it from the full-duplex protocol described in Chapter 5.

The TKP Access protocol provides for the establishment of a LAN connection between the PHY layer, the MAC layer, and the LLC layer.

This chapter's discussion of the TKP Access protocol proceeds as follows:

4.1 History and Design Overview
4.2 Data Structures
4.3 Station Functions
4.4 Support Processes
4.5 Operational Functions
4.6 Error Recovery Functions

4.1 TKP Access Protocol History and Design Overview

This section contains a history and design principle overview of the TKP Access protocol.

4.1.1 TKP Access Protocol History

The TKP Access protocol was developed in the late 1970s and early 1980s as an alternative to the protocols used to control existing Loop network technology and the then-emerging Ethernet network technology (IEEE 802.3), as described in Chapter 1.

The designers of the TKP Access protocol recognized that the Loop and Ethernet technologies failed to handle wiring plant and station failures that often resulted in difficult-to-isolate network outages.

Specifically, the TKP Access protocol provided built-in failure isolation or failure identification mechanisms to handle the following well-known network failures.

- *Attachment failures*: The part of the wiring plant used to attach the station to the network had failures such as shorted data paths and excessive bit error rates due to noise.

- *Streaming station failures*: A station that transmits signal without regard to the network's protocol (e.g., due to a babbling transmitter or continuous transmission of frames) causes the network to become inoperative.

- *Duplicate MAC addresses*: When a station attaches to a network with its MAC address equal to another station's MAC address, difficult-to-isolate failures cause the higher layer protocols to become inoperative or receive messages not intended for the particular higher layer protocol, for example.

- *Station becomes inoperative*: The station is no longer performing the network's protocol due to faults (e.g., stuck transmitters and station buffers which are full), causing the network to be inoperative.

- *Isolation of bit errors on the media*: The media connecting stations is subjected to environmental conditions (e.g., it's a noisy world and the media picks up this noise) and causes bit errors to be introduced at a random rate and frequency.

The TKP Access protocol is designed to handle these well-known failures, thus allowing the network to:

- Accomplish network recovery automatically using the TKP Access Protocol without any customer intervention;
- Accomplish network recovery using intelligent concentrators;
- As a last resort, allow customer intervention.

For more difficult failures, recovery is assisted by information provided by the station, the intelligent concentrator (if present), and network management.

4.1.2   TKP Access Protocol Design Principles

The TKP Access protocol requires each station to support the following functions.

- *Active Monitor's (AM) clocking function:* Provides the crystal for ring clocking.

- *AM's protocol function:* Generates tokens when the lack of frames or tokens is detected, removes frames or priority tokens that circulate the ring more than once, and serves as a starting point of the Neighbor Notification function.

- *Standby Monitor (SM) protocol function:* Checks for the existence and correct operation of the AM.

- *Delineation pattern recognition:* Detection of the beginning and ending of tokens, frames, and abort sequences.

- *Timing:* Provides timers required for the MAC protocols.

- *Access control:* Provides normal and priority token transmit access to the transmission medium as well as protocols to detect token oriented errors.

- *Address recognition:* For the reception (copy) of a frame based on the frame's *destination address* (DA); the IEEE 802.5 station recognizes one unique address (specific address), an all-stations address (broadcast), one or more group addresses, and one or more functional addresses.

- *Code violation/cyclic redundancy check:* Allows detection, counting, and isolation of token and frame errors.

- *Frame control:* Detection of either MAC or LLC frames.

- *Frame status:* Provides the MAC layer with address recognition and frame copy notification.

- *Ring recovery:* Use of the TKP Access Protocol to recover the ring from errors.

The station is also required to handle *attachment* to the network as follows.

1. *Prior* to physically becoming part of the network, the station is required to check the station's portion of the wiring plant by performing the *Lobe media test* (LMT) to ensure the station will not impact the operation of the network.

2. *After* physically becoming part of the network, but *before* becoming part of the network, the station is required to check for the following error conditions to ensure the station will not impact the operation of the network.
   a. Check the station's portion of the wiring plant.
   b. Check for the presence of the station responsible for network clocking and, if not present, provide the necessary clocking.
   c. Ensure the station does not have a MAC address equal to a station already on the ring by performing the duplicate address test function.
   d. Continuously use the TKP Access protocol to check for correct operation of the IEEE 802.5 Token Ring.
   e. Receive network management approval to become an active participant of the network.

3. *After* receiving permission to become part of the network, the station continuously checks for the following types of error conditions:
   a. The station's portion of the wiring plant;
   b. The presence of network clocking and, if not present, invoke the process to select a station to provide ring clocking;
   c. The presence of bit errors in token or frames and reports any of these errors to network management;
   d. Correct operation of the IEEE 802.5 Token Ring protocol.

The TKP Access protocol is divided into the following major areas of responsibility to accomplish the preceding tasks.

1. The *Join* function supports:
   a. Assuring the attachment media is operating at an acceptable bit error rate (LMT);
   b. Detecting the presence or absence of ring clocking (detect monitor presence);
   c. Providing ring clocking if necessary (await new monitor);
   d. Ensuring the station's MAC address is unique on the ring (duplicate address test);
   e. Informing other stations of its presence (Neighbor Notification process);
   f. Requesting permission to become part of the network (request insertion);
   g. Optionally, detecting wiring plant errors (wire fault).

2. The *Transmit* function supports the following:
   a. All stations use *repeat mode* to repeat data on the ring when not transmitting;
   b. Transmission of frames without a token to support the TKP Access protocol ring recovery functions;
   c. Formatting and transmission of frame(s) upon token capture;
   d. Making reservations for future frame transmissions when access is denied;
   e. Recognizing errors requiring termination of frame transmission;
   f. Stripping the frame or frames transmitted from the ring;
   g. Generating normal or priority tokens after frame transmission;
   h. Recognizing lost frame(s).

3. The *Monitor* function supports the following:
   a. All stations use *Neighbor Notification* to notify other stations of its presence and determine the active *upstream neighbor address* (UNA).
   b. The AM station uses *Ring Purge* to provide the ring with a new token when it detects the absence of a token or the presence of a circulating frame or priority token.
   c. All stations use *Claim Token* to detect malfunctions or the absence of an AM and to elect the first or a new AM.
   d. All stations use *Beaconing* to provide a method to locate station or wiring plant failures.

4. The *Error Reporting* function supports the following:
   a. Recognizing and counting token or frame errors;
   b. Recognizing and counting various station or wiring plant errors;
   c. Reporting of any error conditions detected, including the address of the reporting station (source address) and the address of the *nearest active upstream neighbor* (NAUN) to the reporting station to assist network management in the isolation of error conditions.

5. The *Station Interface* function supports the following:
   a. Reports to the MAC sublayer the recognition a frame to be handled by the MAC sublayer;
   b. Reports to the LLC sublayer for recognition of received frames;
   c. Reports to the bridging, routing, and switch layers for recognition of received frames.

6. The *Station Frame Handling* function supports the required frame handling for the MAC layer.

4.2 TKP Access Protocol Data Structures

This section discusses the data structures used by the TKP Access protocol as follows:

- Describes the nomenclature for data as represented in this book;
- Describes the reason for selecting the Manchester encoding scheme;
- Describes the delineation patterns used by the TKP Access protocol;
- Describes the structured messages used by the TKP Access protocol.

4.2.1 Data Nomenclature

The rules for numbering the layout of bits and bytes for control fields and data structures are described in Section 1.3.

In review, all data is transmitted on the ring from high-to-low bit and high-to-low octet order. Thus, the first bit transmitted is bit 0 of octet 0, as illustrated in Figure 4.1.

4.2.2 Manchester-Encoded Data

The Token Ring uses *Manchester-encoded data* as its method of data transmission. This method is utilized for the following reasons:

- Ease of generating structured messages using unique delimiters without requiring bit insertion;
- Improved error detection;
- Lower susceptibility to noise.

The Manchester-encoded data waveforms are covered in Section 3.2.1 and explained in this chapter as it relates to structured messages.

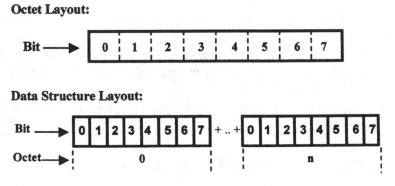

Figure 4.1 Bit and octet layout.

4.2.3 Delineation Patterns for Data Structures

The unique *delineation patterns,* SD and ED, identify the boundaries of the structured messages used by the TKP Access protocol.

4.2.3.1 Starting Delimiter (SD)

The SD is used by the TKP Access protocol to inform the station that a structured message is *starting.* The SD can occur on *any* baud time and, once the beginning of the SD has been recognized, the bit times are defined for the entire structured message until the end of a token, frame, or abort sequence is recognized. The SD is illustrated in Differential Manchester coding by Figure 4.2.

The SD, illustrated in Figure 4.3, is a one-octet field containing a uniquely defined six-bit pattern followed by two bits set to zero indicating the *start* of a token, frame, or abort sequence.

4.2.3.2 Ending Delimiter (ED)

The ED informs the station that a structured message has ended. The ED *always* starts on a multiple of 8 bit times from the *first bit time* of the SD. The ED pattern of "JK1JK1" is recognized on any baud time but, when not on a multiple of 8 bit times from the first bit time of the SD, results in an invalid structured message.

The ED is illustrated in Differential Manchester coding by Figure 4.4.

The ED, illustrated in Figure 4.5, is a one-octet field containing a uniquely defined six-bit pattern and two information bits, the I-bit and the E-bit.

1. The *intermediate frame bit* (I-bit) is used *only* in frames and is set to zero in the last or only frame transmitted by the station. When a station optionally transmits multiple frames, the I-bit is set to one in all frames except the last.

Implementation Note

Even though the I-bit is defined, adapter implementations have made little use of this bit because, in most cases, the performance gain offered by multiple frame transmit is small.

The I-bit is *not used* in the token or abort sequence. The TKP Access protocol does not examine the I-bit.

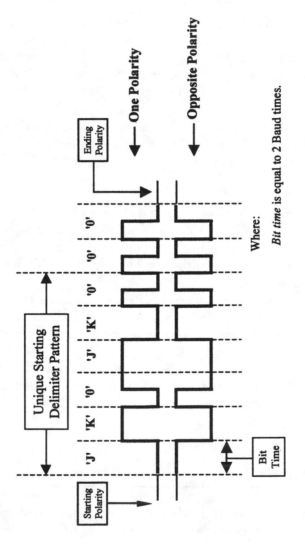

Figure 4.2 SD pattern.

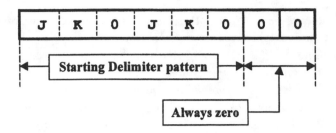

Figure 4.3 SD definitions.

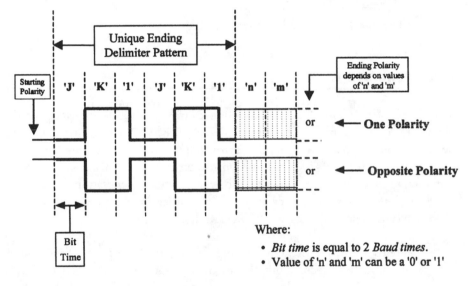

Figure 4.4 ED pattern.

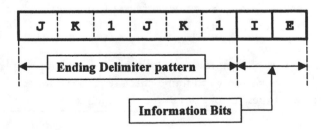

Figure 4.5 ED definitions.

2. The *error detect bit* (E-bit) is set to zero whenever a station transmits a frame or token and is set as follows:
    a. The E-bit is not changed when the frame or token repeated by the station does not contain errors.
    b. The E-bit is *optionally* (see Section 4.3.5) set to one when the token repeated by the station contains a code-violation error. The area of the token covered by the E-bit is illustrated in Figure 4.7.
    c. The E-bit is set to one when the frame repeated by the station contains either a code-violation or a frame check sequence error. The area of the frame covered by the E-bit will be illustrated in Figure 4.9.

The E-bit is not used in the abort sequence.

Implementation Note

The code-violation and frame check sequence error conditions *are not* counted separately by the Standard because the detection of both causes the E-bit to be set to one and may cause the line error counter to be incremented.

However, it is possible for a station to detect a code-violation error, but not a frame check sequence error and vice-versa. The code-violation and frame check sequence error conditions may indicate different types of failures have been detected. The following are two examples.

- Poorly formed Manchester coding patterns may cause a code-violation error to be detected, but the FCS may not detect a bit error. This may indicate an analog type of error (e.g., poor data path between stations).

- Bit-hits may cause an FCS error, but not cause code-violation error. This may indicate a digital type of error (e.g., poorly operating station transmitter or receiver).

Therefore, it might improve error diagnosis if a station implementation recognized the code-violation and frame check sequence error conditions separately and reported these error conditions separately to station management.

4.2.4 TKP Access Protocol Structured Messages

The TKP Access protocol uses three uniquely structured messages to control the operation of the Token Ring, namely:

1. The token (provides authority to transmit data);
2. The frame (used to transmit data to between stations);
3. The abort sequence (used to prematurely terminate frame transmission).

Each ring in a network of Token Rings has its own set of structured messages. Only the data within the LLC frame may be transferred to other IEEE 802.5 Token Rings in the network using bridges or switches or other LAN technologies using routers.

The organization of the structured messages used by the TKP Access protocol is shown in Figure 4.6.

The components of the structured message, illustrated in Figure 4.6, are as follows:

- *SD:* This field indicates the start of a structured message.
- *Null, control, and control and data:* This field has the following forms:
  —*Null:* Only the abort sequence uses this form;
  —*Control:* Only the token uses this form;
  —*Control and data:* Only the frame uses this form.
- *ED:* This field indicates the end of a structured message.
- *Frame fields:* This field is required by the Standard after a frame's ED (see Section 4.2.4.2) and contains the frame status and the IFG fields.

The *Interframe Gap* (IFG) field is used to provide stations enough time between frames, tokens, and aborted frames to prepare to receive the next *structured message.* The size of the *transmitted* IFG is controlled by the speed of the ring and is a minimum of one octet at 4 Mbit/s and five octets at 16 Mbit/s.

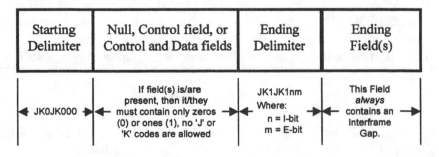

Figure 4.6 Basic structured message.

The size of the IFG after it has been transmitted may be increased or decreased in length by the AM's elastic buffer (see Sections 3.1.2 and 4.3.9.3). However, the AM may *never* decrease the IFG size smaller than one octet for 4 Mbit/s and two octets for 16 Mbit/s. These minimum IFG sizes are required to allow all station implementations to receive back-to-back frames.

IFG Implementation Note

The IFG *size* is an implementation statement of *minimum* requirements. Some implementations can tolerate an IFG smaller than the IFG size specified by the Standard.

4.2.4.1 Token

The *token* is a structured message representing permission for a station to transmit data or, if appropriate, make a reservation to transmit data at a later time. The AM produces the first token on the ring and also replaces the token when the token is lost due to an error. The token is used by the station to gain authority to transmit MAC frames requiring a token for transmission and LLC frames. The station captures the token so that it can transmit data and, upon successful data transmission, normally releases a token to replace the token it captured.

Flow Control Note

The TKP Access protocol has a form of flow control because of its token. Unlike IEEE 802.3 that allows the station to transmit whenever it is ready to transmit and possibly to collide with a message already on the network, the 802.5 Standard compliant station must *wait* for its opportunity to transmit.

Figure 4.7 identifies the token's three required single-octet fields and the location of these fields. A short explanation of each field follows.

The token is three octets in length and consists of the SD, AC, and ED.

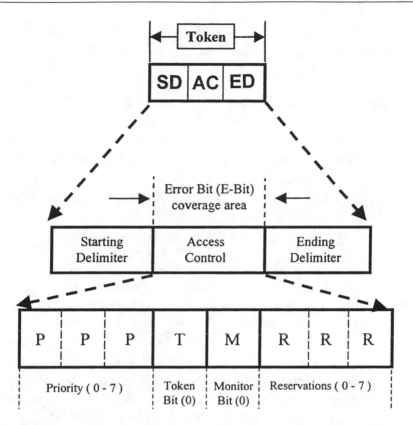

Figure 4.7 Token field descriptions.

Starting Delimiter

The SD is one octet in length and is defined in Section 4.2.3.1.

Access Control (AC)

The token's AC contains the following fields.

The *priority bits* (P-bits) indicate the priority of the token and can have a value of B'000' (0) through B'111' (7). When a station wants to transmit a frame, the frame can be transmitted only if the frame to be transmitted has a priority equal to or greater than the token's priority.

A *token bit* (T-bit) is used to determine which fields follow the AC. As shown in Figure 4.7, the T-bit is a B'0', thus the octet following the AC must be an ED.

Only the AM uses the *monitor bit* (M-bit). The token is always transmitted with the M-bit set to 0. The M-bit is set to 1 by the AM *only* when the M-bit is a 0 *and* the token's access priority is nonzero (B'001' through B'111'). The AM's use of the M-bit is explained in Section 4.4.2.3.

The station uses the value of the *reservation bits* (R-bits) when it wants to transmit a frame but the received token has a priority higher than the frame waiting for transmit. For example, if the token's P-bits are B'011' and the station wants to transmit a frame with an access protocol of B'010', then the station sets the R-bits in the token being repeated as follows.

- If the value of the token's R-bits is less than B'010', then the station sets the R-bits to B'010' (previous lower priority reservation request is lost).

- If the value of the token's R-bits is equal to or greater than B'010', then the station loses its opportunity to make reservations.

Ending Delimiter

The ED, defined in Section 4.2.3.2 and illustrated in Figure 4.8, is a one-octet field containing a uniquely defined six-bit pattern not allowed elsewhere in the token followed by two information bits. The token is invalid if it does not end with this six-bit pattern.

- The first information bit is not used and is always 0.

- The second information bit is the E-bit. The E-bit is set to zero when a station transmits a token. The token's error detection area covered by the E-bit is illustrated in Figure 4.7. The setting of the E-bit is *not changed* when the repeated token does not contain a code-violation error in the AC. The E-bit is set to one when the token contains a code-violation error in the AC.

Interframe Gap

The IFG field provides stations enough time between tokens to detect tokens on a ring whose electrical length is less than a token's length (see Section 4.2.4 for definition).

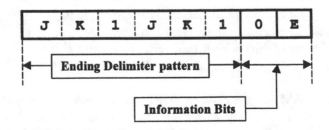

Figure 4.8 Token ending delimiter.

4.2.4.2 Frame

The *frame* is a structured message containing data being transmitted from one station to other station or stations on the ring (including the transmitting station) specified by the DA.

A station having data to transmit prepares a frame containing the FC, DA, SA, an optional RIF, and the frame's information field (frame's data). The station then scans the traffic arriving from the preceding station for an SD indicating the beginning of a token, frame, or abort sequence. When the station recognizes the structured message as a suitable token, it changes the token's AC field by setting the T-bit to B'1' and then transmits the remainder of the frame immediately following the AC field.

Maximum Frame Size

The Standard specifies the maximum frame size for both MAC and LLC frames by defining the maximum *time* allowed to transmit the frame defined in Section 4.2.4.2. The maximum time that can be used to transmit a frame is 9.1 ms and is calculated using the equation

$$\{[\text{Timer, Valid Transmission}] - [\text{Station Delay} + \text{Ring Delay}]\}$$

The components of this calculation are as follows.

- The minimum value of the "Timer, Valid Transmission" is 10 ms.
- The total value of the "Station delay + Ring Delay" is 0.9 ms.

Thus, the maximum time that a frame can be transmitted is 9.1 ms, which the Standard translates into a maximum frame size of 4,550 octets at 4 Mbit/s and 18,200 octets at 16 Mbit/s.

Technical Note About Frame Size

The 9.1-ms time allows a frame size greater than the cited lengths, but these numbers were the limits agreed to by the IEEE 802.5 Committee.

Maximum Frame Size Note

In practice, the maximum frame size is dependent on factors other than station count and ring length delay. Some examples are bridges, routers, and switches, which often support maximum frame sizes of 4,096 octets or less and limit frame sizes going between Token Ring networks, and station device driver implementations and higher layer protocols (e.g., TCP/IP and NetBEUI), which may also limit frame sizes.

Frame Organization

Figure 4.9 illustrates the frame on the ring's data path and identifies the required frame fields and the location of these fields.

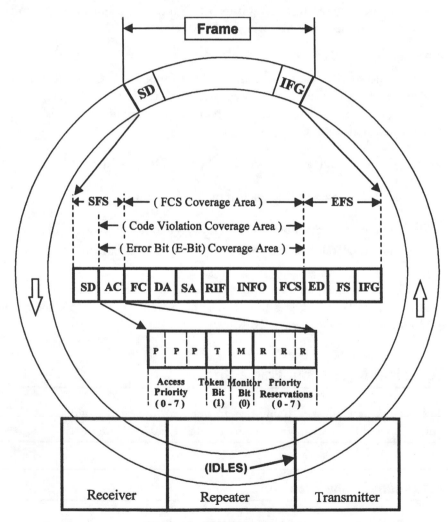

Figure 4.9 Frame field descriptions.

The frame consists of the following fields.

- *Starting frame sequence* (SFS) consisting of the SD and AC;
- *Frame control field* (FC);
- *Destination address* (DA);
- *Source address field* (SA);
- *Routing information field* (RIF)—optional;
- *Information field* (I-field);
- *Frame check sequence field* (FCS);
- *Ending frame sequence* (EFS) consisting of the ED, *frame status field* (FS), and IFG.

These fields are summarized as follows.

Starting Delimiter

The SD is defined in Section 4.2.3.1.

Access Control

The frame's AC contains the following fields.

The *P-bits* indicate the priority of the frame and can have a value of B'000' (0) through B'111' (7).

The *T-bit* is used to determine the fields following the AC. The T-bit, as shown in Figure 4.9, is a B'1' and the fields following the AC make up a frame.

Only the AM uses the *M-bit*. The M-bit is always transmitted in a frame as B'0' and is set to B'1' when repeated by the AM and the T-bit is B'1' (frame). The AM's use of the M-bit is explained in Section 4.4.2.3.

The station sets the frame's *R-bits only* when it has a frame waiting for transmit that has a priority greater than the current value of the token's R-bits. For example, if the station wants to transmit a frame with an access protocol of B'010', then it sets the R-bits in the frame being repeated as follows.

- If the value of the frame's R-bits is less-than B'010', then the station sets the R-bits to B'010' (previous lower priority reservation request is lost).
- If the value of the frame's R-bits is equal to or greater than B'010', then the station loses its opportunity to make reservations.

Frame Control

The FC is one octet in length and contains a two-bit frame type field (bits 0 and 1) and a six-bit control field (bits 2 through 7). The FC is illustrated in Figure 4.10. Following is a summary of supported frame types.

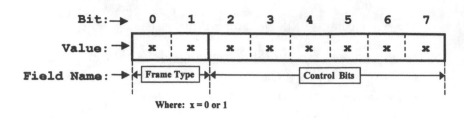

Figure 4.10 FC organization.

1. The frame is a MAC frame if the *frame type* bits are equal to B'00', as shown in Figure 4.11.

   When the frame type bits are B'00', the frame control bits 2 though 7 (B'ZZ-ZZZZ') and the receive treatment provided for the frame are:

   a. B'00-0000': This MAC frame is copied whenever a station has receive buffers available. Thus, it is possible for this type of frame not to be received by a station due to the lack of receive buffers.
   b. B'00-0001': This MAC frame is copied whenever a station has receive buffers available. Thus, it is possible for this type of frame not to be received by a station due to the lack of receive buffers.

B'00-0001' Standards and Implementation Note

The original intention of this value was to allow implementations to change which MAC frames required special receive handling after the IEEE 802.5 Standard had been completed without effecting the TKP Access Protocol. The rationale for which MAC frames actually needed special receive handling was not well understood when the ISO/IEC 8802-5:1985 Standard was generated.

Furthermore, this function was not clearly defined in the IEEE 802.5 Standards dated 1985 through 1992 and was misinterpreted by some implementations. This oversight was corrected by the Standard by allowing the following interpretations.

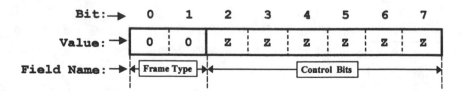

Figure 4.11 Frame control for MAC frames.

Implementations are allowed to handle this frame control type as a frame not requiring special receive handling.

Implementations are allowed to handle this frame control type using special receive buffers when its normal receive buffers are full (not available).

To allow these implementations to coexist, the Clause 3 MAC transmit/ receive tables allow some MAC frames to be transmitted/received with a frame control value of either B'00-0000' or B'00-0001'.

---

   c. B'00-0010' through B'11-1111': These MAC frames require the station to make a best effort to copy. Some implementations handle these frame control types using special receive buffers when the station's normal receive buffers are full (not available). The values assigned by the Standard are as follows.

   (1) B'00-0010': The MAC frame is a Beacon MAC frame.

   (2) B'00-0011': The MAC frame is a Claim Token MAC frame.

   (3) B'00-0100': The MAC frame is a Ring Purge MAC frame.

   (4) B'00-0101': The MAC frame is an Active Monitor Present (AMP) MAC frame.

   (5) B'00-0110': The MAC frame is a Standby Monitor Present (SMP) MAC frame.

   All other values are reserved.

2. The frame is an LLC frame if the *frame type* bits are equal to B'01', as shown in Figure 4.12.

   The LLC frame control bits 2 though 7 (B'ZZZ-ZZZ') are defined as 'rrr-YYY' and have the following meanings.

   a. The "rrr" bits are reserved for future standardization, are transmitted as zero, and may be ignored upon reception.

   b. The "YYY" bits are the LLC frame P-bits and carry the frame's priority assigned by the source LLC entity to the destination LLC entity.

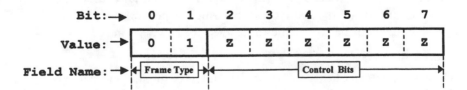

Figure 4.12 Frame control for LLC frames.

The TKP Access protocol does not use the LLC frame priority but may be used by forwarding functions (bridges, routers, and switches) to expedite important frame traffic.

3. Frame types B'10' and B'11' are defined as *undefined frame types* and reserved for future expansion. If these values are defined in the future, the frames using the newly defined frame type *must* follow the rules for undefined frame types specified in the Standard's Clause 3.

    The intention of these rules is to assure that newly defined MAC frame types do not:

    a. Prevent the currently defined TKP Access protocol from operating correctly;

    b. Put new requirements upon a station that supports the TKP Access protocol defined in the Standard, but not the new frame type.

Destination Address

The DA is a six-octet field containing the address of the station that should copy the frame if receive buffers are available. The TKP Access protocol does not allow stations that are not addressed from copying this frame and setting the frame status field's A-bits and C-bits.

Implementation Note

Some product implementations exist that allow all frames to be copied, but these are ring diagnostic tools that do not follow the architecture defined in the Standard. For example, in some cases these "copy-all-frame" stations do not participate in the Neighbor Notification process and thus their presence on the ring cannot be ascertained. Network administration personnel that examine data for performance and service tasks normally use these tools.

High-security installations must take actions to prevent this type of tool from being put on the ring without the knowledge of the network administrator.

The DA may or may not be on the same ring as the station that transmitted the frame. Bridges, routers, or switches can deliver frames with a DA not on the same ring as the transmitter to other rings or networks.

The DA can be any of the forms described in Section 4.3.4.

Source Address

The SA is a six-octet field containing the address of the station used to transmit the frame. It may be the individual address (see Section 4.3.4.2) of the station or the address of the PDU being forward by a bridge, router, or switch.

The high-order bit of addresses normally defines whether the address is an individual or group. However, this definition has no meaning in the SA because bits 1 through 47 always contain the individual address used to transmit and strip the frame. Therefore, the high-order bit of the SA (bit 0) is defined as the *routing information indicator bit* (RII-bit) and is used to indicate the absence or presence of the optional RIF. The RIF is not present if the RII-bit is a zero and is present if the RII-bit is a one.

Routing Information

The *optional* RIF is an even-length field with a minimum of two octets and a maximum of 30 octets that contains *routing control information* used by the source routing function in bridges or switches. The RIF controls the path through the network when a station on one ring transmits a frame to a station on another ring.

ISO/IEC 8802-5 does not use the RIF shown in Figure 4.13 but controls its location within the frame (see Figure 4.9) *and* how to determine its *length* (LTH) so the station can determine where the frame's I-field begins.

The remainder of the explanation of the source routing RIF defines only the format of the RIF; the reader is referred to Annex C of the ISO/IEC 10038 Standard for a detailed explanation of source routing.

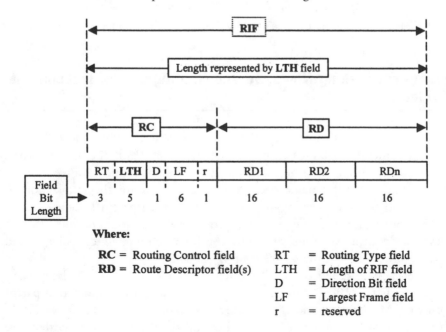

Where:

RC = Routing Control field   RT   = Routing Type field
RD = Route Descriptor field(s)   LTH = Length of RIF field
                                     D   = Direction Bit field
                                     LF   = Largest Frame field
                                     r   = reserved

Figure 4.13 RIF structure. (*Source:* ISO/IEC 10038:1993.)

The format of the RIF shown in Figure 4.13 is controlled by the ISO/IEC 10038 Standard (IEEE 802.1D). Each field has the following general meaning.

RT: Indicates the type of frame forwarding to be done by the bridge through the network:

0xx = The RIF contains a specifically routed frame as per the contents of the RD fields.
10x = The RIF contains a *all routes explorer* (ARE) frame.
11x = The RIF contains a *spanning tree explorer* (STE) frame.

LTH: Indicates the length of the RIF in octets (2 through 30).
D: Indicates the *direction* of frame travel through the network specified by the RD fields:

0 = The frame follows the *forward* route as identified by the RD fields (RD1, RD2, . . . , RDn).
1 = The frame follows the *reverse* route as identified by the RD fields (RDn, . . . , RD2, RD1).

LF: Indicates the maximum length of the frame's information field.
r: Indicates this bit is *reserved.*
RD: Indicates the route to be used by the frame when traversing the network.

Frame Forwarding Notes: Source Routing versus Transparent Bridging

Even though the source routing scheme is more efficient when controlling the path of frames through the network than its Ethernet (IEEE 802.3) counterpart *transparent bridging,* Ethernet (IEEE 802.3) does not support source routing. Ethernet chose this scheme because it did not require the end-station to understand frame forwarding. Those in favor of transparent bridging argued source routing forced an unnecessary burden on the end-station, thus Ethernet never made the move to support source routing (even though technology could have easily supported the source routing scheme).

Because of Ethernet's lack of source routing support, the Bridging Standard (ISO/IEC 10038) was originally written *only* to support transparent bridging. Much later a bridging scheme called *source routing transparent* (SRT) bridging allowing both schemes to be handled by bridges was written (see Annex C of ISO/IEC 10038:1993). This lack of an early description of SRT in the Bridging Standard caused the efficient source routing scheme not to be accepted as the routing scheme of choice. Thus, source routing has been relegated to those networks technologies supporting source routing (ANSI X3T9 and IEEE 802.5 Token Rings).

It should be noted that with the advent of network switches, the lack of an end-to-end frame addressing scheme such as source routing requires implementations to have expensive fast look-up schemes such as *content addressable memory* (CAM). The size of the CAM determines how often transparent bridges have to re-establish end-to-end addressing and therefore has significant impact upon the switches throughput. Source routing on the other hand does not require the switch to have an internal CAM because the forwarding information is contained within the frame's RIF.

Hop Count Change Note

Originally the ISO/IEC 8802-5:1988 Standard limited the number of hops defined by the forwarding information field to seven, thus limiting the RIF to a maximum of 16 octets. The seven-hop limit, which was artificial as far as some implementations were concerned, was set as a reasonable number of bridges to traverse in the network for performance reasons. Even so, the 1988 Standard defined the forwarding information field as shown in Figure 4.13 that allows up to 14 hops. Some implementations, fearing that this hop count would be increased, supported the full 14-hop count supported by the RIF.

As hierarchical networks grew in size, this limit was abandoned in favor of the 14 hops supported by the RIF definition to allow more flexibility in network design.

Information Field

The location of the I-field is shown in Figure 4.9. The minimum I-field is at least four octets for MAC frames and zero or more octets for LLC frames. The exact minimum I-field length for MAC frames is specified by ISO/IEC 8802-5:1998 and LLC frames is specified by ISO/IEC 8802-2:1994.

The maximum I-field for both MAC and LLC frames is a function of ring speed and is equal to

$$\{[\text{Maximum Frame Size}] - [\text{Frame Overhead}]\}$$

where the frame overhead is 22 to 52 at 4 Mbit/s or 26 to 56 octets at 16 Mbit/s and includes the following fields (with associated sizes):

SD (1 octet);
AC (1 octet);
FC (1 octet);
DA (6 octets);
SA (6 octets);
RIF (0 to 30 octets);
FCS (4 octets);
ED (1 octet);
FS (1 octet);
IFG (1 octet at 4 Mbit/s, 5 octets at 16 Mbit/s).

Using the Standard's maximum frame sizes stated earlier, the maximum I-field size at a ring speed of 4 Mbit/s is 4,528 octets without a RIF and 4,498 octets with a maximum 30-octet RIF and at a ring speed of 16 Mbit/s is 18,174 octets without a RIF and 18,144 octets with a maximum 30-octet RIF. An implementation may support any maximum I-field sizes as long as these maximums are not exceeded.

Frame Check Sequence

The FCS is a *cyclic redundancy check* (CRC) that allows the station to detect bit errors in the frame.

The location and the error detection coverage area of the FCS is shown in Figure 4.9. The calculation of the FCS value *does not* include the frame's AC because this field can be changed when repeated by a station since the AM may change the M-bit and *any* station may change the R-bits.

The FCS contains a 32-bit sequence based on a standard generator polynomial of degree 32 as specified by the Standard's Clause 3. This field is transmitted commencing with the coefficient of the highest term.

End Delimiter

The location of the ED is shown in Figure 4.9. The ED, defined in Section 4.2.3.2 and illustrated in Figure 4.14, is a one-octet field containing a uniquely

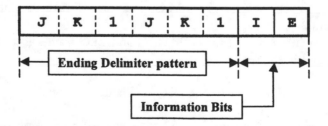

Figure 4.14 Frame ED definitions.

defined six-bit pattern not allowed elsewhere in the frame and two information bits. The frame is invalid if it does not end with this six-bit pattern.

The first information bit is the I-bit. The I-bit is set to zero in the last or only frame transmitted by a station. The I-bit is set to one when a station optionally transmits multiple frames in all frames except the last frame.

The I-bit is not examined by the TKP Access protocol MAC layer but could be used by a receiving station to prepare to receive back-to-back frames.

The second information bit is the E-bit, which is set to zero when a station transmits a frame.

The setting of the E-bit is *not changed* if the frame repeated by the station does not contain a code-violation or frame check sequence error. The E-bit is *set to one* if the frame repeated by the station does contain a code-violation or frame check sequence error. The area of the frame covered by the E-bit is illustrated in Figure 4.9.

The E-bit is used by the TKP Access protocol MAC layer to support error reporting as explained in Section 4.3.7.

Frame Status

The location of the FS is shown in Figure 4.9. The FS field, defined by Figure 4.15, is a one-octet file containing the address recognized bits, frame copied bits, and reserved bits.

The address-recognized bits (A-bits) and frame-copied (C-bits) bits, which are duplicated for reliability, are set as follows.

- Both A-bits are set to 0 when a station transmits the frame. Both A-bits are set to 1 when a station recognizes a valid frame (no errors) with a DA equal to any one of its addresses used to recognize frames.

- Both of the C-bits are set to 0 when a station transmits a frame. Both of the C-bits are set to 1 when a station recognizes a valid frame (no errors) with a DA equal to any one of the its addresses used to recognize frames *and* the station copied the frame.

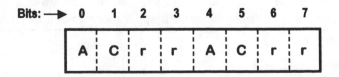

Figure 4.15 FS definitions.

A-bits and C-bits Discussion

The A-bits and C-bits are the subject of much controversy.

The TKP Access protocol uses the A-bits and C-bits for a limited set of MAC functions. The Standard identifies *only* MAC frames when examining these bits (see Appendix D).

The A-bits and C-bits were not to be examined in LLC frames, although their indication was provided to the receiving station's higher layer application software for diagnostic purposes. Higher layer protocols were not supposed to use these bits for *any* purpose since their meaning can be convoluted by bridges. However, the best of intentions have not always been followed. Early bridge implementations used the setting of the A-bits and C-bits to signify whether a station was present on the ring. This resulted in interoperability issues when using bridging in Token Ring. The issue of whether bridges should set the A-bits and C-bits created difficulties with application protocols because these application protocols followed specific bridge implementation interpretation of the A-bits and C-bits. The resulting interoperability issues caused significant problems.

- *A-bit arguments:* It was argued that the A-bits should *only* be set when the bridge recognized the DA. It was also argued the A-bits should also be set *any* time the bridge recognizes a frame as requiring forwarding regardless of the DA.

- *C-bit arguments:* It was argued that the C-bits should be set *any time* the frame was copied even though the bridge did not recognize the DA. It was also argued that the C-bits should *only* be set when the bridge recognized the DA.

- *Resulting conflicts:* This resulted in different implementations of bridges. For example, some bridges set the C-bits, but not the A-bits; some bridges set the A-bits and C-bits; and some bridges set neither the A-bits nor C-bits.

- *Bottom line:* The authors have personally spent countless hours discussing with application layer designers (any level above the MAC layer) why the A-bits and C-bits should not be used and, when they were used, resolving issues with applications caused by their use.

The *reserved* bits (r) are for future standardization and are transmitted as zeros, ignored upon reception, and repeated as received.

Interframe Gap

The Standard requires the IFG field after a frame to provide stations enough time to detect the next structured message (see Section 4.2.4 for a definition). An example for the use of the IFG is back-to-back frames with the same DA.

4.2.4.3 Frame Recognition

Each station on the ring scans each frame at the same time it is repeating the frame, and when the DA matches any of its addresses, the station copies the frame. As it is copying this frame, the station examines the FC byte. If the FC-bits 0 and 1 are equal to B'00', the MAC layer is addressed and handled by the TKP Access protocol MAC Protocol. If the FC-bits 0 and 1 are equal to B'01', the LLC layer is addressed and the frame is passed across the appropriate interface.

This section defines the LLC and MAC frame formats defined by the Standard.

MAC Frame Format

The structure of all MAC frames is shown in Figure 4.16, including the *FC field*, which defines the frame as a MAC frame, and the *vector*, which defines the meaning of the MAC frame. Most, but not all, MAC frame vectors have subvector(s). The structure and use of each of these vectors is explained next.

FC Field

The FC field for the MAC frame has bits 0 and 1 equal to B'00'.

Vector

The *vector*, which identifies the function of a MAC frame, consists of *vector length* (VL); *vector class* (VC); *vector identifier* (VI); and, if present, *subvectors* (SV). The Standard permits only *one* vector in a MAC frame.

The *VL field* is 16 bits in length and contains the vector length in octets including the VL field, VC field, VI field, and all SV (if any). Because only *one* vector is permitted in a MAC frame, it is also the length of the MAC frame's information field.

The *VC field* is 8 bits in length and is divided into two 4-bit fields. Bits 0 through 3 define the MAC frame's *destination class* and bits 4 through 7 define the MAC frame's *source class* as shown in Figure 4.16.

When any layer receives a MAC frame, both the destination *and* the source class must be examined to ensure the MAC frame is valid for the receiving entity. If the destination class and source class are invalid for the receiving entity, the receiving entity takes the following actions.

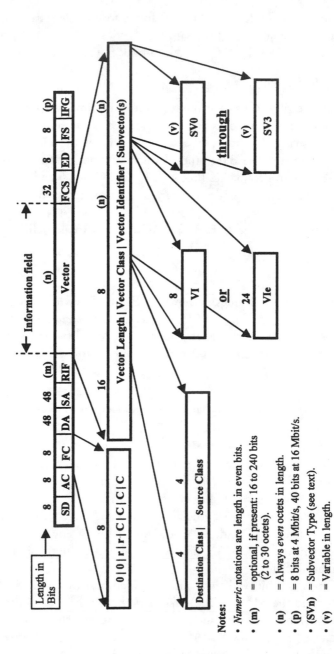

Figure 4.16 General MAC frame format.

- If the destination class is zero (MAC sublayer) receives a MAC frame with an invalid combination of destination class and source class, the frame is handled as follows. If the source class is zero, the MAC frame is ignored and discarded. It is assumed that all MAC frames transmitted by the MAC sublayer are valid; therefore, the MAC sublayer does not handle a negative response.

- If the source class is nonzero, the MAC frame is ignored and discarded and the MAC sublayer responds with a negative response indicating "Inappropriate Source Class."

- If the destination class is nonzero, then a higher layer function receives this MAC frame. The Standard does not specify what action should be taken for this class of MAC frame.

Figure 4.17 illustrates the concept of *destination* and *source class* data flows specified by the Standard (MAC sublayer) and the layers above MAC sublayer (higher layer).

An explanation of the MAC frame flows shown in Figure 4.17 follows.

MAC frame flow 1 uses MAC frames that flow between MAC sublayers in either direction. This flow is used *exclusively* by the MAC sublayer to support the TKP Access Protocol.

MAC frame flow 2 uses MAC frames that flow between the higher layer and the MAC sublayer or between the MAC sublayer and the higher layer. Examples of this flow are as follows.

- The MAC sublayer provides information to the higher layers. Examples of MAC frames using this flow include unsolicited error data from the MAC sublayer or solicited station information from the MAC sublayer as the result of a higher layer request.

- A higher layer requests information from the MAC sublayer. An example of MAC frames using this flow includes requests for MAC sublayer station information.

MAC frame flow 3 is used for communication between the higher layers in either direction. This flow is used *exclusively* by the higher layers and is not defined or used by the Standard. An example of this flow is the two-way communication between the bridge and network management applications.

The *vector classes* used by the TKP Access protocol are as follows.

- B'0000': This value defines the MAC sublayer and can be either the destination or source of a MAC frame. Frames with this *destination*

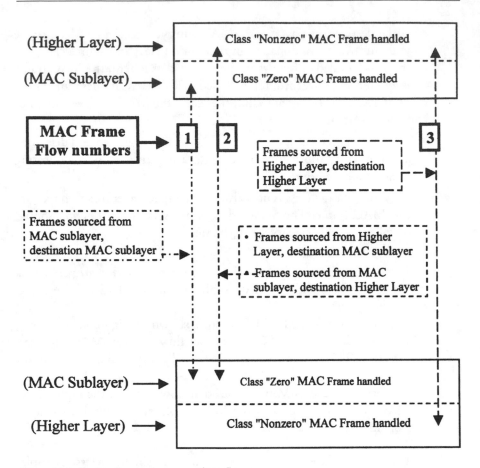

Figure 4.17 Destination and source class flows.

class are handled by the MAC sublayer if recognized, ignored if not recognized, and may result in a negative response being sent to the sourcing station. Frames with this *source* class are transmitted *only* by the MAC sublayer and are handled by the MAC sublayer if recognized and are ignored if not recognized (a response is *never* sent to a source class of B'0000').

- B'0100': This value defines the configuration report server, which is a higher layer application function used by network management.
- B'0101': This value defines the ring parameter server, which is a higher layer application function used by network management.
- B'0110': This value defines the ring error monitor, which is a higher layer application function used by network management.

- B'1000': This value defines the network management, which is a higher layer application function.

The *vector identifier* is used to identify the purpose of the MAC frame. The rules for each vector identifier recognized by the Standard follows.

There are two forms of vector identifier, VI and VIe. The vector identifier, VI, is *one* octet in length and has a value of X'00' through X'FE'. The vector identifiers used or reserved by the Standard are defined in Appendix D. The undefined VI vector codes with a destination or source class of zero (MAC sublayer) are reserved for future expansion.

Standardization Notes

Some MAC frames were used in early implementations and are identified by the Standard as reserved to prevent stations supporting these MAC frames from being nonconformant and new function standardization from using these VI values and causing conflicts between old and new implementations.

For example, the transmit forward function, invented by Ken Wilson (co-author of this book), is defined in Appendix E because of its unique network isolation capabilities. Even though early implementations supported the MAC sublayer's transmit forward function, later decisions made by the IEEE 802.1d Committee (Bridging Standards group) caused its use between multiple rings to be dropped. However, the transmit forward concept was implemented at the LLC layer by some network management implementations.

MAC frames using the *vector identifier extended* (VIe) are not specified by the Standard, but the concept of a VIe is specified for future expansion. No implementations known by the authors use this feature. VIe is *three* octets in length, has a value of X'FF 00 01' through X'FF FF FF', and is defined as shown in Figure 4.18.

Subvector

All data within the vector is contained in subvectors. Vectors have required and optional SVs as defined in Appendix D. Each subvector conveys *one* type of information. The Standard requires the subvectors to be recognized in *any* order.

The components of the SV are *subvector length* (SVL), *subvector identifier* (SVI), *subvector length extended* (SVLe), *subvector identifier extended* (SVIe),

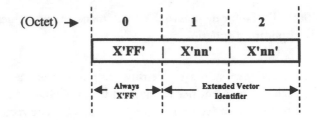

Figure 4.18 VIe definition.

and *subvector value* (SVV). There are four forms of SVs because of the extension process for the subvectors' length and identifier. The extension feature for subvectors follows a complex set of rules, so in order to simplify the understanding of these rules, this book defines four types of SVs: SV0, SV1, SV2, and SV3 as follows.

- SV0 is defined when the value of SVL is one octet in length and has a value of X'04' through X'FE' and the value of SVI is one octet and has a value of less-than X'FF'. SV0 has an SVL field containing the SVL field, which contains the length of SVL, SVI, and SVV. This SV is illustrated in Figure 4.19.

   The subvector SV0 is a minimum of 4 octets in length and contains data associated with the vector identifier. SV0, illustrated by Figure 4.19, contains three fields: the SVL, the SVI, and the SVV.

   This form of SV supports the TKP Access protocol to support an SVI value of less than 255 *and* an SVV length of less than 255 octets. This subvector form is used by most of the MAC frames used by the Standard (see Appendix D).

- SV1 is defined when the value of SVL is one octet in length and has a value of less than X'FF' and the value of SVI is one octet and has a value of X'FF'. The SV1 has an SVL field containing the length of SVL, SVI, SVIe, and SVV. This SV is illustrated by Figure 4.20.

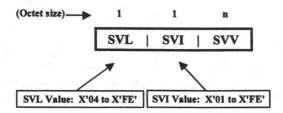

Figure 4.19 SV0 definition.

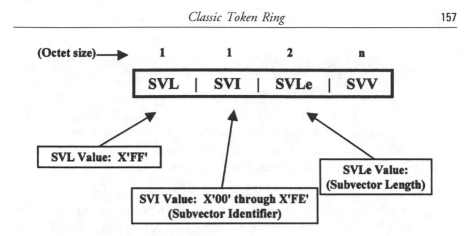

Figure 4.20 SV1 definition.

The subvector SV1 is a minimum of 6 octets in length and contains data associated with the VI. The SV1, illustrated by Figure 4.20, contains four fields: the SVL, the SVI, the SVLe, and the SVV.

This form of SV is supported by the TKP Access protocol to support an SVI value of less than 255 *and* an SVV length of greater than 255 octets.

The LMT MAC frame, which may a data field (SVV) of more than 255 octets, is the only known use of this subvector form (see Appendix D).

- SV2 is defined when the value of SVL is one octet in length and has a value of X'FF' and the value of SVI is one octet and has a value of less than X'FF'. SV2 has an SVL field containing X'FF' and the SVLe field containing the length of SVL, SVI, SVLe, and SVV. This SV is illustrated by Figure 4.21.

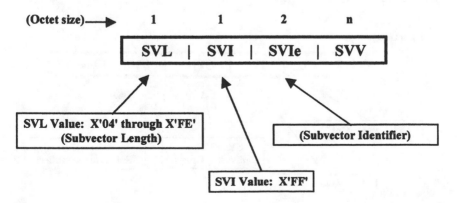

Figure 4.21 SV2 definition.

The subvector SV2 is a minimum of 6 octets in length and contains data associated with the VI. SV2, illustrated by Figure 4.21, contains four fields: the SVL, the SVI, the SVIe, and the SVV.

This form of SV is defined by the Standard to support an SVI value that is greater than 255 *and* has an SVV length of less than 255 octets. The Standard has reserved this form of SV for future definition. No implementations known by the authors use this feature.

- SV3 is defined when the value of SVL is one octet in length and has a value of X'FF' and the value of SVI is one octet and has a value of X'FF'. SV2 has an SVL field containing X'FF' and the SVLe field containing the length of SVL, SVI, SVLe, SVIe, and SVV. This SV is illustrated by Figure 4.22.

The subvector SV3 is a minimum of 8 octets in length and contains data associated with the VI. SV3, illustrated by Figure 4.22, contains five fields: the SVL, the SVI, the SVLe, the SVIe, and the SVV.

This form of SV is defined by the Standard to support an SVI value greater than 255 *and* has an SVV length of greater than 255 octets. The Standard has reserved this form of Subvector for future definition. No implementations known by the authors use this feature.

The SVI definitions used and reserved by the Standard are defined in Appendix D.

MAC frames using the SVIe are not specified by the Standard, but the concept of a SVIe is specified for future expansion. No implementations known by the authors use this feature.

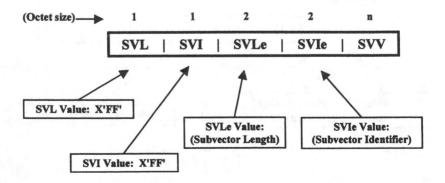

Figure 4.22 SV3 definition.

LLC Frame Format

The format of the LLC frame, illustrated by Figure 4.23, is specified by the Standard.

For information on the contents of the LLC frame, the reader is referred to ISO/IEC 10038:1993.

Abort Sequence

The abort sequence, illustrated in Figure 4.24, is a two-octet structured message consisting of an SD (see Section 4.2.3.1) immediately followed by an ED (see Section 4.2.3.2). The abort sequence is transmitted by a station to terminate the transmission of a frame because of error conditions.

Under normal conditions, the station that transmits an abort sequence does not release a token. However, the Standard supports an option that releases a token when certain transmit under-run error conditions occur (see Section 4.3.5.2).

Any station copying a frame recognizes the abort sequence on any baud boundary. The abort sequence, when recognized by the station that has not yet detected the frame's ending sequence, causes the station to stop copying the frame. No indication of this aborted frame copy is provided to the station's interface(s), and this abort sequence is not counted by the receiving station but is counted by the transmitting station.

The IFG field is used to provide stations enough time between frames to detect the next structured message (see Section 4.2.4.2 for definition).

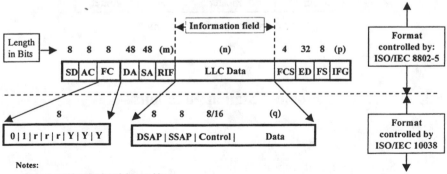

Notes:

- *Numeric* notations are length in even bits.
- **(m)** = optional, if present: 16 to 240 bits (2 to 30 octets).
- **(n)** = Location of LLC Data as required by ISO/IEC 8802-5.
- **(p)** = 8 bits at 4 Mbit/s, 40 bits at 16 Mbit/s.
- **(q)** = 8 bits multiplied by 0 or more.

Figure 4.23 LLC frame.

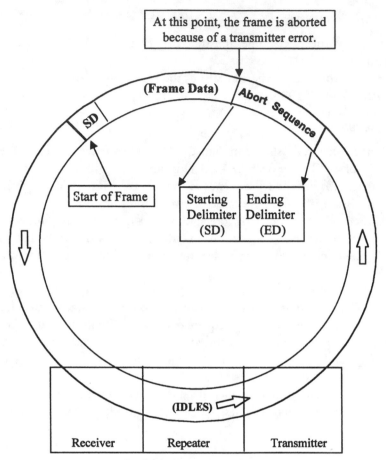

At this point, the frame is aborted because of a transmitter error.

(Frame Data)

Abort Sequence

SD

Start of Frame

Starting Delimiter (SD) | Ending Delimiter (ED)

(IDLES)

Receiver | Repeater | Transmitter

Station's PHY Layer shown

Figure 4.24 Abort sequence.

4.3 TKP Access Protocol Station Functions

This section describes

- Station transmit, repeat, and copy;
- Token Release protocol;
- Priority operation;
- Addressing;
- Station policies;
- Timers;

- Error counters;
- Station Operation Tables;
- Finite state machines.

4.3.1   Station Transmit, Repeat, and Copy Functions

This section discusses how a station *transmits* a frame, *strips* (removes) this frame, and releases a token; *repeats* frames; and *copies* and *repeats* frames. These functions are explained using Figures 4.25 to 4.28.

Figure 4.25 is used to explain how station A, with a frame waiting in its transmit queue for transmission to station C, captures the token and the resulting actions.

Figure 4.26 is used to explain that station A continues to transmit the frame from its transmit queue. Station B continues to repeat the frame transmit-

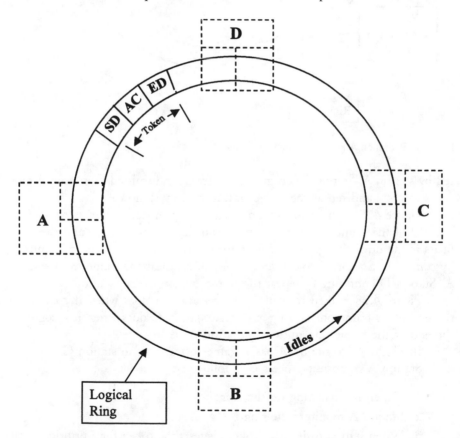

Figure 4.25 Station A waiting for a transmit opportunity.

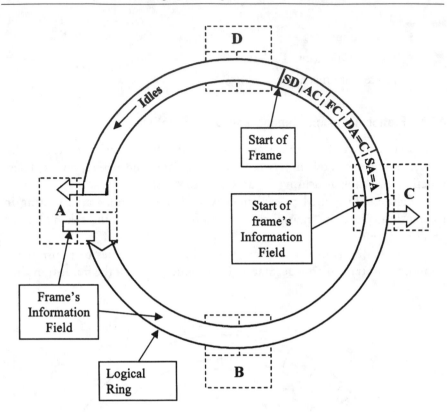

Figure 4.26 Station A starts transmitting frame.

ted by station A. Station C recognizes the frame transmitted by station A and starts to copy and repeat the frame transmitted by station A.

Figure 4.27 is used to explain what station A does when it recognizes the end of the frame being transmitted from its transmit queue and the start of its frame on the ring. Station B continues to repeat the frame transmitted by station A. Station C continues to copy and repeat the frame from station A. Station D continues to repeat the frame transmitted by station A.

Figure 4.28 is used to explain what station A does when it recognizes the end of its frame on the ring and what station C does when it recognizes the end of the frame transmitted by station A.

In Figure 4.25, station A has a frame to transmit to station C. Station A's transmit process is as follows.

1. Token is circulating on the ring.
2. Station A monitors the ring for a token.
3. When station A detects a token, it tests the token for a priority equal to or less than the priority of the frame to be transmitted.

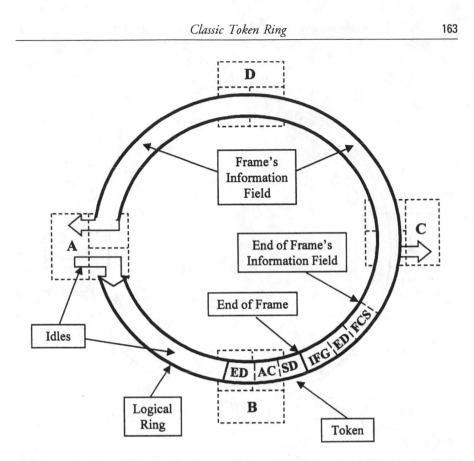

Figure 4.27 Station A strips its frame.

4. If the token has a priority greater than the priority (Pm) of the frame to be transmitted, then station A cannot transmit its frame but may be allowed to make priority reservations (see Section 4.3.3).

5. If the frame to be transmitted has a priority (Pm) equal to or greater than the priority (P) of the token, then station A is allowed to capture the token.

6. Upon capture of the token, station A repeats the token's P-bits (0–3) unchanged, breaks the repeat path between its receiver and transmitter, starts transmitting its frame by setting the T-bit to a 1, sets the M-bit to 0, repeats the P-bits unchanged, and then starts to transmit its frame data.

7. At the same time that station A started to transmit its frame (when the repeat path is broken), its receiver starts to strip (remove) all data from the ring.

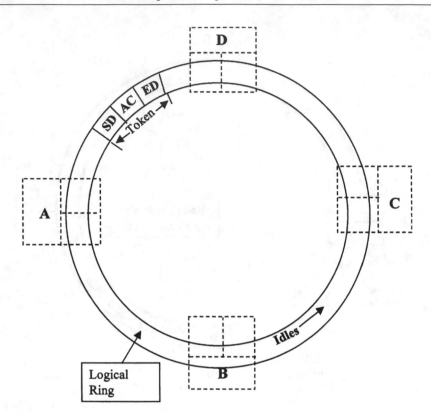

Figure 4.28 Station C completes frame reception and Station A completes strip.

In Figure 4.26, station A continues to transmit its frame to station C and strip data from the ring.

Station A's transmit process continues and the frame is recognized by station C as being a frame it should copy. Station A's transmit process and station C's receive process are as follows.

1. Station A continues to strip the data from the ring and monitors the ring for its transmitted frame by continuously checking for a SD.

2. Station B repeats data on the ring and monitors the ring for frames with its DA. Station B ignores the frame because the frame's DA is not equal to station B's address.

3. Station C repeats data on the ring, monitors the ring for frames, and recognizes the frame is addressed to its station address. To accomplish frame recognition, station C must be capable of buffering at least eight octets of frame data (the AC [1 octet], FC [1 octet], and the DA [6 octets]), so it can make the decision as to whether it should

copy the frame. This buffer allows station C to copy the beginning of the frame's data.

4. Station C detects a frame with a DA equal to its address.

5. Station C starts to copy the frame by transferring the frame's data to its buffer(s).

In Figure 4.27, station A completes transmitting its frame, releases a token followed by idles, and continues to strip its frame from the ring.

Station A's transmit process and station C's receive process proceed as follows.

1. Station A recognizes the end of the frame it is transmitting from the transmit queue and transmits the frame's calculated FCS, EFS (ED, FS, and IFG).

2. Station A releases a token as per the normal or early token release function (see Section 4.3.2) and follows the token with idles.

3. Station A recognizes the start of its frame on the ring and continues to strip the data from the ring.

4. Station B continues to repeat the frame on the ring.

5. Station C repeats data on the ring and, because it has not detected the end of the frame on the ring, continues to *copy* the frame by transferring the frame's data to its buffer(s).

6. Station D continues to repeat the frame on the ring.

In Figure 4.28, station A has completed its transmit strip process and station C has completed copying the frame from station A.

1. Station A completes stripping its transmitted frame and returns to repeating data waiting for the next action to be taken by the TKP Access Protocol or the higher layer application.

2. Station C detects the end of station A's frame and takes one of the following actions.
   a. If the frame is invalid (a code violation or FCS error is detected), it terminates copy of the frame and takes the appropriate action to prevent this frame's reception. It may or may not be able to free the buffers used to receive the frame. No indication of frame reception is provided to any layer (other than to indicate the used receive buffers can be set free). The frame's A-bits and C-bits are repeated unchanged.

    b. If the frame is valid (no code violation or FCS error), it completes copy of the frame by indicating the frame's reception to the appropriate station interface (e.g., if the frame was a MAC frame, to the MAC layer). The frame's A-bits and C-bits are set to 1 indicating the frame was address recognized and copied.

4.3.2   Token Release Protocol

This section explains the protocol responsible for releasing a token after the station has completed transmitting its frame. The TKP Access protocol defines two types of Token Release protocols: Normal Token Release and Early Token Release.

4.3.2.1   Normal Token Release

The early development of Token Ring was for a ring that operated at 4 Mbit/s. One early problem addressed by the Token Ring designers was the ability of important traffic being transmitted before less important traffic. To this end, an access priority scheme was developed (see Section 4.3.6). The control of priority transmission was so important to the developers of TKP Access protocol that they designed the Normal Token Release protocol.

    At 4 Mbit/s, most frames, even short ones, were normally near the logical ring length or longer than the logical ring; thus, very little bandwidth was wasted by stripping a frame until its transmitted SA was detected before releasing a token.

    The Normal Token Release protocol requires the station to detect the SA of the last frame transmitted before it releases a token. This insured that the station would get the latest priority reservations and release a token of the appropriate priority.

4.3.2.2   Early Token Release

At 16 Mbit/s a high percentage of short frames (e.g., LLC response frames) were shorter than the logical ring. Thus, a significant amount of ring bandwidth could be wasted when a high percentage of ring utilization consisted of short frames.

    Preventing this wasted bandwidth became paramount in designer's minds, and the result was Ken Wilson's (co-author of this book) invention of the Early Token Release protocol. This protocol operates only when a frame is shorter than the logical length of the ring plus the frame's length up to its AC and, when active, allows the immediate release of a token after the last field (IFG) of the last or only frame is transmitted.

In his performance paper "Enhancement to the Token Ring Protocol: Early Token Release" presented by Dr. Norman Strole of IBM to the IEEE 802.5 Committee on November 10, 1986, Dr. Strole explains why Wilson's Early Token Release protocol does not seriously interfere with frames needing priority from being transmitted before other less important frames until the ring's bandwidth utilization reached 80 percent or greater.

4.3.2.3 Token Release Protocol Comparison

This section compares the theoretical bandwidth utilization of a 12-KM ring using the Normal Token Release and Early Token Release protocols. The 12-KM ring has an octet length of 31 octets at 4 Mbit/s and 124 octets at 16 Mbit/s.

- A 4-Mbit/s ring with a 27-octet frame (26 octets of frame data plus 1 octet of IFG) on the ring utilizes 87% of the available 4-Mbit/s bandwidth.

- A 16-Mbit/s ring with a 31-octet frame (26 octets of frame data plus 5 octets of IFG) on the ring utilizes 25% of the available 16-Mbit/s bandwidth. This means the Normal Token Release protocol at 16 Mbit/s wastes enough space to hold three more 26 octet frames.

Figure 4.29 illustrates utilization of the 4-Mbit/s and 16-Mbit/s ring when using the Normal Token Release protocol.

Figure 4.30 illustrates utilization of the 16-Mbit/s ring when using Early Token Release. A 16-Mbit/s ring with four 31-octet frames (each consisting of 26 octets of frame data plus 5 octets of IFG) on the ring utilizes 100% of the available 16-Mbit/s bandwidth.

4.3.3 Priority Operation

Access to the ring's bandwidth using the concept of priority is an important feature of the TKP Access protocol. Access priority allows applications to be assigned levels of transmit importance based on the application's need for access to the ring's bandwidth. Unlike Ethernet (IEEE 802.3), which has no access priority, applications needing improved access (high priority) to the ring's bandwidth can override applications not needing improved access (low priority) to the ring's bandwidth.

The TKP Access protocol priority is controlled by the token and frame AC's bits 0 through 2 (P-bits [P]) and bits 5 through 7 (R-bits [R]). The P-bits and R-bits support eight levels of priority (0 through 7), where priority

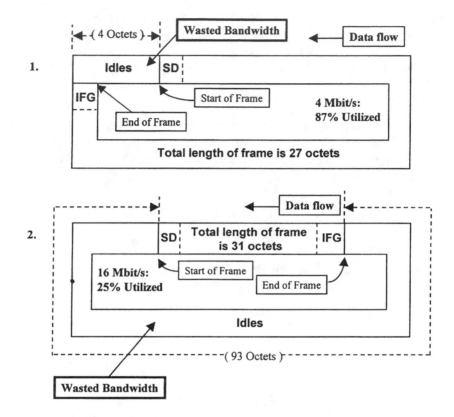

Figure 4.29 Token release comparison 1. Utilization of a 12-KM ring at 4 Mbit/s and 16 Mbit/s by one 26-octet frame (SD through ED).

0 is the lowest priority and 7 is the highest priority. The P-bits and R-bits work together in an attempt to match the highest priority *protocol data unit* (PDU) queued for transmission to the *current ring service priority.* The current ring service priority of the ring is determined by the value of the AC's P-bits within the frame or token currently on the ring. That is, if a token is on the ring and its P-bits are equal to 2, the current service priority is 2. Likewise, if a frame is on the ring and its P-bits are 3, the current service priority is 3.

The TKP Access protocol priority mechanism operates in such a manner that fairness (equal access to the ring) is maintained within a single priority level. However, fairness is not maintained between different levels of priority because a higher priority transmit request takes precedence over a lower priority transmit request. This means high-priority PDUs (e.g., 5) can delay the trans-

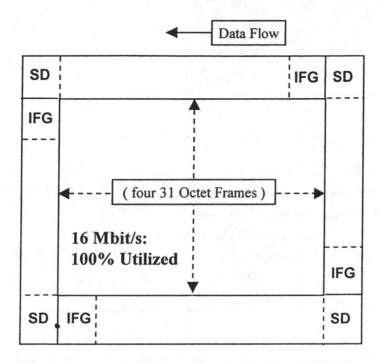

Figure 4.30 Token release comparison 2. Utilization of a 12-KM ring by four 31-octet frames (SD through IFG) at 16 Mbit/s.

mission of lower priority PDUs (e.g., 0 through 4). Because of this priority mechanism, the MAC layer *reserves* priority 7 for its use to prevent interruption of timed TKP Access protocol functions needed to ensure correct operation of the TKP Access protocol (e.g., transmission of the AM's AMP PDU).

MAC Frame Priority Note

The Standard changed the priority scheme for MAC layer MAC frames. Before the ISO/IEC 8802-5:1995 Standard, only the AMP MAC frame was transmitted using priority 7. With the advent of multimedia, it was realized that the MAC functions needed some MAC frames to be transmitted using priority 7. Before the 1995 Standard, some MAC frames were transmitted using priority 0 while others were transmitted with a priority of 3. Thus, it is possible to see different priorities being used by stations, some manufactured before the 1995 Standard and some manufactured after the 1995 Standard.

The following frames may have a priority of 0 or 7:

- Response MAC frame;
- Report station address MAC frame;
- Report station state MAC frame;
- Report station attachments MAC frame.

The following frames may have a priority of 3 or 7:

- SMP MAC frame;
- Duplicate address test MAC frame;
- Report new AM MAC frame;
- Report SUA change MAC frame;
- Report neighbor notification incomplete MAC frame;
- Report AM error MAC frame;
- Report error MAC frame.

The operation of the priority mechanism is explained in the following two subsections. Section 4.3.3.1 identifies the terms used to explain TKP Access protocol priority, and Section 4.3.3.2 explains how the station evaluates the frame or token P and R values to control priority using a stacking table (STACK_TABLE). The text and tables used in this section use the following operators:

- &, denotes "and";
- |, "or";
- <, "less than";
- =, "equal to";
- ≥, "greater than or equal to";
- ≤, "less than or equal to."

4.3.3.1 Priority Terms

Table 4.1 defines all the terms used by Section 4.3.3.2 to explain TKP Access protocol control of the STACK_TABLE.

4.3.3.2 Priority Stacking Table Operations

Any station that releases a *new* priority token [TK_AC(P > 0)] is required to remember the priority of the frame [FR_AC(P)] or token [TK_AC(P)] received

Table 4.1
Definition of Priority Terms

Term	Meaning of Term
Ring service priority	The priority value of the current token or frame *on* the ring.
STACK_TABLE	A push-down/push-up table containing up to four two-paired entries, one entry containing the old ring service priority (Sr) and one entry containing the new ring service priority (Sx). When frame or token evaluation occurs, only the top entry is examined even though the STACK_TABLE may have up to three more pairs of entries (with lower priority values).
Active STACK_TABLE	A STACK_TABLE that has one or more active entries as the result of previous priority transmissions.
Sr	The top entry of the STACK_TABLE containing the old ring service priority
Sx	The top entry of the STACK_TABLE containing the new ring service priority
FR(P < Sx)	A frame is received with a current ring service priority (P) less than the station's new ring service priority (Sx).
FR_AC(P)	Value of the received frame's priority field (bits 0–2)
FR_AC(R)	Value of the received frame's reservation field (bits 5–7)
TK_AC(P)	Value of the received token's priority field (bits 0–2)
TK_AC(R)	Value of the received token's reservation field (bits 5–7)
Pm	Priority value of the frame queued for transmission
Pr	A register containing the most recently received FR_AC(P) [always equal to the current ring service priority]
Rr	A register containing the most recently received reservation. Equal to one of the following: • Transmit machine is in the transmit fill state then the *last* received FR_AC(R); • If the transmit machine is in the repeat state then TK_AC(R).

Px	A station priority value representing the higher value of either: (1) the received reservation (Rr) or (2) the priority value of the frame queued for transmission (Pm) when greater than the received reservation (Rr). Equal to one of the following:
	If (PDU_QUEUED & Pm > Rr), then Px = Pm.
	If (PDU_QUEUED & Pm ≤ Rr), then Px = Rr.
	If QUE_EMPTY, then Px = Rr.

CLEAR_STACK	A STACK_TABLE process that clears all entries of the STACK_TABLE
POP	A STACK_TABLE process that removes the Sr and Sx values from the top of the STACK and causes all other STACK_TABLE entries (if any) to move up one position
RESTACK	A STACK_TABLE process that replaces the top entry of STACK_TABLE values (Sr and Sx). Values already in the other positions of the STACK_TABLE (if any) remain *unchanged.*
STACK	A STACK_TABLE process that pushes a value into the top of the STACK_TABLE and moves values already in the STACK_TABLE (if any) down one position

as the old ring service priority (Sr) and the priority of the new token released as the new ring service priority (Sx) in its STACK_TABLE.

The station evaluates an *active* STACK_TABLE (table has at least one entry) when it is:

- In any Transmit mode and receives a frame or token with P < Sx (see Table 4.2);

- In Transmit Repeat mode and receives a token with P = Sx (see Table 4.3);

- In Transmit Fill mode and has met the conditions for releasing a new token (see Table 4.4).

The *only* entry evaluated in the STACK_TABLE is the *top* entry.

Reception Handling of a Frame or Token with P < Sx

The station in any Transmit mode (Transmit Repeat, Transmit Data, Transmit Fill, or Transmit Fill and Strip) takes the action defined in Table 4.2 when it has an active STACK_TABLE with a new ring service priority (Sx) less than the priority value of the received frame [FR_AC(P < Sx)] or token [TK_AC(P < Sx)].

This action is used, for example, when a lost token is detected by the AM and causes the transmission of a Ring Purge MAC frame that has a priority of 0. This action causes the STACK_TABLE to be cleared (CLEAR_STACK) and set inactive.

Using an example, Figure 4.31 illustrates the actions defined in Table 4.2 when a token is received with a priority value less than the value saved by the station as the *new ring service priority.*

This same action also occurs if a frame is received with a priority value less than the value the station has saved as the new ring service priority [FR_AC(P < Sx)].

Table 4.2
Repeating Station Control of STACK_TABLE When P < Sx

Frame/Token Received	STACK_TABLE Control	Frame/Token Change
If FR_AC(P < Sx)	then CLEAR_STACK	and frame is not changed
If TK_AC(P < Sx)	then CLEAR_STACK	and token is not changed

Example STACK_TABLE Operation when TK_AC(P<Sx)

The *top entry* of the STACK_TABLE contains the priority value of the latest *new ring service priority* (Sx=4) released by this Station and the priority value of the *old ring service priority* (Sr=2).

The Token received [TK_AC(P<4;R=x)] has a priority *less-than* the Station's *new ring service priority* (Sx=4) meaning the *current ring service priority* (P<4) has been lowered (e.g., by the Active Monitor), thus the STACK_TABLE is no longer valid. Thus, the STACK_TABLE is CLEARED and set inactive.

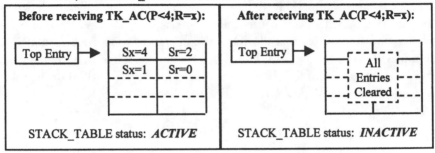

Figure 4.31 Reception of a token with P < Sx.

Standards Note for the FR_AC(P < Sx) Event

This event is marked *optional-i* in the Standard. The rationale for this optional statement is as follows.

ISO/IEC 8802-5:1992 had many incomplete and confusing definitions; as a result, some companies implemented FR_AC(P < Sx) while others did not. When the 1992 Standard was updated, the IEEE 802.5 committee was not allowed to make changes that would make these implementations nonconformant (existing implementations are protected against future Standards changes by the IEEE 802 "grandfather clause"). The IEEE 802.5 committee agreed to create the optional-i statement as the method to be used in the updated Standard (now ISO/IEC 8802-5:1995) to recommend a function be included in all new station implementations.

STACK_TABLE Implementation Note

Some implementations use a STACK_TABLE *pointer* to indicate which entries in the STACK_TABLE are in use. These implementations reset the table pointer and set the STACK_TABLE *inactive* (only an active STACK_TABLE is examined by the TKP Access protocol). This has the same effect as *clearing* all STACK_TABLE entries.

Reception Handling of a Token with P = Sx

The station in the transmit repeat mode with an active STACK_TABLE takes the actions defined in Table 4.3 when it detects the priority token originally released by the station [TK_AC(P = Sx)]. The entry number included in column 1 is used to explain each table entry usage. Each entry number is followed by an example that illustrates the operation supported. Although all examples illustrate the use of the frame's reservation field, it must be remembered that Px is the greater of the station's queued frame's priority (Pm) or the token or frame reservation field (Rr), as defined in Section 4.3.3.1.

- *Entry number 1:* This action is used when a request has been made (Px) to change (RESTACK) the current ring service priority. It should be remembered that Px is the greater of the station's queued frame's priority (Pm) or Rr. The value of the old ring service priority (Sr) remains unchanged. The example in Figure 4.32 illustrates this action.

- *Entry number 2:* This action is used to decrease (POP) the current ring service priority [TK_AC(P = Sr)] because the value of Px is equal to or less than the old ring service priority. This action causes the top entry of the STACK_TABLE to be removed and other entries (if any) to be moved up one position in the STACK_TABLE. The example in Figure 4.33 illustrates this action.

Handling of a Frame Received in Transmit Fill Mode

The station in the Transmit Fill mode (station is completing the transmission of a frame) with an active or inactive STACK_TABLE takes the actions defined in Table 4.4 when it receives a priority frame that it originally released (Pr = Sx) or when a current ring service priority is greater than its last released

Table 4.3
Repeating Station Control of STACK_TABLE Upon Token Reception

Entry Number	Token Received	STACK_TABLE Control	Token Change
1	If TK_AC(P = Sx) & Sr < Px	then RESTACK(Sx = Px)	and TK_AC(P = Px & R = 0)
2	If TK_AC(P) = Sx & Sr ≥ Px	then POP(Sx & Sr)	and TK_AC(P = Sr & R = Px)

Example STACK_TABLE Operation when TK_AC(P=Sx;R>Sr)

The *top entry* of the STACK_TABLE contains the priority value of the latest *new ring service priority* (Sx=4) [released by this Station] and the priority value of the *old ring service priority* (Sr=2) [priority elevated by this Station].

The Token received [TK_AC(P=4;R=3)] has a priority *equal-to* the *current ring service priority* (this Station release original priority Token). The Token's reservation field (R=3) causes the Station to RESTACK the top entry of the STACK_TABLE by replacing the Sx=4 with Sx=3.

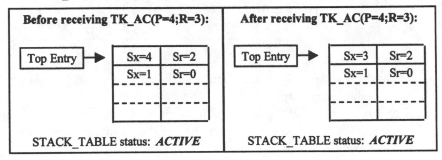

Figure 4.32 Repeat mode handling of TK(P = Sx; R > Sr).

Example STACK_TABLE Operation when TK_AC(P=Sx;R<=Sr)

The *top entry* of the STACK_TABLE contains the priority value of the latest *new ring service priority* (Sx=4) [released by this Station] and the priority value of the *old ring service priority* (Sr=2) [priority elevated by this Station].

The Token received [TK_AC(P=4;R=1)] has a priority *equal-to* the *current ring service priority* (this Station release original priority Token). The Token's reservation field (R=1) causes the Station to POP the top entry of the STACK_TABLE by removing Sx=4 and Sr=2 and moving other entries up one position.

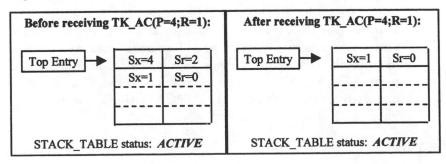

Figure 4.33 Repeat mode handling of TK(P = Sx; R <= Sr).

Table 4.4
Transmitting Station Control of the STACK_TABLE for Token Release

Entry Number	Frame Test	STACK_TABLE Control	Released Token
1	If Pr = Sx & Sr < Px	then RESTACK(Sx = Px)	and FR_AC(P = Px & R = 0)
2	If Pr = Sx & Sr ≥ Px	then POP(Sx & Sr)	and FR_AC(P = Sr & R = Px)
3	If Pr > Sx & Pr < Px	then STACK(Sx = Px & Sr = Pr)	and FR_AC(P = Px & R = 0)
4	If Pr > Sx & Pr ≥ Px	then not changed	and FR_AC(P = Pr & R = Px)

priority token (Pr > Sx). The entry number included in column 1 is used to explain each table entry usage.

Each entry number in Table 4.4 is first explained and then followed by an example that illustrates how the station changes the STACK_TABLE when the frame test condition is met.

Entry Numbers 1 and 2

Table 4.4 entry numbers 1 and 2 are used when the station receives a frame with the priority equal to the last priority token released by this station. Although all examples illustrate the use of the frame's reservation field (R), it should be remembered that Px is the greater of the station's queued frame's priority (Pm) or R and therefore could also be the result of Pm being greater than Rr.

- *Entry number 1:* This action is used when a request has been made (Px) to change (RESTACK) the current ring service priority. The value of the old ring service priority (Sr) remains unchanged. The example in Figure 4.34 illustrates this action.

- *Entry number 2:* This action is used to decrease (POP) the current ring service priority [FR_AC(P = Sr)] because the value of Px is equal to or less than the old ring service priority. This causes the top entry of the STACK_TABLE to be removed and other entries (if any) to be moved up one position in the STACK_TABLE. The example in Figure 4.35 illustrates this action.

Entry Numbers 3 and 4

Table 4.4 entry numbers 3 and 4 are used when the station receives a frame with the priority equal to the last priority token released by another station (P > Sx). This condition occurs when this station has a frame queued for

Example STACK_TABLE Operation when FR_AC(P=Sx;R>Sr)

The *top entry* of the STACK_TABLE contains the priority value of the latest *new ring service priority* (Sx=4) [released by this Station] and the priority value of the *old ring service priority* (Sr=2) [priority elevated by this Station].

The Frame received [FR_AC(P=4;R=3)] has a priority *equal-to* the *current ring service priority* (this Station release original priority Token). The Frame's reservation field (R=3) causes the Station to RESTACK the top entry of the STACK_TABLE by replacing the Sx=4 with Sx=3.

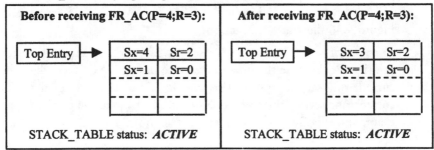

Figure 4.34 Transmit Fill mode handling of FR_AC(P = Sx; R > Sr).

Example STACK_TABLE Operation when FR_AC(P=Sx;R<=Sr)

The *top entry* of the STACK_TABLE contains the priority value of the latest *new ring service priority* (Sx=4) released by this Station and the priority value of the *old ring service priority* (Sr=2).

The Frame received [FR_AC(P=4;R<=?)] has a priority *equal-to* the Station's *new ring service priority* (Sx=4) and the requested Px is *equal-to* or *less-than* the *old ring service priority* (Px<=2). The top entry of the STACK_TABLE is removed (POP) and the other entries (if any) are moved up one position.

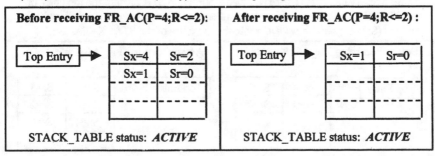

Figure 4.35 Transmit Fill mode handling of FR_AC(P = Sx; R ≤ Sr).

transmission with a priority value (Pm) that was equal to or greater than the priority of the token released by the other station.

- *Entry number 3:* This action is used to create a new ring service priority (STACK) because the new ring service priority (Sx) is the current ring service priority (Pr). This causes a new entry to be added to the top entry of the STACK_TABLE and the other entries (if any) to be moved down one position in the STACK_TABLE. The example in Figure 4.36 illustrates this action.

- *Entry number 4:* This action is used to pass the current ring service priority (Pr) because the new ring service priority (Sx) is less than the current ring service priority (Pr) and Px is less than or equal to the current ring service priority. No STACK_TABLE action is taken. The example in Figure 4.37 illustrates this action.

4.3.4 Addressing

The TKP Access protocol frames have two types of addresses: the SA, which identifies the station that transmitted the frame, and the DA, which identifies the address of the station or stations that are to receive the frame.

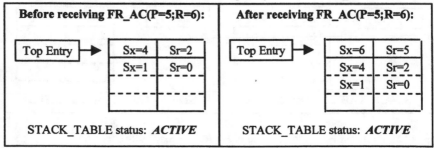

Example STACK_TABLE Operation when FR_AC(Pr>Sx & Pr<Px)

The *top entry* of the STACK_TABLE contains the priority value of the latest *new ring service priority* (Sx=4) [released by this Station] and the priority value of the *old ring service priority* (Sr=2) [priority elevated by this Station].

The Frame received [FR_AC(P=5;R=6)] has a priority *greater-than* the Station's *new ring service priority* (another Station released this priority Token). The Frame's reservation field (R=6) causes the Station to STACK into the top entry of the STACK_TABLE by moving down all entries (if any) and putting the new (Sx) and old (Sr) *ring service priorities* into the top entry of the STACK_TABLE.

Before receiving FR_AC(P=5;R=6):			After receiving FR_AC(P=5;R=6):		
Top Entry ▶	Sx=4	Sr=2	Top Entry ▶	Sx=6	Sr=5
	Sx=1	Sr=0		Sx=4	Sr=2
				Sx=1	Sr=0
STACK_TABLE status: *ACTIVE*			STACK_TABLE status: *ACTIVE*		

Figure 4.36 Transmit Fill handling of (Pr > Sx & Pr < Px).

Example STACK_TABLE Operation when FR_AC(Pr>Sx & Pr>=Px)

The *top entry* of the STACK_TABLE contains the priority value of the latest *new ring service priority* (Sx=4) [released by this Station] and the priority value of the *old ring service priority* (Sr=2) [priority elevated by this Station].

The Frame received [FR_AC(P=5;R=3)] has a priority *greater-than* the Station's *new ring service priority* (another Station released this priority Token). The Frame's reservation field (R=3) causes the Station leave the STACK_TABLE *unchanged* and passes the reservation (R=3) in the Token (so this reservation is not lost).

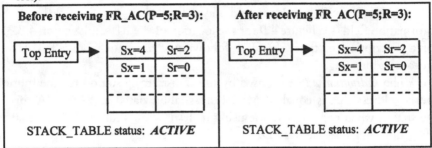

Figure 4.37 Transmit Fill handling of (Pr > Sx & Pr ≥ Px).

This section describes the types of IEEE 802.5 on-ring addresses used for the delivery of and the removal of IEEE 802.5 Token Ring frames on a single LAN.

All addresses are explained using the most significant bit (MSB) notation unless otherwise noted (see Section 1.3 for a discussion of MSB versus least significant bit (LSB)).

Addressing Note

The delivery of some frames may be to other LANs. These frames are sent from one LAN to another LAN using either the transparent bridging or the source routing method of delivery described in ISO/IEC 10038.

- The Transparent Bridging Forwarding protocol uses the DA within the IEEE 802.5 frame and tables in the forwarding mechanism to forward frames to the destination LAN.

- The Source Routing Forwarding protocol uses the RIF contained within the IEEE 802.5 frame to forward frames to the destination LAN. It is important to realize that the RIF is independent of and should not be confused with on-ring addressing.

The frame has its DA in the 3rd through the 8th octets and its SA in the 9th through the 14th octets after the SD, as shown in Figure 4.38.

The DA and SA are considered single 48-bit binary fields for decoding purposes and are numbered bits 0 through 47. However, in the SA field, bit 0 is not part of the actual SA. For addressing purposes, bit 0 of the SA is always considered to be zero, indicating an individual address.

Therefore, bit 0 of the SA is assigned as the *routing information indicator* (RII) bit. When the RI bit is set to 0, there is no RIF. When the RII bit is set to a 1, the RIF is present and located between the SA field and the information field (see Figure 4.9 for the location of the RIF and Section 4.3.3.2 for a description of the RIF). The following is a summary of TKP Access protocol addressing.

The *individual address* is used by the transmitting station for the frame's source address (SA is equal to My Address, abbreviated as SA = MA) or by the station when copying a frame using it destination address (DA is equal to My Address, abbreviated as DA = MA). There are two types of individual addresses, namely, universally administered and locally administered. It should be noted that when bridges, routers, and switches forward frames from another ring or network, the SA will not, for example, be the bridge's station individual address. In this case, the SA is equal to the originating station's individual address.

The *group address* is a special type of broadcast address used in DA by applications that need to address a particular function assigned to one or more stations. All stations that have this particular group address set as being one of the addresses the station recognizes will, if possible, copy the frame. A group address can be any of several types: universally administered, locally administered, or special group addresses that may be either functional or broadcast.

The *null address* is used only within the DA and is not copied by any station. The null address is used for testing the ability of the ring to circulate a frame without errors and not be recognized or copied by any station.

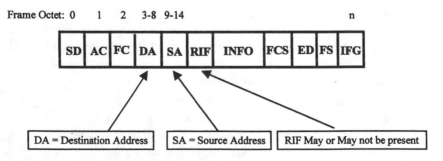

Frame Octet: 0   1   2   3-8 9-14   n

SD AC FC DA SA RIF INFO FCS ED FS IFG

DA = Destination Address   SA = Source Address   RIF May or May not be present

Figure 4.38 Frame address locations.

4.3.4.1 Universally and Locally Administered Addresses

This section describes the format and characteristics of the *universally administered address* (UAA) and *locally administered address* (LAA) used in the frame. Bit 1 of the DA and SA fields defines which type of address is being used, as shown in Figure 4.39.

Universally Administered Addresses

The UAA is a 48-bit address controlled by a universal authority administered by the Secretary General, ISO Central Secretariat in Genève, Switzerland. This group assigns to manufacturers one or more universal address sets. The UAA has the form shown in Figure 4.40.

The manufacturer is assigned by the universal authority one or more *Manufacturer's ID* (MFID) values for the address bits 2 through 23, and the manufacturer is responsible for ensuring that bits 24 through 47 are unique between each station manufactured.

Locally Administered Addresses

The LAA is a 48-bit address administered by a local authority (usually the customer's network administrator). The LAA has the form illustrated in Figure 4.41.

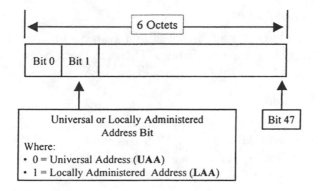

Figure 4.39 Universal or local administered address control.

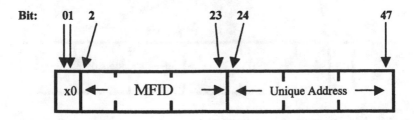

Figure 4.40 Universal administered address form.

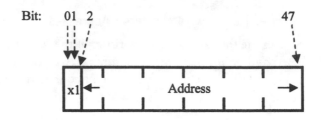

Figure 4.41 Locally administered address form.

The network administrator allowing LAA type addresses to be used within his or her networks is responsible for ensuring that each station is unique.

- At a minimum, all stations on a single ring are required to have unique addresses and the TKP Access protocol enforces this requirement.
- It is recommended that all stations within the networks controlled by the network administrators be unique (unique addresses between rings *are* not enforced by the TKP Access protocol).

Now that the meaning and identification of UAAs and LAAs have been explained, each of these types of address is either an individual or group address.

4.3.4.2   Individual Address

The individual address is used by the station as its SA for transmit and as its receive address when copying frames. There are two types of individual addresses: the UAA individual address and the LAA individual address.

UAA Individual Address

The UAA individual address is a 48-bit address characterized by the bits 0 and 1 being equal to B'00', bits 2 through 23 containing the MFID, and bits 24 through 47 containing the unique station address. The UAA individual address is illustrated by Figure 4.42.

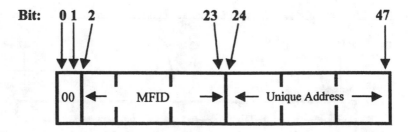

Figure 4.42 UAA individual address form.

LAA Individual Address

The LAA individual address is a 48-bit address characterized by bits 0 and 1 being equal to B'01' and bits 2 through 47 containing the unique station address. The LAA individual address is illustrated by Figure 4.43.

Local individual addresses are administered by an authority local to each LAN. The administrator of the LAA must assure that each station has a unique address for at least a ring. It is advisable to have LAAs that are unique to the network to prevent stations from receiving undesired frames.

Address Uniqueness versus Interoperability Note

The point of address uniqueness cannot be overstressed. Many customers using LAA have faced severe duplicate address problems when, for example, the administrator decides to increase the size of the Token Ring LAN by collapsing two rings into one and the same station individual address existed on both rings (only possible with LAA addresses). The most common problem is the failure of a station to insert into the ring due to the failure of the TKP Access protocol's duplicate address test. Recovering from this duplicate address problem (which did not exist before the reconfiguration) is time consuming because the customer must setup each station detecting a duplicate address test problem. This problem is complicated by application software that "learn" a station's individual address and then use this address for future communications.

4.3.4.3 Group Addresses

There are several different types of group addresses defined for the TKP Access protocol: universally administered group addresses and locally administered group addresses, which can be further divided into locally administered group addresses, functional addresses, and broadcast addresses.

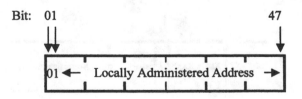

Figure 4.43 LAA individual address form.

Universally Administered Group Address

The universally administered group address is a 48-bit address characterized by bits 0 and 1 being equal to B'10'. The universally administered group address is illustrated by Figure 4.44.

The universal group address is intended to be unique among all stations in conformance with the Standard. Bits 2 through 23 of the address contain a constant MFID, which is administered by a central authority so as to be unique among manufacturers of all stations used in the IEEE 802 standards.

There are a set of universal group addresses that have been defined by the IEEE 802 committee for the exclusive use of specific networking protocol needs (e.g., TCP/IP and bridges). These universal group addresses as well as the addresses in the LSB notation are listed in Technical Report ISO/IEC TR 11802-2:1995(E).

Locally Administered Group Addresses

There are two types of locally administered group addresses: the group address and the functional address.

- *Group address:* The locally administered group address is a 48-bit address characterized by bits 0 and 1 being equal to B'11', bit 16 being equal to B'1', and bits 17 through 47 containing the locally administered group address, which is illustrated by Figure 4.45.

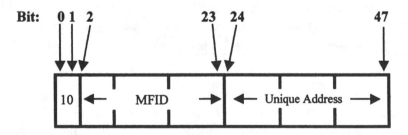

Figure 4.44 Universally administered group address form.

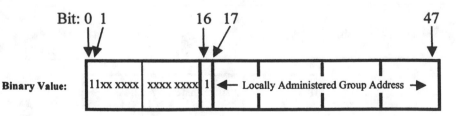

Figure 4.45 Locally administered group address.

The group address is administered by an address administrator for each IEEE 802.5 Token Ring LAN and is used to address multiple stations with a single frame.

- *Functional address:* The functional address is a 48-bit address characterized by bits 0 and 1 being equal to B'11', bit 16 being equal to B'0', and bits 17 through 47 being bit significant addresses. A station can have as many functional addresses set as supported by the station. The functional addresses is illustrated by Figure 4.46.

  A '1' in the corresponding position of the bit-significant address field independently indicates each of the 31 functional addresses. Decoding a functional address is a bit-for-bit binary comparison of bits 0 through 16, followed by an AND of bits 17 through 47, of the frame destination address with the corresponding bits of the assigned station functional address. This allows a station to assume any or all of the 31 functional addresses with minimized hardware requirement.

  A functional address uniquely identifies a group of one or more stations that have a functional association with a particular functional address.

  — *Standardized functional addresses:* The functional addresses, identified in Table 4.5, have been defined by the Standard for exclusive use by the TKP Access protocol functions.

  Other functional addresses, not used by the TKP Access protocol, have also been defined and are listed in the Technical Report

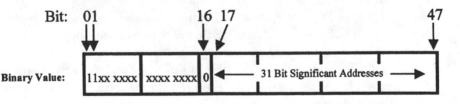

Figure 4.46 Functional address form.

Table 4.5
TKP Access Protocol Functional Address Assignments

Name of Function	Address (MSB)	Address (LSB)
AM	X'C0 00 00 00 00 01'	X'03 00 00 00 00 80'
Ring Parameter Server	X'C0 00 00 00 00 02'	X'03 00 00 00 00 40'
Ring Error Monitor	X'C0 00 00 00 00 08'	X'03 00 00 00 00 10'
Configuration Report Server	X'C0 00 00 00 00 10'	X'03 00 00 00 00 08'

ISO/IEC TR 11802-2:1995(E). This technical report also lists all the addresses in the LSB notation shown in Table 4.5.

— *Customer-assigned functional addresses:* The functional addresses X'C0 00 00 08 00 00' through X'C0 00 40 00 00 00' have been reserved for use by the customer (bits 17 through 28 shown with bit 16 always equal to 0).

Implementation Note

It is possible for different application software packages to assign conflicting meanings to the same bit-significant functional address. Care must be taken to prevent these types of application software packages from operating at the same time.

The problems caused by conflicting functional addresses are difficult to resolve because the cause of the failure is not usually obvious.

- *Broadcast address:* Every station irrespective of other addresses it may have been assigned shall recognize the broadcast address, also known as the All Stations Address. All stations on the ring shall attempt to copy. The broadcast address is either of the two forms shown in Figure 4.47.

   The address X'C000,FFFF,FFFF' address form is the result of early TKP Access protocol definitions and should not be used for other than MAC frames.

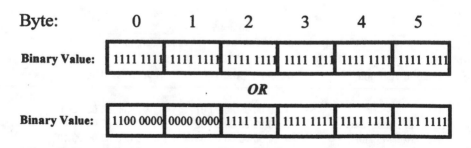

Figure 4.47 All stations broadcast addresses.

Note About Early Addressing

Early in the development of the ISO/IEC 8802-5 Standard it was thought that ring addressing was necessary and the station address should be broken into two parts: ring addressing and station addressing. The ring address was located in bits 2 through 15, and the station address was located in bits 16 through 47. Later in the development of the Standard, the ring address was abandoned in favor of a full 46-bit address in bits 2 through 47 for station addressing and the use of source routing for ring addressing. However, some implementations were unable to react to this Standard change, and the functional address remained assigned to bits 17 through 47 with bit 16 indicating the address was either a functional or group address.

Thus, earliest versions of the ISO/IEC 8802-5 Standard proposed that the DA be divided into two segments, one for ring addressing and one for station addressing. These two segments were as shown in Figure 4.48.

Bit 0 was used to indicate a ULL or LAA. Bit 1 was used to indicate individual or group addressing. Bits 2 though 15 were the ring address, and bits 16 through 47 were the station address.

Later in the development of the ISO/IEC 8802-5 Standard, ring addressing was abandoned in favor of source routing.

Early implementations were caught in this change and could not respond to this change. The effect was that bits 2 through 15 were required to be 0.

4.3.5 Station Policies

Station operation is controlled by flags specified by the Standard. Flags are used to remember important conditions and are state independent, although they may only be tested in particular states.

These flags are divided into the two categories: *station policy flags,* specified in clause 3, and *MAC protocol flags,* specified in clause 4.

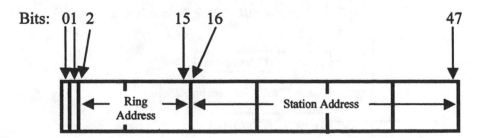

Bits: 01 2   15 16   47

Ring Address   Station Address

Figure 4.48 Obsolete ring addressing.

The MAC protocol flags, which control *how* the Standard's station operation tables operate, are not covered in this book because their definition would not add anything to the understanding of the TKP Access protocol.

The station policy flags control station policy and are option type flags set before the station starts the Join process. The TKP Access protocol does not change these flags. The naming concept for the station policy flags is shown in Figure 4.49.

Station policy flags are divided into functional flags and methodology flags.

4.3.5.1 Functional Flags

Functional flags specify information that determines how the Station Operation Tables operate and come in various formats.

The *claim contention option* method (FCCO) determines the operation of the Claim Token process (see Section 4.4.2.4) when the station receives a Claim Token MAC frame with a SA lower than the station's individual address. The station contender option is described in Section 4.4.2.2.

The *early token option* (FETO) determines whether the station supports the Normal Token Release or the Early Token Release protocols described in Section 4.3.2.

The *medium rate option* (FMRO) determines whether the station is supporting 4-Mbit/s or 16-Mbit/s medium rate speed.

The *reject remove option* (FRRO) determines whether the station accepts or rejects the Remove Ring Station MAC frame. For example, using this option flag, servers can be configured to reject the Remove Ring Station MAC frame to prevent shutting down the customer's server function.

4.3.5.2 Methodology Flags

ISO/IEC 8802-5:1992 had several functions that were imprecisely defined. As a result, these functions were implemented using slightly different methodologies.

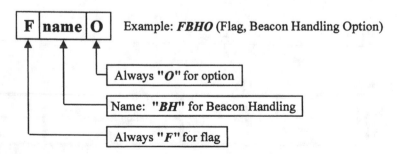

Figure 4.49 Station policy flags.

The IEEE 802.5 committee agreed upon an acceptable set of Station Operation Table functions to accommodate these different methodologies when the committee created ISO/IEC 8802-5:1995 as a replacement for the 1992 Standard.

The purpose of the methodology flags are summarized as follows and are detailed in the Standard's clause 3.

The *beacon handling option* (FBHO) is used by the Station Operation Tables to determine when the station starts handling the beacon process during the Join process (see Section 4.5.1.1).

The *error counting option* (FECO) is used by the Station Operation Tables to determine how the station uses the *timer error report* (TER) (see Section 4.3.6) to report errors.

The *good token detection method* (FGTO) is used by the Station Operation Tables to determine how the station detects the presence of a valid token.

Token error option (FTEO) is used by the Station Operation Tables to determine whether the station supports the detection of code violations (see Section 3.1.1) within the E-bit coverage area of the token (see Section 4.2.4.1).

The *token handling option* (FTHO) is used by the Station Operation Tables to determine how the station uses the timer valid transmission (TVX, see Section 4.3.6) to detect the absence of a token.

ISO/IEC 8802-5:1995 had many errors updated by the ISO/IEC 8802-5:1998 document. The IEEE 802.5 committee agreed to add one new optional function to accommodate implementations that experienced excessive underrun conditions when transmitting (called a Station Error in the Standard). The rationale for this change is the impact upon ring's throughput when station error recovery is accomplished as specified by the Standard. Recovery of the station error causes the Ring Purge process to operate, which takes about 11 ms and causes the ring to be inoperative during this time period. The flag below was defined to optionally allow transmit underrun to release a token when the station underrun condition occurred, thus reducing the throughput impact of the underrun condition.

The *transmit underrun behavior option* (FTUBO) is used by transmit Station Operation Table to handle the transmit underrun conditions that do not impact the operation of the TKP Access protocol.

4.3.6 Timers

The TKP Access protocol uses timers to determine when events should occur or have failed. Timers are explained by first discussing the basis of the time interval for the timers and then discussing each timer specified by the Standard.

4.3.6.1   Timer Basis

The following assumptions were used in the definition of time periods for the timers that time ring events and were generated theoretically using the maximums for station count, station delays, and ring length and its associated ring delay. These times do not represent actual configurations, but rather the maximums for a worst case timing situation.

The maximum *ring length* is 182 km. This number consists of the maximum number of stations or repeaters, which is 260; and the maximum average distance between stations, which is 700m.[1] These numbers were the foundation of the values for the timers identified by the Standard. The Standard, however, specifies less distance between stations. These timer values are designed to allow concentrators to have active retiming repeat functions envisioned but not available in 1985. Concentrator ports with active retiming repeat functions are now commonplace.

The total *ring delay* is the sum of the bit delay of the active stations on the ring and the media delay for 182 km. The wire or fiber media signal speed is 199.86 km/ms (approximately two-thirds that of light) for a maximum ring delay of 0.911 ms. The maximum total station delay is implementation and ring speed dependent but consists of the following components:

- Supported ring speeds are 4 or 16 Mbit/s.
- Each station has approximately 40 ns of delay in the receive/transmit path for a total delay of 0.011 for 260 stations.
- One AM station with a delay of up to 32 bits.
- 252 SM stations with a delay of approximately 2.5 bits (4.75 baud) each for a total of 630 bits of delay.
- A maximum of 7 stations with a protocol fairness delay of 9.5 bits each or 66.5 bits.

Table 4.6 summarizes the ring delay numbers calculated using these assumptions and used to develop the original values of the timers as specified in the ISO/IEC 8802-5:1985 Standard (the first approved Token Ring Standard).

Changes in this Standard have increased some of these values, but the original values of the timers have been maintained.

---

1. The 700m between stations far exceeds the recommended drive distance of normal stations but provides delay enough for a variety of concentrator connection mechanisms (such as fiber between the station and the concentrator).

Table 4.6
Ring Speed versus Ring Delay

Ring Speed	Media Delay +	Station Delay =	Total Ring Delay
4 Mbit/s	0.911 ms	0.193 ms	1.104 ms
16 Mbit/s	0.911 ms	0.057 ms	0.968 ms

Timer values and tolerances are not included in this section but are specified in the definition for each timer description in clause 3 of the Standard.

4.3.6.2 Timer Definitions

The following timers, listed alphabetically, ensure correct operation of the TKP Access protocol. Also included is an insight to some implementations that use equivalent timers but with different timer names.

Timer, AM (TAM)

Used by the AM to periodically start the neighbor notification process, this timer is also known as the "ring poll timer" or "poll timer" in some implementations. TAM expiration causes the AM station to start the neighbor notification process by queuing the AMP MAC frame for transmission on the next available token.

Timer, Beacon Repeat (TBR)

Used by all stations in the beacon repeat mode to detect the absence of Beacon MAC frames, this timer is also known as the "escape timer" in some implementations. TBR expiration causes the station in the Beacon Repeat mode to start the Claim Token process by transmitting a Claim Token MAC frame without waiting for a token.

Timer, Beacon Transmit (TBT)

Used by all stations in the Beacon Transmit mode to determine that the station in the Beacon Repeat mode has been unable to recover the ring, TBT expiration causes the station in the Beacon Transmit mode to start the beacon test process.

Timer, Claim Token (TCT)

Used by all stations to detect the failure of the Claim Token process, this timer is also known at the "monitor contention timer" in some implementations. TCT expiration causes the station in the Claim Token process to start the Beacon process by transmitting a Beacon MAC frame without waiting for a token.

Timer, Error Report (TER)

Used by all stations after Join has completed to report error conditions, this timer is also known as the "soft error timer" in some implementations. TER expiration causes the station that has detected and counted errors (see Section 4.3.7) to queue the Report Error MAC frame for transmission on the next available token.

Timer, Insert Delay (TID)

Used by all stations when inserting into the ring to assure the station has entered the physical ring before making decisions about frames it receives, TID expiration allows the station to examine received MAC frames and take the appropriate actions.

Timer, Join Ring (TJR)

Used by all stations during the Join process to detect the failure of the Join process, TJR expiration causes the station to exit the Join process and enter the bypass state.

Timer, No Token (TNT)

Used by the SM station to detect the absence of a normal priority token (P = 0) or a priority token (P > 0) followed by a frame (known as good token in the Standard), this timer is also known as the "good token timer" in some implementations. TNT expiration causes the SM station to enter the Claim Token process by transmitting a Claim Token MAC frame without waiting for a token.

Timer, Queue PDU (TQP)

Used by all stations to pace the transmission of Beacon, Claim Token, and SMP MAC frames, this timer is also known as the "pacing timer" in some implementations when transmitting the Beacon and Claim Token MAC frames. This timer is also known as the "receive poll" timer in some implementations when transmitting the SMP MAC frames. TQP expiration causes the:

- Station in the Beacon Transmit mode to transmit the Beacon MAC frame without waiting for a token;
- Station in the Claim Token Transmit mode to transmit the Claim Token MAC frames without waiting for a token;
- SM station to queue the SMP MAC frame for transmission on the next available token.

Timer, Remove Hold (TRH)

Used by all stations to delay deinsertion until the concentrator has removed the station from the ring, TRH expiration causes the station to exit the Join process and enter the bypass state.

Timer, Request Initialization (TRI)

Used by all stations during the Join process to wait for a response to its last Request Initialization MAC frame, TRI expiration either causes the request initialization process to continue or the station to fail the Join process and enter the bypass state.

Timer, Ring Purge (TRP)

Used by the AM to detect the failure of the Ring Purge process, TRP expiration causes the station to start the Claim Token process by transmitting a Claim Token MAC frame without waiting for a token.

Timer, Return to Repeat (TRR)

Used by a transmitting station to detect the loss of its transmitted frame, this timer is also known as the "physical trailer timer" in some implementations. TRR expiration causes the station in the transmit fill mode or the transmit fill and strip mode to return to the repeat mode. This timer in some implementations causes the AM station in the Ring Purge process to transmit the Ring Purge MAC frame without waiting for a token.

Timer, Remove Wait (TRW)

Used by all deinserting stations to remain in the repeat mode until the concentrator has removed the station from the ring (reduces the deinserting station's interruption of the ring until the concentrator switches the station out of the ring), TRW expiration causes the station to exit the Join process and enter the bypass state.

Timer, Signal Loss (TSL)

Used by all stations to detect the presence of a steady state of the PHY condition "signal loss." TSL expiration causes the station to start the Claim Token process by transmitting a Claim Token MAC frame without waiting for a token.

Timer, Standby Monitor (TSM)

Used by all stations to detect the absence of an AM or to detect token streaming (via the absence of an AMP MAC frame), TSM expiration causes the station to start the Claim Token process by transmitting a Claim Token MAC frame without waiting for a token.

Timer, Valid Transmission (TVX)

Used by the AM to detect the absence of any frame (SD followed by a AC with P = n, T = 1, and M = 0) or any token (SD followed by a AC with P = n, T = 0, and M = 0), this timer is also known as the "any token timer" in some implementations. TVX expiration causes the AM station to start the Ring Purge process by transmitting a Purge MAC frame without waiting for a token.

Timer, Wire Fault (TWF)

Used by all stations to sample the wire fault condition once the timer TWFD (follows) allows sampling, TWF expiration causes the station to examine the wire fault condition and, if active, to enter the bypass wait state in preparation for deinsertion from the ring.

Timer, Wire Fault Delay (TWFD)

Used by all stations to prevent the sampling of the wire fault condition during the Join process or when inserting after exiting the beacon test state, TWFD expiration allows the station to sample the wire fault condition using the timer TWF.

4.3.7 Error Counters

This section defines the TKP Access protocol error counters. Before defining these error counters, the concept of fault domain is presented because all of the TKP Access protocol error counters relate to the domain of the error condition detected.

4.3.7.1 Fault Domain

The TKP Access protocol error counters are incremented by station error detectors and are contained in all station configurations. These station error detectors isolate failures to either a basic or a complex fault domain.

Basic Fault Domain

The basic fault domain is identified when a station's error detector isolates an error condition to a set of configurable elements as follows.

- *Ring errors:* Ring errors (data or signal) are detected and counted only by the first detecting station. Other stations also detect some of these errors but are prevented from counting the error condition by the E-bit in the ED. Thus, ring error detectors provide isolation to a transmitting station, a receiving station, and the components between these two stations (e.g., wire and wiring concentrators).

- *Station errors:* Station errors are detected and counted only by the station. Thus, station error detectors provide isolation to the failing station.

Figure 4.50 illustrates the error domains of ring errors and station errors.

Complex Fault Domain

The *complex fault domain* is identified when a station detects an error condition that could have been caused by any station on the ring and is illustrated by Figure 4.51.

For example, if station 3 was the AM and detected a token error, any station on the ring could have caused the error condition.

4.3.7.2   Isolating error counters

The following isolating error counters track error conditions that identify the "basic fault domain" reported in the Report Error MAC frame:

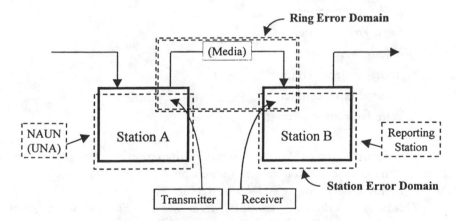

- **Ring Error Domain error conditions**
    - Line Error (any station, receive detection)
    - Burst Error (any station, receive detection)
- **Station Error Domain error conditions**
    - Abort Delimiter Error (any station, transmit detection)
    - Internal Error (any station, internal detection)
    - Wire Fault Error (any station, internal detection)
    - A/C Set Error (any station, internal detection)

Figure 4.50   Basic fault domain.

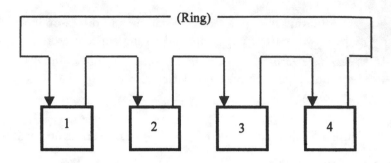

- Lost Frame Error (any station)
- Frame Copied Error (any station)
- Receive Congestion Error (any station)
- Frequency Error (any Standby Monitor station)
- Token Error (Active Monitor)

Figure 4.51 Complex fault domain.

- Line error counter;
- Burst error counter;
- A/C set error counter;
- Abort sequence counter;
- Internal error counter;
- Wire fault error counter.

The Report Error MAC frame identifies the reporting station as downstream to the error condition via the frame's SA and identifies the UNA via the Report Error MAC frame's UNA subvector.

This book uses the more descriptive term *nearest active upstream neighbor* (NAUN) rather than UNA because it includes the word *active*.

Line Error Counter

The line error counter is incremented once per frame for each occurrence of the following:

- The detection of a FCS error and the received frame's E-bit is set to 0;
- The detection of a code violation within the covered frame fields and the received frame's E-bit is set to 0.

Burst Error Counter

The burst error counter is incremented when the station detects the burst five condition (five baud times without a Differential Manchester transition). Normally, only one station detects the burst five condition because the station that detects a burst four condition (four baud times without a Differential Manchester transition) conditions its transmitter to transmit idles if the burst five condition is detected. However, because of ring jitter, it is possible that the burst four condition eventually lengthens to become a burst five condition.

A/C Set Error Counter

The incrementing of the ARI/FCI set error counter is conditioned by the reception of the first neighbor notification cycle after a Ring Purge. It is incremented when a station receives more than one AMP MAC frame or SMP MAC frame with the A-bits and C-bits set to B'0000' without first receiving an intervening AMP MAC frame. This counter indicates that the upstream station is unable to set the A-bits and C-bits in a frame that it must recognize and copy.

Abort Error Counter

The abort error counter is contained in any station configuration and is incremented when any condition causes the transmitter to release an abort sequence. When the station releases an abort sequence, the token is not released unless the "transmit underrun behavior option" indicates a token should be released (see Section 4.3.5.2). This counter isolates the error condition to the station causing the abort sequence to be released. Excessive abort errors have a serious impact upon ring performance because each time the absence of a token is detected a recovery time of approximately 11 ms is required.

Internal Error Counter

The internal error counter is contained in any station configuration that supports the detection of a recoverable internal error condition that does not cause an abort sequence to be transmitted. Implementations that do not support the recoverable internal error condition cause the station to deinsert from the ring and are not required to increment the internal error counter.

Wire Fault Error Counter

The wire fault error counter is contained in all stations. This counter is incremented when the wire fault condition is detected but does not last long enough to cause deinsertion (short-term wire fault conditions are not detected because the wire fault condition must be continuously present for between 5s and 10s). Therefore, this error counter identifies short-term wire fault conditions—

meaning the Wire Fault Detection process was started but reset prior to the time-out that would have caused the Wire Fault Detection process to force deinsertion.

4.3.7.3 Nonisolating Error Counters

The following nonisolating error counters track error conditions that identify the complex fault domain that are reported in the Report Error MAC frame:

- Lost frame error counter;
- Frame copied error counter;
- Receive congestion error counter;
- Frequency error counter;
- Token error counter.

Lost Frame Error Counter

The lost frame error counter is contained in all stations. This counter is incremented when a station, in the transmit mode, fails to strip the frame it transmitted before the *return to repeat* timer (TRR) expires. No token is released when this error occurs.

The AM detects the absence of the token; causes a Ring Purge; and upon successful completion of Ring Purge, releases a new token. Excessive lost frame errors can have serious impact upon ring performance since each detection of the absence of a token requires a recovery time of approximately 11 ms (10 ms to detect token's absence, approximately 1 ms for Ring Purge).

Frame-Copied Error Counter

The frame-copied error is incremented when a station receives a frame addressed to its specific address but finds the A-bits and C-bits set to B'1111'. This error indicates a possible line hit, a duplicate address, or a station copying unauthorized frames.

Receive Congestion Error Counter

The receive congestion error counter is incremented when a station recognizes a frame that it should copy but is unable to copy the frame due to congestion.

Frequency Error Counter

The frequency error counter, found only in SM, is incremented when a station detects an excessive difference between the ring data frequency and its internal crystal frequency.

Token Error Counter

The token error counter, found only in AMs, is incremented when the AM detects an error with the token protocol as follows:

- A circulating priority token is detected (a token with a nonzero priority with the M-bit set to 1).
- A circulating frame is detected (a frame with the M-bit set to 1).
- The TVX expires, indicating the absence of a token.

4.3.7.4 Assured Delivery Process

The Report Error MAC frame is used to report important station and ring error information to the ring error monitor function. The TKP Access protocol provides the Report Error MAC frame with an Assured Delivery process in an attempt to assure this frame is delivered to the on-ring ring error monitor function. The rationale for providing Assured Delivery is the impact that the loss of this frame could have upon the ring error monitor function's ability to do accurate error analysis of station or ring errors.

Assured Delivery Note

The Report SUA Change MAC frame also uses the Assured Delivery process described in this section.

Requirements

The Assured Delivery process is designed to fulfill the following frame delivery requirements.

- Significantly improve the probability of the delivery of certain on-ring MAC frames. However, delivery of frames cannot be guaranteed since the MAC layer does support two-way communication between entities.
- A delivery mechanism with a high probability of preventing duplicate copies of the frames to the designated function.
- Retransmission as a result of bit-hits effecting the ED or frame status fields must be low. This is accomplished by providing an interpretation of the bits within the ED and the frame status field that accounts for possible bit-hits.
- The delivery mechanism must not be destination class sensitive.

The Assured Delivery process examines the MAC frame transmitted and, if it appears the destination station did not copy the frame, then a controlled retransmission of the frame may occur.

Report Error MAC Frame Assured Delivery Note

The station transmitting the Report Error MAC frame is faced with the possibility of error counters being incremented between, for example, the first transmission of this frame and subsequent retransmission. The Standard requires the station to send the Report Error MAC frame with the latest error counts and the reset of the error counters to occur after the Assured Delivery process has terminated.

Assured Delivery Operational Note

The Assured Delivery process is designed to prevent duplicate copies of frames. However, error conditions can occasionally cause duplicate frames to be received by the designated functions (such as the ring error monitor or configuration report server). These functions must be designed to handle an occasional duplicate frame.

Assured Delivery Basis

The basis of the Assured Delivery process for the Report SUA Change and Report Error MAC frames depends on the setting of the E-bit from the ED and the A-bits and C-bits from the frame status field. The value of the E-bit from the ED and the A-bits and C-bits from the frame status field are combined to make the *transmit frame delivery value* (TFDV), as shown in Figure 4.52.

Table 4.7 explains two examples using the algorithm shown in Figure 4.52.

When a frame is transmitted requiring Assured Delivery, the TFDV is created from the stripped frame, examined, and the action defined in Table 4.8 is taken.

Table 4.8 identifies all the possible TFDVs and specifies the action taken by the station for each of these values and the rationale for this action. The TFDVs of B'01010', B'10000', and B'11010' are the only values causing the Assured Delivery process to "retry" the transmission of the frame. All other TFDVs terminate the Assured Delivery process.

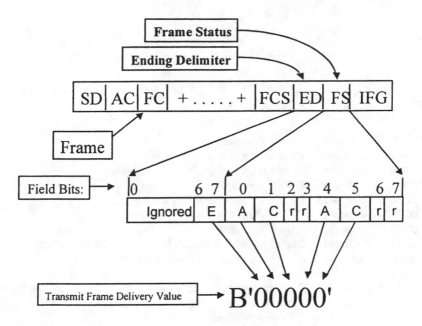

Figure 4.52 Basis of the TFDV.

Table 4.7
Examples of TFDV Interpretation

TFDV	Meaning
B'01010'	Destination station is congested since A = 1, C = 0, and no frame errors were detected by any station on the ring.
B'11010'	Destination station is congested since A = 1, C = 0, and a frame error has been detected by some station(s) on the ring.

The "Retry" action in Table 4.8 indicates the frame is retransmitted for MAC frames requiring Assured Delivery. The TFDVs of B'01010', B'10000', and B'11010' were selected because they offer the most accurate support of the second and third Assured Delivery requirements specified in Section 4.3.7.4. The station retransmits the frame for up to a total of four times. If all retries are unsuccessful, the station terminates the Assured Delivery process.

4.3.7.5 Customer Error Views

Each of the isolating error counters identified in Section 4.3.7.2 and nonisolating error counters identified in Section 4.3.7.3 can be seen by the customer from different points of view, as shown in Figure 4.53.

Table 4.8
TFDV Analysis

TFDV	Assured Delivery Action	Rationale for This Action
B'00000'	Terminate	DA is not present
B'00001'	Terminate	DA is not present or TFDV is unreliable
B'00010'	Terminate	DA is not present or TFDV is unreliable
B'00011'	Terminate	TFDV is unreliable
B'00100'	Terminate	DA is not present or TFDV is unreliable
B'00101'	Terminate	TFDV is unreliable
B'00110'	Terminate	TFDV is unreliable
B'00111'	Terminate	Destination station received frame or TFDV is unreliable
B'01000'	Terminate	DA is not present or TFDV is unreliable
B'01001'	Terminate	TFDV is unreliable
B'01010'	Retry	Destination station is congested and could not copy the frame
B'01011'	Terminate	Destination station received frame or TFDV is unreliable
B'01100'	Terminate	TFDV is unreliable
B'01101'	Terminate	Destination station received frame or TFDV is unreliable
B'01110'	Terminate	Destination station received frame or TFDV is unreliable
B'01111'	Terminate	Destination station received frame
B'10000'	Retry	Station may be present, but did not recognize the frame
B'10001'	Terminate	TFDV is unreliable
B'10010'	Terminate	TFDV is unreliable
B'10011'	Terminate	TFDV is unreliable
B'10100'	Terminate	TFDV is unreliable
B'10101'	Terminate	TFDV is unreliable
B'10110'	Terminate	TFDV is unreliable
B'10111'	Terminate	Destination station received frame or TFDV is unreliable
B'11000'	Terminate	TFDV is unreliable
B'11001'	Terminate	TFDV is unreliable
B'11010'	Retry	Destination station is congested and could not copy the frame
B'11011'	Terminate	Destination station received frame or TFDV is unreliable
B'11100'	Terminate	TFDV is unreliable
B'11101'	Terminate	Destination station received frame or TFDV is unreliable
B'11110'	Terminate	Destination station received frame or TFDV is unreliable
B'11111'	Terminate	Destination station received frame

Customer View 1

This error view allows the customer to identify only those isolating and nonisolating errors detected by station 1 without knowledge of errors being detected by other stations on the ring. It should be note, however, that the TKP Access protocol requires any error detected by station 1 to be reported to the ring error monitor function in station 3.

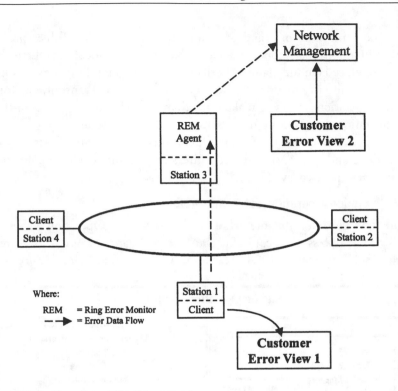

Figure 4.53 Customer error views.

Customer View 2

This error view allows the customer to view the isolating and nonisolating errors reported by any station on the ring (including station 3). Therefore, the ring error monitor function in station 3 can analyze these error reports and provide the customer with error status for the entire ring including stations 1 to 4.

4.3.8 The Station Operation Tables

The TKP Access protocol defined by the Standard is supported by the following six Station Operational Tables:

- Join Station Operation Table;
- Transmit Station Operation Table;
- Monitor Station Operation Table;
- Error Handling Station Operation Table;
- Interface Handling Station Operation Table;
- Miscellaneous Frame Handling Station Operation Table.

The Join, Transmit, and Monitor Station Operation Tables processes contain functions that support the base TKP Access protocol. The Error Handling, Interface Handling, and Miscellaneous Frame Handling Station Operation Tables support functions that do not cause the Join, Transmit, or Monitor Station Operation Tables to make state changes but rather support on-the-fly type operations required by these Station Operation Tables.

This section covers the Station Operation Tables by first illustrating their interactions and relationship to the finite state machines that implement the major functions used by the TKP Access protocol.

4.3.8.1   Station Operation Table Interactions

Figure 4.54 illustrates the interaction of the six Station Operation Tables used by the TKP Access protocol defined in the Standard's clause 4 as corrected by the ISO/IEC 8802-5:1998. The Join, Transmit, and Monitor *finite state*

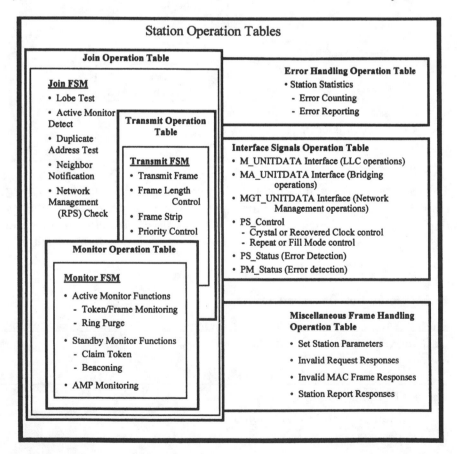

Figure 4.54   Station Operation Tables.

machines (FSM) implement the major functions of the TKP Access protocol and are part of the Join, Transmit, and Monitor Station Operation Tables, respectively.

The Join Station Operation Table, illustrated in Figure 4.54, is the *master* Station Operation Table while the other five tables are its *slaves*. Once the Join Station Operation Table is started, the Monitor and Transmit Station Operation Tables are activated by the Join Station Operation Table at the appropriate time to allow the TKP Access protocol to operate. The Error Handling, Interface Signals, and Miscellaneous Frame Handling Tables are used to support the Join, Transmit, and Monitor Station Operation Tables.

4.3.8.2 Station Operation Table Overviews

The following is an overview of the functions performed by the Station Operation Tables.

Join Station Operation Table

The Join Station Operation Table is used by the TKP Access protocol:

- To start the TKP Access protocol;
- For the LMT;
- For AM detection;
- To provide an AM if not present on the ring;
- For a Duplicate Address Test;
- For Neighbor Notification;
- To Request Initialization;
- For Join completion.

Transmit Station Operation Table

The Transmit Station Operation Table is used by the TKP Access protocol:

- For token recognition;
- To make reservations;
- For frame transmission;
- To strip transmitted frames;
- To release token after frame transmission;
- For recognition of transmit error conditions.

Monitor Station Operation Table

The Monitor Station Operation Table is used by the TKP Access protocol for the following functions:

- Repeat;
- Ring Purge;
- Claim Token;
- Beacon Transmit;
- Beacon Repeat;
- Beacon Test.

Error Handling Station Operation Table

The Error Handling Station Operation Table is used by the TKP Access protocol to count error conditions and to report error conditions.

Interface Signals Station Operation Table

The Interface Signals Station Operation Table is used by the TKP Access protocol:

- To recognize MAC and LLC frames to be reported to a station's LLC protocol (MA_UNITDATA.indication), a station's Management protocol (MGT_UNITDATA.indication), or a station's Bridge protocol (M_UNITDATA.indication);
- To set the station into either the repeat or fill mode;
- To set the station to transmit using either the station's crystal clock or the ring's recovered clock;
- To set the frame copy bits;
- To set the address recognize bits;
- To queue frames for transmission;
- To recognize the signal loss condition;
- To recognize wire fault condition (see Section 2.3);
- To count the frequency error condition;
- To recognize a token.

Miscellaneous Frame Handling Station Operation Table

The Miscellaneous Frame Handling Station Operation Table is used by the TKP Access protocol:

- To set appropriate station parameters;
- To respond to the Change Parameter or Initialize MAC frames;
- To respond to frame's detected with syntax error conditions;
- To respond to invalid remove station requests;
- To respond to requests made for station information.

4.3.9 Finite State Machines

The Join, Monitor, and Transmit FSMs are used by the Standard to explain the operation of the TKP Access protocol by identifying the state changes required for the various TKP Access protocol functions. The Join, Transmit, and Monitor Station Operation Tables have many events that do not cause state changes and, therefore, are not part of the FSMs. However, these events and actions—specified by Error Handling, Interface Handling, and Miscellaneous Frame Handling Station Operation Tables—are required for correct operation of the TKP Access protocol.

4.3.9.1 Join Ring Finite State Machine

The Join Ring process controls the station's insertion into the ring as follows.

1. Assures the media (wire or fiber) attaching the station to the ring has an acceptable bit error operating range.
2. Assures the ring has an AM.
   a. When the inserting station is the first station on the ring it temporarily performs the duties of the AM until the Join process is complete. Whether the station becomes the AM or SM depends on the insertion timing of other stations on the ring.
   b. When the inserting station is not the first station to join the ring it continues the Join process and becomes the SM only upon successful completion of the Join process.
3. Assures the inserting station does not have an address equal to another station on the ring.
4. Participates in the Neighbor Notification process.
   a. When the station is not the first station on the ring it waits until it receives an AMP or an SMP MAC frame and then transmits the SMP MAC frame.
   b. When the station is the first station on the ring it transmits the AMP MAC frame.
5. Requests permission to complete insertion by transmitting the Request Initialization MAC frame to the *ring parameter server* (RPS).

6. Starts a Wire Fault process to ensure the media attaching the station to the ring has acceptable electrical characteristics.

4.3.9.2   Transmit Finite State Machine

The TKP Access protocol uses the Transmit FSM to transmit all frame data on the ring, to remove the transmitted frame from the ring, and to control token priority (see Section 4.3.6) during the removal of this frame.

The Transmit FSM supports two types of transmission: Transmit Immediate and Transmit on Token. After discussing these transmission schemes, the Transmit function is discussed.

Transmit Immediate

The Transmit Immediate function allows the station to transmit a frame without waiting for a token and, as used by the TKP Access protocol (token passing), always transmits using the station's crystal as the transmit clock.

Transmit on Token

The Transmit on Token function allows the station to transmit a frame only when the station detects a token of a priority of equal to or less than the priority of the frame queued for transmission. The station transmits frames using the station's crystal clock when the station is an AM and using the clock derived from the ring's signal when the station is an SM.

Transmit Functional Operation

The Transmit function is explained using Figure 4.55 without regard to the station's clocking scheme because all transmissions are required to follow the same procedures.

Station 1 in Figure 4.55 transmits a frame and stations 2 and 3 take the following actions.

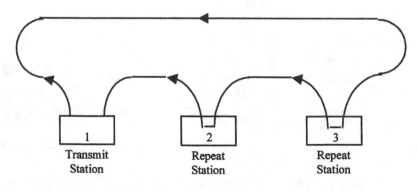

Figure 4.55 Transmit operation.

1. The frame's format is defined in Section 4.3.3.2.

2. Station 2 repeats the frame. If the frame has a code violation or a FCS error, then station 2 sets the E-bit to 1; or if the frame does not have a code violation or a FCS error and is addressed to station 2, then station 2 attempts to copy the frame. If station 2 can copy the frame, then station 2 sets the A-bits and C-bits to 1; or if station 2 cannot copy the frame (congested), then station 2 sets the A-bits to 1 and does not set the C-bits to 1.

3. Station 3 repeats the frame. If the frame has a code violation or a FCS error, then station 3 sets the E-bit to 1; or if the frame does not have a code violation or a FCS error and is addressed to station 3, then the station 3 attempts to copy the frame. If station 3 can copy the frame, then station 3 sets the A-bits and C-bits to 1; or if station 2 cannot copy the frame (congested), then station 2 sets the A-bits to 1 and does not set the C-bits to 1.

4. Station 1 removes (strips) its frame from the ring. If the frame has a code violation or a FCS error, then station 1 ignores the frame; or if the frame does not have a code violation or a FCS error and is addressed to station 1, then the station 1 attempts to copy the frame. If station 1 can copy the frame, then station 1 sets the A-bits and C-bits to 1; or if station 1 cannot copy the frame (congested), then station 1 sets the A-bits to 1 and does not set the C-bits to 1.

5. Station 1 completes the strip of this frame and returns to repeat.

Standards Note About "Loop Back"

The TKP Access protocol allows station 1 to transmit a frame to itself. This capability is called the "Loop Back" function, which is useful when the station's MAC layer transmits a frame to, for example, the REM functional address. Because the MAC layer does not know where the REM is located, using the Loop Back function the station does not need to understand whether this function is located within this transmitting entity. The Loop Back function is a unique TKP Access protocol operation since most LAN architectures do not support this function in their MAC layer but rather at the layers above MAC.

4.3.9.3   Monitor Finite State Machine

Because the token is key to the operation of the Token Ring network, a token monitor feature is included in the form of either an AM or an SM. The TKP

Access protocol first establishes an AM and then all other stations (if any) become SMs. Any station can act as an AM or an SM, and a Claim Token process (see Section 4.4.2.4) establishes an AM if there is none. Figure 4.56 illustrates a typical ring with one AM and *n* SMs.

1. The AM, shown in Figure 4.56 as station 1, has two distinct functions: *Ring Control* and *Protocol.*
   - Station 1 is responsible for three types of Ring Control functions, as described in Chapter 3.
     — *Ring clocking:* Station 1 supplies the ring with a crystal clock. The data being transmitted by station 1 uses the station's crystal clock and is either data received from the ring (the repeat path) or data being transmitted by station 0.
     — *Ring jitter:* Station 1 supplies a variable length elastic buffer to ensure the ring is always an integral number of bits in length.
     — *Ring length:* Station 1 supplies a 24-bit latency buffer to ensure the ring is long enough to contain at least one token.

       Station 1 always receives data using a receive clock derived from the ring's signal and always transmits using a clock derived from its crystal. All other stations use a clock derived from the ring's signal for both receive and transmit.
   - Station 1 provides Protocol functions that detect the following types of token and frame error conditions as defined in this section:
     — Absence of an AM;
     — Absence of any token or frame;
     — Presence of a circulating priority token;
     — Presence of more than one token;
     — Presence of more than one AM;
     — Presence of a circulating frame (a frame not removed by its originator).
2. The SM, shown in Figure 4.56 as stations 2 through *N*, also has two distinct functions.
   - *Ring Control:* Stations 2 through *N* use a recovered clock derived from the ring's signal to repeat received data (repeat path) to transmit station data on the ring.
   - *Protocol:* Stations 2 through *N* provide the ring with a Protocol function to detect the absence of an AM or the absence of any good token.

A station already inserted into the ring can either be an AM or an SM, while a station just inserting into the ring can never be the AM *unless* it is the only station on the ring.

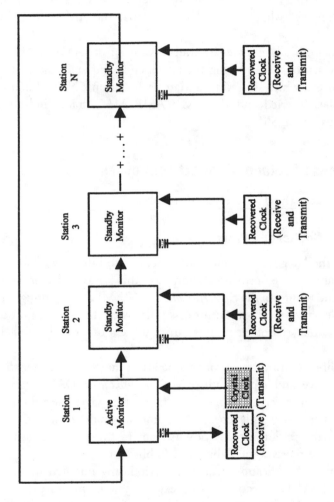

Figure 4.56 Active and standby monitors.

The Claim Token process (see Section 4.4.2.4) is always used to select which station becomes the AM even if only one station exists on the ring. Once the Claim Token process has selected an AM, all other stations (if any) become an SM.

The newly elected AM uses the Ring Purge process to release a token and to return the ring to a known state. The AM ensures normal token operation on the ring, while each SM ensures that the AM is functioning correctly.

The AM periodically initiates by transmission of the AMP MAC frame informing other stations on the ring of its presence and start the Neighbor Notification process, which is used to update each station's NAUN address if a new station has inserted and the AM's AMP MAC frame indicates its operational status to the SM.

4.4 TKP Access Protocol Support Processes

This section discusses the TKP Access protocol.

4.4.1 Neighbor Notification Process

The purpose of the Neighbor Notification process is to inform *each* station of the SA of its nearest active upstream station, save this SA as its *stored upstream address* (SUA), and to inform the *ring error monitor* (REM) of ring configuration changes. The Neighbor Notification process starts with the AM and proceeds around the ring, one station at a time, until all stations on the ring have participated.

The Neighbor Notification process operates using the characteristics of how stations receive and repeat frames with a *broadcast* DA (all stations' address). Each station receiving a broadcast frame first saves the value of the A-bits and C-bits and then, as it repeats the frame, it sets the A-bits and C-bits to 1. Thus, *only* the station directly downstream to the station transmitting the broadcast frame receives the A-bits and C-bits as 0.

The Neighbor Notification process is periodically initiated by the AM when its timer TAM (see Section 4.3.6) expires and it queues the priority seven AMP MAC PDU for transmission. The AM has two special conditions that it tracks.

1. When the AM station detects its *own* AMP MAC frame with the A-bits and C-bits equal to 0, this is a single station ring and causes the AM to terminate the Neighbor Notification process and inform its station management of the single station ring condition.

2. When the AM station detects the expiration of timer TAM, indicating it should restart the Neighbor Notification process, it checks to see if it has received either an AMP or SMP MAC frame with A-bits and C-bits equal to 0. If it has not detected this condition, then the AM has detected that the Neighbor Notification process was not completed successfully, causing it to transmit the Neighbor Notification Incomplete MAC frame to REM. The following can cause this condition.

- The AMP or SMP MAC frame is lost due to an error condition either on the ring or in a station.
- The ring's bandwidth is so heavily utilized that the Neighbor Notification process is prevented from completing before the timer TAM expired.

The station receiving the AMP or SMP MAC frame with the A-bits and C-bits equal to 0 saves the SA of this frame (if necessary) as its SUA. As the process continues around the ring, each station (if any) that receives either an AMP MAC frame or SMP MAC frame with the A-bits and C-bits equal to 0 paces the transmission of the SMP MAC frame.

Figure 4.57 illustrates stations 1 through 3 transmitting an AMP MAC frame or an SMP MAC frame; how the stations determine these frames are transmitted by the upstream station using the setting of the A-bits and C-bits and the value of the SUA saved by these stations.

Using Figure 4.57, the operation the Neighbor Notification process proceeds as follows.

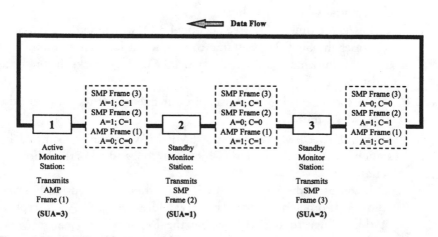

Figure 4.57 Neighbor Notification process.

1. Station 1 transmits the AMP MAC frame to the broadcast DA (all stations' address).

2. Station 2 receives the AMP MAC frame transmitted by station 1 with the A-bits and C-bits equal to 0 and takes the following actions.

   a. It compares the SA of the AMP MAC frame to its currently saved SUA. If different, it saves this new address as its SUA and reports this change to the REM using the Report SUA Change MAC frame; or if the same, continues the Neighbor Notification process.

   b. Sets the A-bits and C-bits to 1.

   c. Resets the timer TQP (see Section 4.3.6), which, upon expiration, causes the SMP MAC frame to be queued for transmission.

   d. Resets the timer TSM (see Section 4.3.6), which tracks the presence of the AM.

3. Station 3 receives the AMP MAC frame transmitted by station 1 with the A-bits and C-bits equal to 1. It sets the A-bits and C-bits to 1 even though both the A-bits and C-bits are already set to 1 and resets the timer TSM (see Section 4.3.6), which tracks the presence of the AM.

4. When station 2's TQP timer expires, it queues for transmission the SMP MAC frame.

5. Station 3 receives the SMP MAC frame transmitted by station 2 with the A-bits and C-bits equal to 0 and takes the following actions.

   a. It compares the SA of the AMP MAC frame to its currently saved SUA. If different, it saves this new address as its SUA and reports this change to the REM using the Report SUA Change MAC frame; or if the same, continues the Neighbor Notification process.

   b. Sets the A-bits and C-bits to 1.

   c. Resets the timer TQP (see Section 4.3.6), which, upon expiration, causes the SMP MAC frame to be queued for transmission.

   d. Resets the timer TSM (see Section 4.3.6), which tracks the presence of the AM.

6. When station 3's TQP timer expires, it queues for transmission the SMP MAC frame.

7. Station 1 (the AM) receives the SMP MAC frame transmitted by station 3 with the A-bits and C-bits equal to 0 and takes the following actions.

   a. It compares the SA of the AMP MAC frame to its currently saved SUA. If different, it saves this new address as its SUA and reports this change to the REM using the Report SUA Change MAC frame; or if the same, continues the Neighbor Notification process.

    b. Sets the A-bits and C-bits to 1.
    c. Resets the timer TSM (see Section 4.3.6), which tracks the presence
       of the AM.
    d. Terminates the Neighbor Notification process.

This procedure results in the Neighbor Notification process proceeding one station at a time until each station on the ring has learned its nearest active upstream station's SA. Finally, using timer TSM (see Section 4.3.6), each station times the occurrence rate of the Neighbor Notification process. If this timer expires, then at least two Neighbor Notification process cycles have been missed, indicating the absence of a functional AM and the station starts the Claim Token process (see Section 4.4.2.4).

Pacing Note

There are two pacing actions in the Neighbor Notification process.

1. The AM *starts* the Neighbor Notification process by transmitting the AMP MAC frame using the 7-s timer AM (TAM; see Section 4.4.2).
2. The SMs *continue* the Neighbor Notification process after receiving an AMP MAC frame or a SMP MAC frame with its A-bits and C-bits equal to 0 but delays the transmission of the SMP MAC frame using the 20-ms timer Queue PDU (TQP; see Section 4.4.2).

The SMP frame *pacing* was added to prevent lock out of stations by the high-priority AMP and SMP MAC frames when the Neighbor Notification process is operating. The time duration of the AM's timer TAM (7 sec) is designed to allow one Neighbor Notification process cycle to complete before the AM starts the next Neighbor Notification process cycle with a heavily loaded ring and a maximum station count (260).

Neighbor Notification Process Operational Note

It is possible for *more than one* Neighbor Notification process cycle to be operating at one time. Consider the following example.

When an error condition causes the AM to execute its Ring Purge process, a Neighbor Notification process cycle currently under way is *suspended* until the Ring Purge process successfully completes. At this time, the *interrupted* Neighbor Notification process cycle *restarts* with the *last* station to receive the AMP or SMP MAC frame. However, as part of the Ring Purge process, the AM starts *another* Neighbor Notification process cycle. Thus, it is possible for *multiple* Neighbor Notification process cycles to be operating at the same time.

4.4.2 Ring Recovery Process

The TKP Access protocol attempts to restore the ring to normal token operation using the Ring Recovery process when ring or station error conditions are detected. In the following subsections, Ring Recovery is explained first, followed by discussions of the functions used by the Ring Recovery process, of the Ring Purge process, of the Claim Token process, and of the Beacon process.

4.4.2.1 Token Ring Recovery Overview

Figure 4.58 illustrates that Ring Recovery is supported by Token-Oriented or Non-Token Oriented TKP Access protocols; by the recovery flows for the Ring Purge, Claim Token, and Beacon processes; and by which Ring Recovery functions handle the hard and soft errors (see shading).

The key points in Figure 4.58 are as follows.

- The TKP Access protocol goes *up* the recovery flow when recovery is successful, *down* the recovery flow when recovery fails.
- *Token-Oriented* protocols recover from errors using functions that *use* tokens, whereas *Non-Token-Oriented* protocols recover from errors using functions that *do not use* tokens.
- *Hard Error Recovery* handles errors that *prevent* the TKP Access protocol from operating correctly, whereas *Soft Error Recovery* handles errors that *do not prevent* the TKP Access protocol from operating correctly.

4.4.2.2 Ring Recovery Functions

The following functions are used by the Ring Purge, Claim Token, and Beacon processes:

- M-bit function;
- Station Contender function;
- Signal Loss function;
- Beacon Type codes;
- Beacon Repeat Count function;

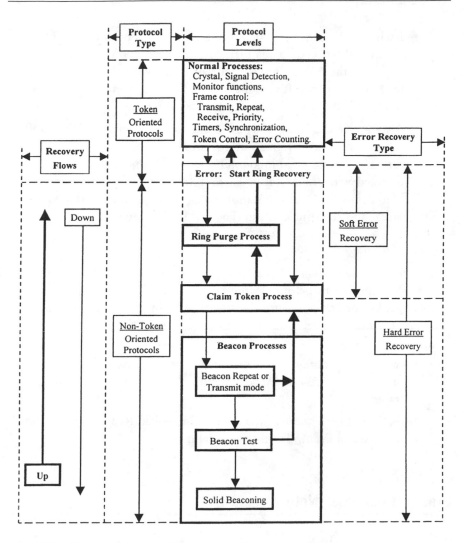

Figure 4.58 Ring Recovery overview.

- Beacon Repeat Escape function;
- Beacon Repeat C-bit function;
- Beacon Test function.

M-bit Function

The M-bit function (bit 4 of the AC) detects circulating frames and token. The M-bit is set to 0 when a station transmits a frame or token. The TKP Access protocol has two different uses for the M-bit.

- Each time the AM function repeats a frame or a priority token (Access Control P-bits greater than 0), the M-bit function sets the M-bit to 1. When the AM detects a frame or token with the M-bit set equal to 1, it enters the Ring Purge process in the Ring Purge Transmit mode and transmits a Ring Purge MAC frame.

- A station in the Beacon Repeat mode that is *downstream* (NADN) to the Beacon MAC frame's transmitting station activates the M-bit function when its C-bit counter goes to 0. Any Beacon MAC frame detected with the M-bit equal to 1 (indicating a circulating Beacon MAC frame) causes the station to enter the Claim Token Transmit mode and transmit a Claim Token MAC frame.

Station Contender Function

The *Station Contender function* is an option setup (see FCCO in Section 4.3.5) by station management before the station starts the Join process and cannot be changed by the station. Once the station has completed the Join process, the station handles the reception of the Claim Token MAC frame as follows.

- A *Contender station* enters or remains in the Claim Token Transmit function when its address is higher than a received Claim Token MAC frame SA.

- A *Noncontender station* enters the Claim Token Repeat function when it receives a Claim Token MAC frame.

Station Contender Note

The Station Contender function operates *only* when the station receives a Claim Token MAC frame and is independent of error conditions requiring the station to enter the Claim Token Transmit function.

Signal Loss Function

The Signal Loss function is a timed error condition used to detect the absence or presence of a ring signal at the station's receiver. Once the station detects the Signal Loss condition, the timer Signal Loss (TSL) begins timing the Signal Loss condition. If the Signal Loss condition has been active for the *entire* period of TSL, then the station sets a flag (FSL) to *remember* the condition. The station then takes the action called for in the process currently active as follows.

- The station enters the Claim Token process in the Claim Token Transmit mode, transmits a Claim Token MAC frame and remembers the Signal Loss condition.

- The station remembers the Signal Loss condition, but continues the Claim Token process in the Claim Token Transmit function.

- The station takes the action defined under Beacon Type.

Signal Loss Note

Because the Signal Loss condition is often an intermittent signal, the flag that remembers the Signal Loss condition (FSL) can bet set to 0 and 1 over time. This results in the Beacon process changing the Beacon MAC frame's Beacon Type to indicate the *presence* of the Signal Loss condition (Beacon Type 2) or *absence* of the Signal Loss condition (Beacon Type 3).

Beacon Type

The Beacon MAC frame contains a *Beacon Type* subvector that identifies the *reason* the station entered the Beacon process. This subvector identifies the meaning of the Beacon MAC frame.

- X'0001' Set Recovery Mode (highest priority): This Beacon Type is not used by the TKP Access protocol other than to recognize a Beacon MAC frame with this Beacon Type. A Beacon MAC frame with a Beacon Type of 1 is used by a station supporting the "IEEE 802.5C Supplement for Dual Ring Architecture" and is originated by the IEEE 802.5C station acting as a Recovery Node with function class zero authorization. The station receiving a Beacon MAC frame with a Beacon Type of 1 enters the Beacon Repeat mode.

- X'0002' Signal Loss: This Beacon Type is transmitted when the *Claim Token Timer* (TCT) expires and the Claim Token Transmit function is entered because of a Signal Loss condition or when a station is transmitting a Beacon MAC frame with a Beacon Type of 3 or 4 *and* the station detects a Signal Loss condition.

- X'0003' Bit Streaming or Frame Streaming: This Beacon Type is transmitted when the TCT expires in Claim Token Transmit mode and *no* Claim Token MAC frames were received or when a station is transmitting a Beacon MAC frame with a Beacon Type of X'0002' (Signal Loss) and the Signal Loss condition disappears.

- X'0004' Claim Token Streaming (lowest priority): This Beacon Type is transmitted when the TCT expires in Claim Token Repeat mode or Claim Token Transmit mode *and* a Claim Token MAC frame has been received.

If the station in the Beacon Transmit function receives a Beacon MAC frame that was not transmitted by this station with a Beacon Type subvector value of equal or higher value than the Beacon Type subvector value being transmitted by this station, the station enters the Beacon Repeat function.

Beacon Repeat Count Function

The *Beacon Repeat Count function* is a method of detecting a Beacon MAC frame from the station's *downstream* station. Once the station receives eight Beacon MAC frames from its downstream station (NADN), the fault domain has been established between two stations and the station exits the Beacon Repeat function and enters the Beacon Test process.

Beacon Repeat Escape Function

The *Beacon Repeat Escape function* is used by the station's Beacon Repeat function to detect the absence of Beacon MAC frames. The station in the Beacon Repeat function resets the *Beacon Repeat Timer* (TBR) every time a Beacon MAC frame is received. When TBR expires, the station in the Beacon Repeat function enters the Claim Token Transmit function by transmitting a Claim Token MAC frame.

Beacon Repeat C-bit Function

The *Beacon Repeat C-bit function* is a method of preventing a circulating Beacon MAC frame (caused by a station transmitting a Beacon MAC frame deinserting from the ring but leaving the Beacon MAC frame on the ring). The C-bit function is activated when a station detects a Beacon MAC frame from its *upstream* station (NAUN).

The station entering the Beacon Repeat Mode resets the TBR, the Beacon Repeat frame counter is set to eight, and the C-bit counter is initialized to 2 and takes the following actions.

- If a Beacon MAC frame is received from the nearest upstream station, the C-bit counter is decremented.
- If the C-bit counter reaches zero, the station enables the Circulating Frame Detection function (the same function used by the AM for a circulating frame or token). When the station detects a Beacon MAC

frame with the AC M-bit equal to 1, a Circulating Beacon MAC frame has been detected, the station enters the Claim Token Transmit mode, and transmits a Claim Token MAC frame.

If the Beacon MAC frame is not from the nearest upstream station, the C-bit counter is set to 2 and disables the circulating frame functions.

Beacon Test Function

The *Beacon Test* function is a method of testing the station for errors. The station deinserts from the ring and runs station tests (type of test is not defined by the Standard) and ends its test by executing the equivalent of the Join LMT (see Section 4.5.1.1). If the station's tests are:

- *Successful,* the station inserts into the ring and returns to the Beacon process in the same mode it entered the Beacon Test (either Beacon Transmit mode or Beacon Repeat mode);

- *Unsuccessful,* the station remains off-ring and notifies station management of this condition.

The station *must* return to the normal TKP Access protocol operation *before* the Beacon Test function is allowed to be executed again.

4.4.2.3 Ring Purge Process

Figure 4.59 illustrates what causes the Ring Purge function to be entered and its success and failure conditions.

The following is an explanation of the numbered boxes in Figure 4.59.

1. The AM's *Timer, Valid Transmissions* (TVX) expires, indicating the absence of a token or frame on the ring; enters the Ring Purge Transmit mode; and transmits a Ring Purge MAC frame.

2. The AM detecting a circulating frame or token enters the Ring Purge Transmit mode and transmits a Ring Purge MAC frame. It detected the circulating frame or token based on the following:
   a. A frame's AC M-bit is equal to 1.
   b. A token's AC P-bits are greater-than 0, the M-bit is equal to 1, *and* a frame does not follow the token.

3. A station in the Claim Token Transmit mode receives a Claim Token MAC frame with an SA equal to the station's address and its AC M-bit equal to 0, enters the Ring Purge Transmit mode, and transmits a Ring Purge MAC frame.

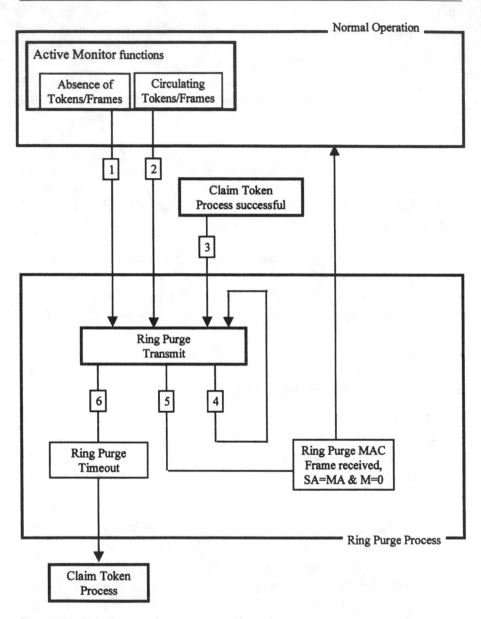

Figure 4.59 Ring Purge process.

4. A station in the Ring Purge Transmit mode detects the need to transmit another Ring Purge MAC frame. How the Ring Purge function retransmits the Ring Purge MAC frames is not specified by the Standard, but the frame is transmitted without using a token. Once the Ring Purge MAC frame is transmitted, the frame is followed by

idles and the station strips the ring until the next frame is transmitted or is terminated by the Ring Purge process.

Ring Purge Transmit Note

Even though not specified in the Standard, most implementations pace the transmission of the Ring Purge MAC frame using the station's TRR or approximately every 4 ms.

5. The station's Ring Purge Transmit function detects the successful completion of the Ring Recovery process by reception of a Ring Purge MAC frame with the SA equal to the AM's station address (MA) *and* an AC M-bit equal to 0. The AM releases a new token at the priority detected in the last or only Ring Purge MAC frame and queues the AMP MAC frame (restarts the Neighbor Notification process).

6. The station's *timer Ring Purge* (TRP) expires, indicating the Ring Purge process has failed in its attempt to recover the ring's token, continues the Ring Recovery process by entering the Claim Token process in the Claim Token Transmit mode, and transmits a Claim Token MAC frame.

4.4.2.4 Claim Token Process

Figure 4.60 illustrates what causes the Claim Token process to be entered and its success and failure conditions.

The following is an explanation of the numbered boxes in Figure 4.60.

1. The Ring Purge process failed in its attempt to restore the ring's token and enters the Claim Token Transmit function by transmitting a Claim Token MAC frame.

2. The Beacon process in the Beacon Transmit function (see Section 4.4.2.5) was successfully completed and enters the Claim Token Transmit function by transmitting a Claim Token MAC frame.

3. Any station that has completed Join *or* an AM station that has not completed Join detects the absence of an AM by failing to receive an AMP MAC frame before its *timer SM* (TSM) expires. When TSM expires, the station enters the Claim Token Transmit function by transmitting a Claim Token MAC frame.

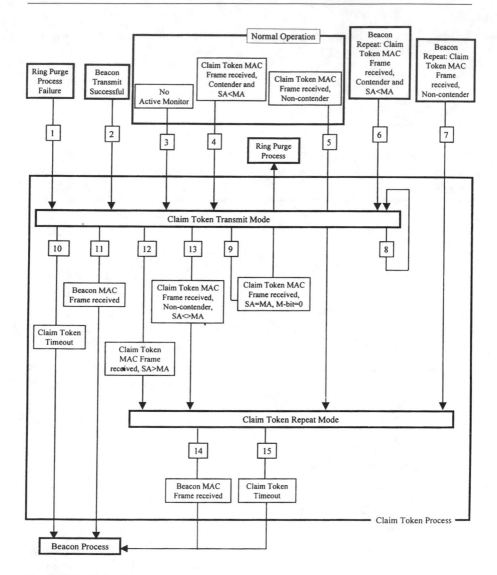

Figure 4.60 Claim Token process.

4. A Contender station in the Repeat function receives a Claim Token MAC frame with an SA that is less than the value of its address and enters the Claim Token Transmit function by transmitting a Claim Token MAC frame.

5. A Noncontender station in the Repeat function receives a Claim Token MAC frame and enters the Claim Token Repeat function.

6. A Contender station in the Beacon Repeat function receives a Claim Token MAC frame with an SA that is less than the value of its address

and enters the Claim Token Transmit function by transmitting a Claim Token MAC frame.

7. A Noncontender station in the Beacon Repeat function receives a Claim Token MAC frame and enters the Claim Token Repeat function.

8. The station in the Claim Token Transmit function detects the need to retransmit the Claim Token MAC frame. The Claim Token Transmit function paces the transmission of the Claim Token MAC frame using the *timer Queue PDU* (TQP) at 10- to 30-ms intervals (TQP is a 20-ms timer) without waiting for a token. Once this frame is transmitted, it is followed by idles. The station strips the ring until the next frame is transmitted or is terminated by the associated function.

9. The station in the Claim Token Transmit function successfully *completes* the Claim Token process when it receives a Claim Token MAC frame with an SA equal to its station address, the AC's M-bit is equal to 0, and enters the Ring Purge process in the Ring Purge Transmit function by transmitting a Ring Purge MAC frame.

10. The station in the Claim Token Process *fails* to complete the Claim Token Transmit mode before its *timer Claim Token* (TCT) expires and enters the Beacon Transmit function by transmitting a Beacon MAC frame.

11. The station in the Claim Token Transmit function receives a Beacon MAC frame and enters the Beacon Repeat function.

12. The station in the Claim Token Transmit function receives a Claim Token MAC frame with an SA greater than its station address (SA > MA) and enters the Claim Token Repeat function.

13. A Noncontender station in the Claim Token Transmit function receives a Claim Token MAC frame with an SA not equal to its station address (SA <> MA) and enters the Claim Token Repeat function.

14. The station in the Claim Token Repeat function *fails* the Claim Token Process when its TCT expires and enters the Beacon Transmit function by transmitting a Beacon MAC frame.

15. The station in the Claim Token Repeat function receives a Beacon MAC frame and enters the Beacon Transmit function by transmitting a Beacon MAC frame.

4.4.2.5 Beacon Process

Figure 4.61 illustrates what causes the Beacon process to be entered and its success and failure conditions.

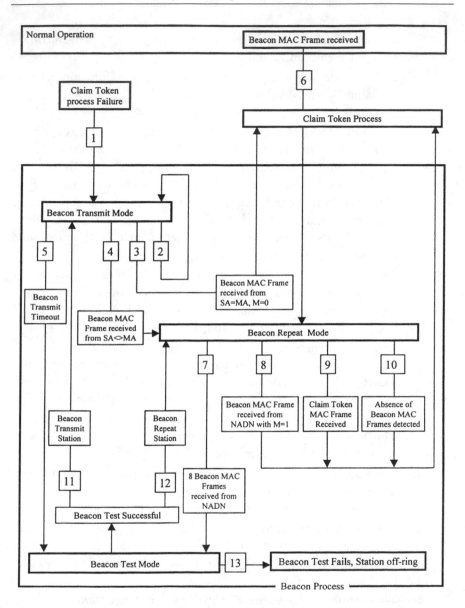

Figure 4.61 Beacon process overview.

The following is an explanation of the numbered boxes in Figure 4.61.

1. The station enters the Beacon process because the TCT expired in the Claim Token Transmit or Claim Token Repeat modes, resets its *timer Beacon Transmit* (TBT), enters the Beacon Transmit mode,

and transmits a Beacon MAC frame with the appropriate Beacon Type code (see Section 4.4.2.2).

2. The station in the Beacon Transmit mode detects the need to retransmit the Beacon MAC frame. The Beacon Transmit function paces the transmission of the Beacon MAC frame using the TQP at 10- to 30-ms intervals (TQP is a 20-ms timer) without waiting for a token. Once this frame is transmitted, it is followed by idles. The station strips the ring until the next frame is transmitted or is terminated by the Beacon process.

3. The station in the Beacon Transmit mode detects the successful completion of the Beacon process when it receives the Beacon MAC frame that it transmitted with the M-bit equal to 0 (see Section 4.4.2.2) and a SA equal to its address, enters the Claim Token process in Claim Token Transmit mode, and transmits a Claim Token MAC frame.

4. The station in the Beacon Transmit mode receives the Beacon MAC frame with a SA not equal to its address and a Beacon Type code (see Section 4.4.2.2) equal to or less than the Beacon Type being used by the station and enters the Beacon Repeat mode.

5. The station in the Beacon Transmit mode detects the expiration of its TBT and enters the Beacon Test function (see Section 4.4.2.2).

6. The station in the normal TKP Access protocol mode (not in Ring Recovery) receives a Beacon MAC frame and enters the Beacon Repeat mode. If the Beacon MAC frame is from its *upstream* station (NAUN), then the station starts its Beacon Repeat Count function (see Section 4.4.2.2).

7. The station in the Beacon Repeat mode detects that it has received eight Beacon MAC frames from its *downstream* station and enters the Beacon Test function (see Section 4.4.2.2).

8. The station in the Beacon Repeat mode detects a possible successful completion of the Beacon process when it receives a circulating Beacon MAC frame (AC M-bit is equal to 1; see Section 4.4.2.2), enters the Claim Token Process in the Claim Token Transmit mode, and transmits the Claim Token MAC frame.

9. The station in the Beacon Repeat mode detects a possible successful completion of the Beacon process when it receives a Claim Token MAC frame, enters the Claim Token Process in the Claim Token Transmit mode, and transmits a Claim Token MAC frame.

10. The station in the Beacon Repeat mode detects a possible successful completion of the Beacon process when its TBR expires (absence of Beacon MAC frames; see Section 4.4.2.2), enters the Claim Token Process in the Claim Token Transmit mode, and transmits a Claim Token MAC frame.

11. The Beacon Test function (see Section 4.4.2.2) is successful, and the station entered the Beacon Test function from the Beacon Transmit mode. The station enters the Beacon Transmit mode.

12. The Beacon Test function (see Section 4.4.2.2) is successful, and the station entered the Beacon Test function from the Beacon Repeat mode. The station enters the Beacon Repeat mode.

13. The Beacon Test function (see Section 4.4.2.2) is unsuccessful, and the station remains off-ring and notifies station management.

4.5 TKP Access Protocol Operational Functions

The Join, Transmit, and Monitor State Operation Tables identified in clause 4 of the Standard are a method of specifying the TKP Access protocol's operational function.

4.5.1 Join Process

This review of the Join process starts with a look at the functions used by Join. Then an overview of the Join process includes a detailed explanation of the success and failure of each Join process function.

4.5.1.1 Join Function Definitions

The Join process uses the following functions:

- LMT;
- Bypass Wait;
- Detect Monitor Present;
- Await New Monitor;
- Duplicate Address Test;
- Neighbor Notification;
- Request Initialization;
- Join Complete.

Lobe Media Test

The LMT function ensures the media connection between the station and the wiring plant (concentrator or switch) is able to sustain data traffic with an acceptable bit error rate (BER). Annex P of the ISO/IEC 8802-5:1998 Amd.1 defines a method of assuring BER.

The Standard does not specify *how* the station performs the LMT but does specify:

- A definition of the repeat path provided by the attachment unit in support of the station's LMT;
- That the acceptable BER of the lobe is *better than* 10^{-5};
- That the station must use standard MAC frames and is not allowed to use invalid transmissions to test the lobe.

Most station implementations transmit the Lobe Test MAC frame enough times to ensure the BER of the link between the station and the attachment unit is better than 10^{-5} (e.g., assuring no more that 1 error in 1,000,000 bits).

Early Implementation Note

Early implementations used an LMT that transmitted over 2,000 MAC frames each over 1,000 octets in size. Although this required the lobe to have a BER of over 10^{-5}, it was soon found that this test also rejected lobes running at a nominal BER as high as 10^{-9}. These early tests were designed to have a bell-shaped acceptance rate centered at 10^{-7} (supposedly assuming the acceptance of lobes running greater than 10^{-6}). These implementations were quickly changed to have a bell-shaped acceptance rate centered at 10^{-6} (assures the acceptance of lobes running greater than 10^{-5}).

Unfortunately, the Standard required the LMT to operate only during the process of joining the ring. Later it was found that a more acceptable process was to require the Beacon Test function (see Section 4.4.2.5) to execute an equivalent of the LMT. Fortunately, this shortcoming was not present in all implementations. During the development of the TXI Access protocol (see Chapter 5), the Standard was changed to require the LMT during the Beacon Test process. However, due to the IEEE grandfather clause protecting past implementations, the TKP Access protocol still does not require the Beacon Test function to execute the LMT.

Bypass Wait

This function is used by the station *after* inserting into the ring to keep its repeat path active until the concentrator or switch has enough time to remove the station from the ring. The purpose of this function is to cause less interruption to the ring while the station is deinserting from the ring.

Detect Monitor Present

This function examines the ring for the presence of the AMP, Ring Purge, or SMP MAC frames.

Await New Monitor

This function executes when the joining station did not detect the presence of an AM. It uses the Claim Token and Ring Purge processes to activate an AM, which releases a token and queues the AMP PDU for transmission on the next available token (starts the Neighbor Notification process).

Duplicate Address Test

The station uses this function to ensure the station's individual address is unique on the ring. The Duplicate Address Test MAC frame is transmitted to a Duplicate Address equal to the individual address of the station joining the ring. If the frame comes back with the A-bits equal to 1 more than twice, then the station has an address equal to another station on the ring, deinserts from the ring, and provides this information to its station management.

Neighbor Notification

The station uses this function to inform its NADN that it has entered the ring. The function uses the Neighbor Notification process defined in Section 4.4.1.3.

Request Initialization

The station uses this function to request permission from the Ring Parameter Server to insert into the ring as follows.

- If the Ring Parameter Server *is* present on the ring, then the station must receive a response to its request for insertion. Failure to receive this response causes the station to deinsert from the ring and provide this information to its station management.

- If the Ring Parameter Server *is not* present on the ring, then the station is allowed to remain on the ring and continues the Join process by entering Join Complete.

Join Complete

The station uses this function to complete the Join process. This state allows the station to communicate with other stations on the ring by activating its ability to receive and transmit non-MAC frames.

4.5.1.2 Join Function Overview

Figure 4.62 illustrates what causes the Join process to be entered and its success and failure conditions.

The following is an explanation of the numbered boxes in Figure 4.62.

1. The station is requested by station management to start joining the ring via the "Connect.MAC" interface signal and enters the LMT function.

2. The station's LMT function *is completed* because it passed its LMT, inserts into the ring, resets the *timer Join Ring* (TJR), and enters the Detect Monitor Present function.

3. The station's LMT function *fails* by detecting its LMT failed, deinserts from the ring, notifies its station management, and enters the Bypass function.

4. The station's Detect Monitor Present function *is completed* since an AM *is* present on the ring by reception of the AMP, Ring Purge, or SMP MAC frames; resets the TJR; and enters the Duplicate Address Test function.

5. The station Detect Monitor Present function *is completed* by determining an AM is not present on the ring or is being established by another station on the ring given the expiration of the TJR or the reception of a Claim Token MAC frame. These two events cause the station to enter the Await New Monitor function.

6. The Detect Monitor Present function *fails* by detecting any of the following conditions:
   - The detection of a "Disconnect.MAC" interface signal from station management;
   - The reception of a Beacon MAC frame;
   - The reception of a Remove MAC frame from an authorized server and the station is configured to allow recognition of the Remove MAC frame;

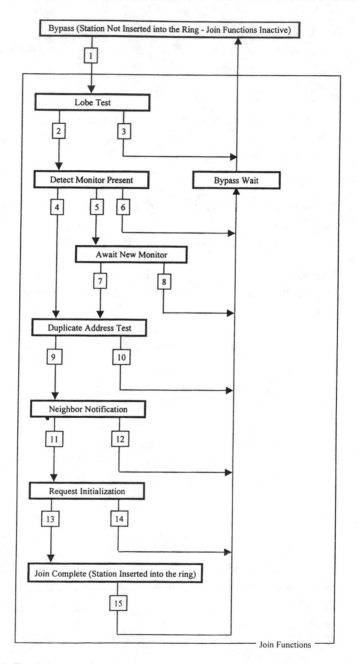

Figure 4.62 The Join process.

- The detection of an unrecoverable station internal error;
- The detection of the PHY signal indicating a ring speed error;
- The optional detection of the presence of a Wire Fault condition.

All of these conditions cause the station to deinsert from the ring and enter the Bypass Wait function.

7. The Await New Monitor function *is completed* by detecting that the AM has been successfully established, resets the TJR, and exits to the Duplicate Address Test function.

8. The Await New Monitor function *fails* by detecting any of the following conditions:
   - The detection of a "Disconnect.MAC" interface signal from station management;
   - The detection of an uncorrectable transmit station error;
   - The reception of a Beacon MAC frame;
   - The reception of a Remove MAC frame from an authorized server and the station is configured to allow recognition of the Remove MAC frame;
   - The detection of an unrecoverable station internal error;
   - The detection of a Claim Token process failure;
   - The detection of a Ring Purge process failure;
   - The optional detection of the presence of a wire fault condition.

All of these conditions cause the station to deinsert and enter the Bypass Wait function.

9. The Duplicate Address Test function *is completed* by detecting the station's individual address is unique to the ring, resets the TJR, and enters the Neighbor Notification function.

10. The Duplicate Address Test function *fails* by detecting any of the following conditions:
    - The detection of a "Disconnect.MAC" interface signal from station management;
    - The detection of an uncorrectable transmit station error;
    - The reception of a Beacon MAC frame;
    - The detection of a Duplicate Address Test failure;
    - The reception of a Remove MAC frame from an authorized server and the station is configured to allow recognition of the Remove MAC frame;
    - The detection of the Signal Loss condition when not the AM;
    - The detection of an unrecoverable station internal error;
    - The detection of a Claim Token process failure;
    - The expiration of the TJR;

- The detection of a Ring Purge process failure;
- The optional detection of the presence of a Wire Fault condition. All of these conditions cause the station to deinsert and enter the Bypass Wait function.

11. The Neighbor Notification function *is completed* by detecting either its own AMP MAC frame or another station's SMP MAC frame; resets the TJR; and enters the Request Initialization function.

12. The Neighbor Notification function *fails* by detecting any of the following conditions:
    - The detection of a "Disconnect.MAC" interface signal from station management;
    - The detection of an uncorrectable transmit station error;
    - The reception of a Beacon MAC frame;
    - The reception of a Remove MAC frame from an authorized server and the station is configured to allow recognition of the Remove MAC frame;
    - The detection of the Signal Loss condition when not the AM;
    - The detection of an unrecoverable station internal error;
    - The detection of a Claim Token process failure;
    - The expiration of the TJR;
    - The detection of a Ring Purge process failure;
    - The optional detection of the presence of a Wire Fault condition.
    All of these conditions cause the station to deinsert and enter the Bypass Wait function.

13. The Request Initialization function *is completed* successfully by detecting the absence of the Ring Parameter Server or receiving the Initialize or Set Parameters MAC frames. These two events cause the station to complete the Join process by entering the Join Complete function.

14. The Request Initialization function *fails* by detecting any of the following conditions:
    - The detection of a nonresponse from the Ring Parameter Server function;
    - The detection of a "Disconnect.MAC" interface signal from station management;
    - The detection of an uncorrectable transmit station error;
    - The reception of a Beacon MAC frame and the station is not yet handling the Ring Recovery process;
    - The reception of a Remove MAC frame from an authorized server and the station is configured to allow recognition of the Remove MAC frame;

- The detection of an unrecoverable station internal error;
- The expiration of the TJR;
- The detection of a Ring Recovery process failure and the station is not yet handling the Ring Recovery process;
- The optional detection of the presence of a Wire Fault condition.

All of these conditions cause the station to deinsert and enter the Bypass Wait function.

15. The Join Complete function *fails* by detecting any of the following conditions:
    - The detection of a "Disconnect.MAC" interface signal from station management;
    - The detection of an uncorrectable transmit station error;
    - The reception of a Remove MAC frame from an authorized server and the station is configured to allow recognition of the Remove MAC frame;
    - The detection of an unrecoverable station internal error;
    - The detection of the presence of a Wire Fault condition and the station is not executing the Beacon Test function.

    All of these conditions cause the station to deinsert and enter the Bypass Wait function.

Note About Join and a Beaconing Ring

Some station implementations provide "hooks" to allow a station to insert into the ring even if the ring is Beaconing. This function is useful because it allows the service technician easy access to the *addresses* of the two stations (the Fault Domain; see Section 4.3.7.1) involved in the Beacon process (see Section 4.4.2.5).

4.5.2 Transmit Process

This subsection discussing the functions used by the Transmit process and provides an overview of the Transmit process.

4.5.2.1 Transmit Function Definitions

The functions used by the Transmit process are Repeat, Transmit Data, Fill, and Strip.

Repeat

The Repeat function is the normal condition of the transmit process when not transmitting or stripping frames and is responsible for:

1. Responding to the Ring Recovery process requests for Transmit Immediate. Only the Beacon, Claim Token, and Ring Purge MAC frames are allowed.

2. Responding to the MAC and LLC layers request for PDU Transmit.
   a. If the PDU exceeds the maximum frame length allowed, the station remains in the Repeat function without transmitting the frame (transmit request is lost).
   b. If the PDU does not exceed the maximum frame length allowed, the station examines the incoming token priority for Transmit as follows.
      (1) When the PDU priority is equal to or greater than the priority of the token, the station starts transmitting data by inserting the FC field, the DA, and the SA fields; resets the TRR; and then enters the Transmit Data function.
      (2) When the PDU priority is less than the priority of the token, the station remains in the Repeat function waiting for the next token but attempts to make reservations by setting the token's Access Control Priority Reservation bits (RRR).
         (a) The station makes reservations when the RRR bits are less than the priority of the PDU by setting the RRR field to the priority of the PDU.
         (b) The station loses it chance to make reservations when the RRR bits are equal to or greater than the priority of the PDU and repeats the RRR field unchanged.
   c. If the PDU does not exceed the maximum allowed frame length, the station examines the incoming frame's Priority Reservation field (RRR) as follows.
      (1) The station makes reservations when the RRR bits are less than the priority of the PDU by setting the RRR field to the priority of the PDU.
      (2) The station loses it chance to make reservations when the RRR bits are equal to or greater than the priority of the PDU and repeats the RRR field unchanged.

Transmit Data Function

The Transmit Data function is responsible for the transmission of data. An overview of its responsibilities include the following.

1. Detection of the following error conditions occur *any* time the Transmit Data function is operating.

a. Assure the third octet of the token is an ED and if not, increment the Token Error counter and enter the Repeat function (transmit request remains outstanding);

b. The detection of any error condition requiring the transmission of the Abort Sequence, incrementing the Abort Sequence counter, and returning to the Repeat State (transmit request is lost);

c. The detection of a station error requiring the station to deinsert by setting a *Force Bypass conditional flag* (FBPF) to 1 (recognized by the Join process) and entering the Repeat function;

d. Optionally detecting a transmit overrun condition, transmitting an Abort Sequence, and entering the Fill function.

2. Transmitting the requested frame.
   a. Calculate the frame's FCS as the frame is transmitted.
   b. Once the first frame's data is transmitted, take one of the following actions:
      (1) If transmitting *one* frame or the *last* of multiple frames, then end the current frame by transmitting the appropriately set ED, Frame Status, and IFG fields and entering the Fill function.
      (2) If optionally transmitting *multiple* frames, then assure the next frame does not exceed the maximum total frame length allowed for *all* frames.
         (a) If it would exceed the maximum total frame length allowed, then do not transmit the next frame but end the current frame by transmitting the appropriately set ED, Frame Status, and IFG fields and entering the Fill function.
         (b) If it would not exceed the maximum total frame length allowed, then end the current frame by transmitting the appropriately set ED, Frame Status, and IFG fields; then transmit the next frame's SD, AC (same value as the first frame), FC, DA, and SA; and restart the Transmit function at the beginning of this item.

Fill Function

The Fill function strips the transmitted frame or frames and is responsible for the release of the token. If the conditions for transmitting a token *are* detected, then the station sets the priority fields PPP and RRR as defined in Priority Operations (see Section 4.3.3) and enters the Strip function. If the conditions for transmitting a token *are not* detected before the TRR (see Section 4.4.2) expires, then the station increments the Lost Frame error counter (see Section 4.4.4.3) and enters the Repeat function *without* releasing a token.

Strip Frame and Fill Function

The Strip Frame and Fill function strips the remainder of the transmitted frame or frames by searching the incoming data stream for the last or only frame transmitted and takes one of the following actions. At the same time it is searching for this ending condition, the station is transmitting idles (Fill).

- When the conditions for ending the Strip function *occur* before the TRR (see Section 4.4.2), the station has successfully transmitted its frame or frames and stops stripping the ring and transmitting idles (Fill) by entering the Repeat function.

- When the conditions for ending the Strip function *do not occur* before the TRR expires, the station may or may not have successfully transmitted its frame or frames but has lost the frame or frames it transmitted. The station increments the Lost Frame error counter (see Section 4.4.4.3) and stops stripping the ring and transmitting idles (Fill) by entering the Repeat function *without* releasing a token.

4.5.2.2 Transmit Function Overview

Figure 4.63 illustrates what causes the Transmit functions to be entered and their success and failure conditions.

The following is an explanation of the numbered boxes in Figure 4.63

1. The station in the Repeat function detects a token with an *AC Priority field* (PPP) greater than the priority of the PDU, sets the Priority Reservation field (RRR) as defined in the Repeat function, and remains in the Repeat function waiting for the next token.

2. The station detects a token with the PPP equal to or less than the priority of the PDU, takes the action defined by the Repeat function, and enters the Transmit function.

3. The station has completed transmitting its frame or frames, takes the action defined for the Transmit Frame function, and enters the Strip Frame and Release Token function.

4. The station detected an error condition, takes the action defined for the Transmit function, and enters the Repeat function.

5. The station successfully detects the conditions necessary for token release, sets the priority fields PPP and RRR as defined in Priority Operations (see Section 4.3.3), and enters the Fill function.

6. The station detected an error condition, takes the action defined for the Strip Frame and Release Token function (see Section 4.5.2.1), and enters the Repeat function.

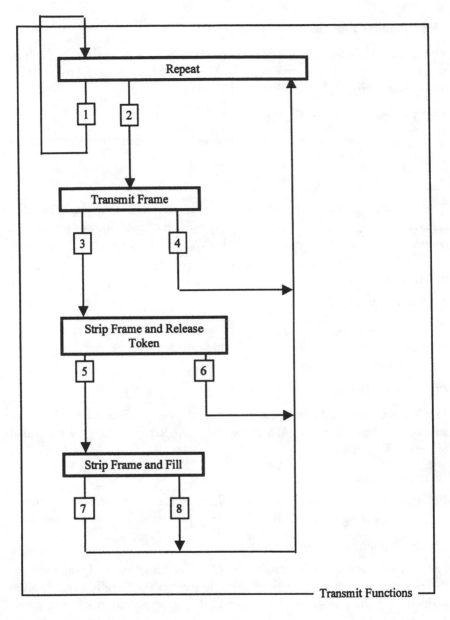

Figure 4.63 The Transmit process.

7. The station successfully detected the necessary conditions for ending the Strip Frame and Fill function and enters the Repeat function.

8. The station detected an error condition, takes the action defined for the Strip Frame and Fill function (see Section 4.5.2.1), and enters the Repeat function.

4.5.3   Monitor Process

This discussion of the Monitor process covers the functions used by the Monitor process and provides an overview of the Monitor process relying on the details of the Ring Purge, Claim Token, and Beacon processes.

4.5.3.1   Monitor Function Definitions

The functions used by the Monitor process are Repeat, Ring Purge, Claim Token, and Beacon.

Repeat Function

The Repeat function is activated by the Join function once the Join LMT function is complete. This is the normal condition of the Monitor process and allows the station to transmit and receive frames. Whether the station is the AM or SM determines which functions TKP Access protocol monitoring functions are active.

Ring Purge Process

The AM station uses the Ring Purge process defined in Section 4.4.2.3 in an attempt to put a new token on the ring using the Ring Recovery process defined in Section 4.4.2.

Claim Token Process

The station uses the Claim Token Process defined in Section 4.4.2.4 in an attempt to establish a new AM using the Ring Recovery process defined in Section 4.4.2.

Beacon Process

The station uses the Beacon process defined in Section 4.4.2.5 in an attempt to recover the ring using the Ring Recovery process defined in Section 4.4.2.

4.5.3.2   Monitor Function Overview

Figure 4.64 illustrates what causes the Monitor functions to be entered and their success and failure conditions. This figure uses the Ring Recovery process defined in Section 4.4.2 and is followed by an explanation of each of the numbered boxes.

The following explanation of the numbered boxes in Figure 4.64 uses the Monitor functions defined in Section 4.5.3.1 and the Ring Recovery process defined in Section 4.4.2.

  1. The AM station detects the *need* for executing the Ring Purge process defined in Section 4.4.2.3.

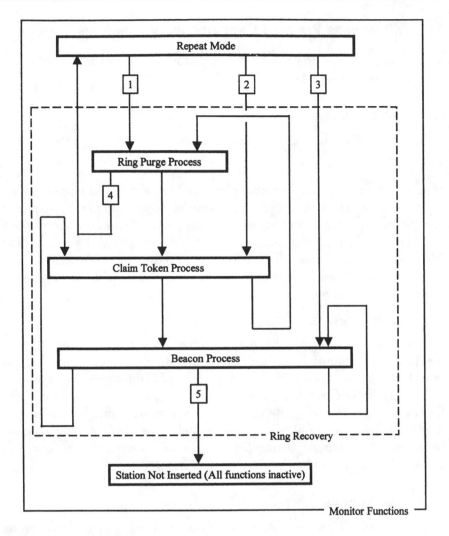

Figure 4.64 The Monitor process.

2. The station detects the *need* for executing the Claim Token Process defined in Section 4.4.2.4.

3. The station detects the *need* for executing the Beacon process defined in Section 4.4.2.5.

4. The AM station detects the *successful* completion of the Ring Purge process and enters the Repeat function.

5. The station detects the *failure* of the Beacon Test function defined in Section 4.4.2.5 and remains off-ring.

4.6 TKP Access Protocol Error Recovery Functions

This section discusses how the TKP Access protocol recovers from hard errors and soft errors.

4.6.1 Types of Errors

Failures in the Token Ring are classified as either hard errors or soft errors.

4.6.1.1 Hard Errors

Hard errors prevent the TKP Access protocol from restoring normal Token Ring operation. Hard errors include bit and frame streaming error, frequency error, signal loss error, and internal error conditions. Some hard errors allow the Ring Recovery process to automatically recover normal token operation while others require manual intervention by the customer, with older unpowered concentrators, and ring recovery by the concentrator, with newer powered and intelligent concentrators.

4.6.1.2 Soft Errors

Soft errors either allow the TKP Access protocol to operate normally or allow the TKP Access protocol to restore the Token Ring to normal operation but cause higher layer performance degradation (e.g., LLC throughput degradation). Soft errors include line error (frame and token bit errors), lost tokens and frames, lost monitor (e.g., by deinsertion or powered off), lost delimiters, multiple tokens, circulating priority tokens and frames, and multiple monitors (often caused when multiple rings are joined either by manual reconfiguration and intelligent concentrator recovery procedures).

4.6.2 Hard Error Recovery

The discussion of Hard Error Recovery will proceed according to the following outline:

- Definition of the hard error types used in the explanation of the two Hard Error Recovery time lines illustrated in Figures 4.65 and 4.66;
- Definition of reconfiguration used in the explanation of these Hard Error Recovery time lines;
- Definition of the Hard Error Recovery time lines to explain Hard Error Recovery as a function of time;
- Definition of the Beacon Remove functions that attempt to accomplish automatic reconfiguration used in the Solid Hard Error Recovery time lines illustrated in Figure 4.65.

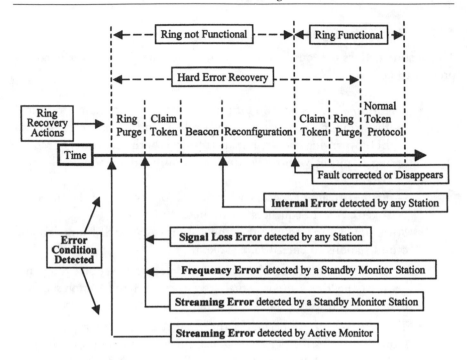

Figure 4.65 Solid Hard Error Recovery time line.

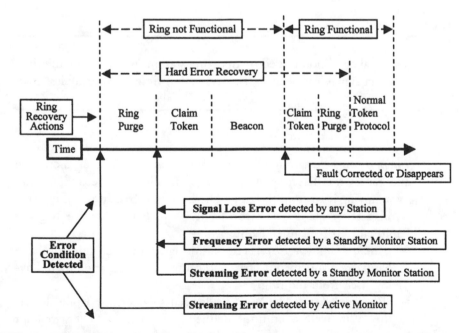

Figure 4.66 Intermittent Hard Error Recovery time line.

4.6.2.1   Hard Error Types

This section defines the four hard error types that cause Hard Error Recovery process to be invoked.

- The *streaming error* has two forms: bit streaming and frame streaming.
  - Bit streaming removes (destroys) tokens and frames (e.g., SD and ED) by writing over (repeat and transmit both occurring) or replacing (transmit without waiting for transmit opportunity) the ring data.
  - Frame streaming occurs when a station continuously transmits tokens, Abort Sequences, or frames.
  Streaming is detected by either the station's AM or SM function and causes stations to enter Claim Token.
- A *frequency error* occurs when a station detects an excessive difference between the ring's input signal frequency and the station's crystal frequency and causes stations to enter Claim Token.
- The *signal loss error* occurs as the result of a broken ring, receiver saturation (excessive signal), a transmitter malfunction, or a receiver malfunction and causes a station to eventually cause reconfiguration.

Note

The loss of signal condition caused by inserting stations is *masked* by the time it takes to detect signal loss because relay transition time is generally less than 3 ms.

- *Internal errors* are detected by the station's hardware and/or microcode and causes the detecting station to eventually cause reconfiguration by removing itself from the ring. There are four basic types of internal error: Station Hardware Error Fault, Wire Fault, Receive Register Fault, and Strip Register Fault.
  - *Station Hardware Error:* The station encounters an unrecoverable hardware or microcode error and causes reconfiguration by removing itself from the ring.
  - *Wire Fault Error:* The Wire Fault occurs when the adapter detects the "Wire Fault condition" defined in Section 2.3. The wire fault detection capability of Token Ring is the first commercial attempt

to identify media problems between the connection mechanism (e.g., concentrator) and the station and was an important step forward in fault diagnosis of media problems.

— *Receive Address Register Fault:* The Receive Address Register is used to compare addresses on the ring to the address of the station. If the station's Receive Register cannot make a comparison between a frame's address equal to its address, then no frames are copied. This fault is not detected by the TKP Access protocol; thus, the station needs, for example, a background test that tests for this condition (test should operate when the station is not busy).

— *Strip Address Register Fault:* The Strip Register Fault is not detected by the TKP Access protocol and thus needs the assistance of the station, such as an internal test that operates when the TRR expires. When TRR expires, the station should test the Strip Address Register against the *transmitted* frame's SA (may be different than the station's individual address). If these registers compare equal, then the station can assume the frame was lost due to a line hit.

If these registers do not compare equally, then the station should assume a hardware error and cause reconfiguration by removing from the ring.

4.6.2.2 Hard Error Recovery Time Lines

Hard Error Recovery is invoked as the result of fault conditions that are either *solid* (fault is long term and causes reconfiguration) or *intermittent* (fault is short term and disappears before reconfiguration is required). Figure 4.65 illustrates Hard Error Recovery as a function of time for a *solid* error, and Figure 4.66 illustrates Hard Error Recovery as a function of time for an *intermittent* error.

The time lines for both of these figures operate as follows. Each identified error condition enters the time line at the point shown and then the TKP Access protocol executes the *processes* going in the direction of the time line. For example, Figure 4.65 illustrates the error condition "Streaming Error Detected by the AM" as causing the Ring Purge process to begin; if this process fails, the Claim Token process begins and, if this process fails, the Beacon process begins.

Solid Hard Error Recovery Time Line

The following is an explanation of the highlighted error condition terms in Figure 4.65.

Detection of the *streaming error* by the AM causes the station to enter the Ring Purge and Claim Token processes, which both will fail, causing the station to enter Beacon Transmit mode and eventually causing reconfiguration.

Detection of the *streaming error* by the SM causes the station to enter the Claim Token process, which fails, causing the station to enter Beacon Transmit mode and eventually causing reconfiguration.

Detection of the *frequency error* by the SM causes the station to enter the Claim Token process, which fails, causing the station to enter Beacon Transmit mode and eventually causing reconfiguration.

Detection of the *signal loss error* by any station causes the station to enter the Claim Token process, which fails, causing the station to enter Beacon Transmit mode and eventually causing reconfiguration.

Detection of an *internal error* by any station causes the station to configure itself out of the ring. It is possible for this error to cause other stations to enter the Ring Purge, Claim Token, or Beacon processes or to have no impact to the 802.5 Token Ring protocol.

Reconfiguration is started when the Beacon process settles (see Beacon Recovery process).

Intermittent Hard Error Recovery Time Line

The following is an explanation of the highlighted error condition terms in Figure 4.66.

Detection of the *streaming error* by the AM causes the station to enter the Ring Purge and Claim Token processes, which both fail, causing the station to enter Beacon Transmit mode when the fault disappears.

Detection of the *streaming error* the SM causes the station to enter the Claim Token process, which fails, causing the station to enter Beacon Transmit mode when the fault disappears.

Detection of the *frequency error* by the SM causes the station to enter the Claim Token process, which fails, causing the station to enter Beacon Transmit mode when the fault disappears.

Detection of the *signal loss error* by any station causes the station to enter the Claim Token process, which fails, causing the station to enter Beacon Transmit mode when the fault disappears.

Reconfiguration is started when the Beacon process settles (see Beacon Recovery process).

Note

In the preceding scenario, both Ring Purge and Claim Token failed. However, it is possible for the Ring Purge or the Claim Token processes to execute when the "Ring Is Not Operational" to start functioning correctly if the fault disappears to skip directly to either the Ring Purge or the Claim Token process in the "Ring Operational" phase.

4.6.2.3 Beacon Recovery Process

The Beacon Recovery process *starts* when the first station transmits a Beacon MAC frame and *ends* when the TKP Access protocol is restarted. This explanation of the Beacon Recovery process procedes according to the following outline.

- How the TKP Access protocol *establishes* where the fault exists;
- How *reconfiguration* is used to remove the fault;
- How the TKP Access protocol is *restarted* when the fault is removed.

The Beacon Recovery process is illustrated in Figure 4.67 when automatic reconfiguration is successful and in Figure 4.68 when automatic reconfiguration is unsuccessful.

The components of Beacon Recovery process illustrated in Figures 4.67 and 4.68 are defined as follows.

Fault Domain Establishment

The Beacon process is designed to first detect hard errors and then to establish the fault domain (see Section 4.3.7.1). The fault domain is established by a Beacon process designed to "walk" the ring backward (against the ring's data flow) toward the fault domain containing the detected hard error.

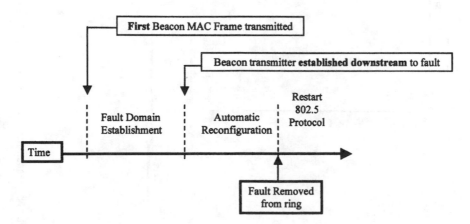

Figure 4.67 Beacon Recovery automatic reconfiguration successful.

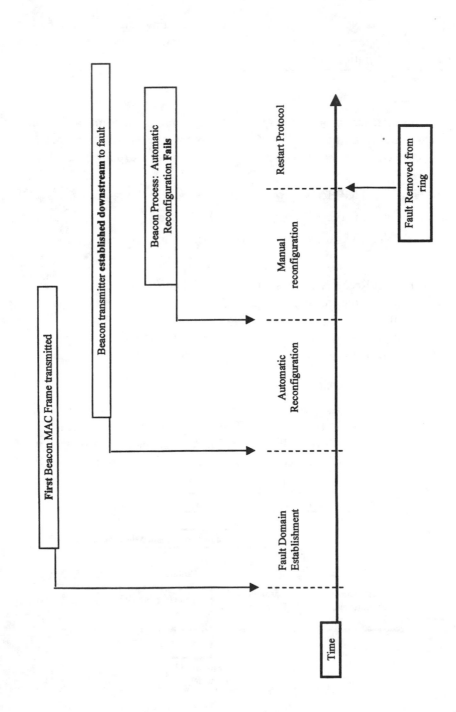

Figure 4.68 Beacon Recovery automatic reconfiguration unsuccessful.

Once the station downstream to the hard error is the Beacon MAC frame transmitter, the Beacon Repeat and Beacon Transmit functions are activated. These functions operate in a given sequence to test their stations as described by reconfiguration in the next subsection.

Reconfiguration

Using reconfiguration, the station in *Beacon Repeat* (BNR) and the station in *Beacon Transmit* (BNT) attempt to recover the ring, each at their own particular time on the reconfiguration time line (see the BNR and BNT time lines in Figure 4.69). The TKP Access protocol assumes that the BNR station is *upstream* to the source or is the cause of the hard error; or the BNT station is *downstream* or is the cause of the hard error. If neither the BNR station nor the BNT station is the source of the hard error, automatic reconfiguration using the TKP Access protocol will be unsuccessful. Reconfiguration using the TKP Access protocol is illustrated in Figure 4.69, followed by an explanation of this figure's key points.

The following key points about Figure 4.69 need to be understood.

A: The BNR station deinserts after receiving eight Beacon MAC frames from its *downstream* station (the Beacon MAC frame received by the BNR station contains the UNA subvector with its individual address), successfully executes its internal self-test, reinserts, remains inserted, and waits for the TKP Access protocol recovery to occur.

B: The BNT station must wait long enough to allow the BNR station that is not the source of the hard error to deinsert, test itself, and reinsert *before* the BNT station starts executing its station tests.

C: The BNT station deinserts after waiting for 18 to 26s (depends on when implementation occurred), successfully executes its internal self-test, reinserts, remains inserted, and waits for the TKP Access protocol recovery to occur.

At this point, reconfiguration using the TKP Access protocol has failed to automatically recover the ring because neither the BNR station nor the BNT station is the source of the hard error (see Figure 4.68).

R1: The BNR station has deinserted and its self-test is unsuccessful. The source of the hard error has been removed from the ring and the BNT station starts recovery when it receives its own Beacon MAC frame (see Figure 4.67).

R2: The BNT station has deinserted and its self-test is unsuccessful. The source of the hard error has been removed from the ring and the station downstream to the BNT station recognizes that the BNT station has removed itself and starts recovery (see Figure 4.67).

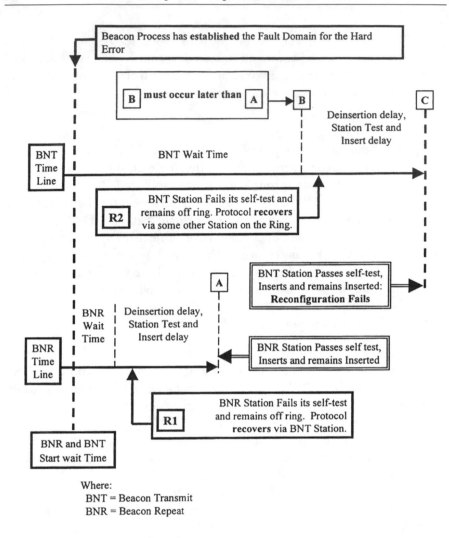

Figure 4.69 Automatic reconfiguration.

Manual Reconfiguration

The customer must take manual reconfiguration steps, or reconfiguration supported by intelligent concentrators must be used to recover the TKP Access protocol if time "C" in Figure 4.69 is reached using the TKP Access protocol.

Restart TKP Access Protocol

The TKP Access protocol is restarted when a station recognizes that it is either receiving its own Beacon MAC frame (point "R1" in Figure 4.69) or another station recognizes the absence of Beacon MAC frames. These stations enter the Claim Token process to start the restoration of the TKP Access protocol (see Figures 4.65 and 4.66).

4.6.2.4 Hard Error Recovery Example

This section covers the Ring Recovery sequence that takes place when a station detects a solid Signal Loss condition to illustrate the Beacon Recovery process described in Section 4.6.2.3.

The following three cases of Signal Loss detection by the Beacon process are possible.

- Signal Loss detected due to a *ring fault* between stations 114 and 185, as illustrated in Figure 4.70.

- Signal Loss detected due to a *receiver fault* in station 185 as illustrated in Figure 4.71.

- Signal Loss detected due to *transmitter fault* in station 114 as illustrated in Figure 4.72.

These three cases are all handled using the *same* process; that is, the *receiver* detects the existence of a fault by detecting the *signal loss condition*. The end result differs, however, because of *where* the fault exists. For example, the TKP Access protocol cannot recover from the media fault illustrated in Figure 4.70. On the other hand, faults in the *receiver* or *transmitter* illustrated in Figures 4.71 and 4.72, respectively, may cause the TKP Access protocol to remove the failing station. When the failing station removes itself, those stations remaining on the ring accomplish Ring Recovery.

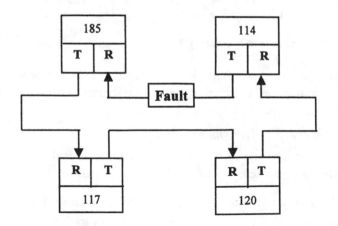

Figure 4.70 Ring fault between stations 114 and 185.

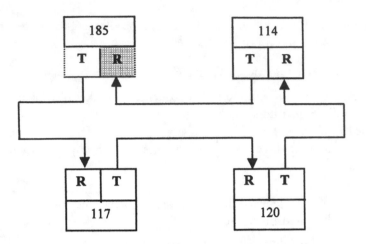

Figure 4.71 Receiver fault in station 185.

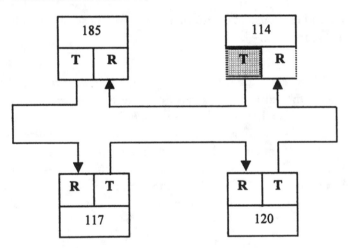

Figure 4.72 Transmitter fault in station 114.

Because all of these examples use the same recovery process, only the transmitter fault is used to describe the step-by-step process used by the Beacon Recovery process described in Section 4.6.2.3.

Detailed Recovery of Transmitter Fault

Figure 4.72 illustrates station 114 that has a transmitter fault causing the Signal Loss condition to be detected by its downstream station (185).

1. Station 185 detects a Burst 4 error caused by station 114's malfunction.
2. Station 185 detects a Burst 5 error caused by station 114's malfunction.
3. States of stations 185, 117, 120, and 114 will vary depending upon which station's Token protocol timers expire first.

    a. When the AM's TVX timer expires (no tokens or frames detected in 10 ms), it enters the Purge Transmit mode until its TPR timer expires (1 to 2s). When TPR expires, the station enters the Claim Token Transmit mode and transmits a Claim Token MAC frame with either error code X'0003' (no Claim Token MAC frames detected) or X'0004' (Claim Token MAC frames detected).

    b. If the SM's timer for no good tokens expires (2.6 to 5.2s), they enter the Claim Token Transmit mode and transmit a Claim Token MAC frame with either error code X'0003' (no Claim Token MAC frames detected) or X'0004' (Claim Token MAC frames detected).

4. When station 185 detects the Signal Loss condition (250 to 500 ms), it enters the Claim Token Transmit mode and transmits a Claim Token MAC frame with either error code X'0003' (no Claim Token MAC frames detected) or X'0004' (Claim Token MAC frames detected). Stations 114, 117, and 120 enter the Claim Token Repeat mode upon receiving the station 185 Claim Token MAC frame. When the station 185 Claim Token timer expires, station 185 enters Beacon Transmit mode by transmitting a Beacon MAC frame with error code of X'0002' (meaning the station entered Claim Token because of a signal loss error).

5. If stations 114, 117, and 120 have entered either the Claim Token process or the Beacon process, they enter the Beacon Repeat mode upon receiving the station 185 Beacon MAC frame.

6. Station 114 is the station upstream to the Beacon Transmitter and starts counting Beacon MAC frames. When station 114 has counted eight Beacon MAC frames, it removes from the ring and enters the Bacon Test process, which does an internal test of the station. Because the Transmitter is malfunctioning, station 114 stays off-ring.

7. Station 185 remains in Beacon Transmit mode until station 114 removes itself to perform the Beacon Test process. When station 114 is no longer on the ring, station 185 detects its own Beacon MAC frame and starts the Claim Token process.

8. The Claim Token process is completed successfully because the faulty station 114 has been removed from the ring. A new AM is selected that produces a new token causing the successful completion of Ring Recovery.

4.6.3 Soft Error Recovery

This discussion of Soft Error Recovery first addresses the definition of soft error types and then that of the Soft Error Recovery time line, which explains the Soft Error Recovery process as a function of time.

4.6.3.1 Soft Error Types

Four types of soft errors are defined.

- Type 1 errors require no Ring Recovery to be executed.
- Type 2 errors require the Ring Purge function to be executed by the AM to restore a token on the ring.
- Type 3 errors require the Claim Token function to restore an AM for the ring and the Ring Purge function to be executed by the newly elected AM to restore a token on the ring.
- Type 4 errors require the Beacon function to establish the Fault Domain, the Claim Token function to restore an AM for the ring, and the Ring Purge function to be executed by the newly elected AM to restore a token on the ring.

Type 1 Soft Errors (SE_T1)

Type 1 soft errors consist of four error conditions: line error, short-term burst error, A/C set error, and multiple AM error.

- *Line error* is detected as defined in Section 4.3.7.2.
- *Short-term burst error* is detected when the Burst Error, defined in Section 4.3.7.2, occurs over a short enough period of time that prevents a frame or token from being removed from the ring. For example, it is possible that within one frame, both a burst error and line error are detected.
- *A/C set error* is defined in Section 4.3.7.2.
- *Multiple monitor error* is detected when the AM station receives an AM Present MAC frame that it did not transmit. The AM detecting this error condition queues for transmission the Report Monitor Error MAC frame with an error code of X'0002' and becomes an SM (this may cause the ring to be without an AM).

These SE_T1 error conditions may cause frames and tokens to be lost.

Type 2 Soft Errors (SE_T2)

Type 2 soft errors consist of seven error conditions: long-term burst error, lost frame error, multiple tokens, corrupted token error, lost token error, circulating priority token error, and circulating frame error. All of these errors cause the AM to execute its Ring Purge function (exceptions noted).

- *Long-term burst error* is detected when the burst error defined in Section 4.3.7.2 occurs over a long enough period of time to cause a frame or token to be removed from the ring.

- *Lost frame error* is detected as defined in Section 4.3.7.3 and leaves the ring without a token. The AM station detects the absence of a token, executes its Ring Purge function, and then releases a new token.

- *Multiple token error* is detected indirectly by the AM. Multiple tokens allow two stations to start transmitting and neither station can strip the frame they transmitted, so both stations strip the ring of all data until the TRR forces them to quit transmitting and return to repeat without releasing a token. The AM station detects the absence of a token, executes its Ring Purge function, and then releases a new token.

- *Corrupted token error* is detected by a station when it detects what appears to be a token. Upon detecting that the third octet of the token is not an ED, the station stops transmitting by releasing an Abort Sequence and returning to repeat without releasing a token. The AM station detects the absence of a token, executes its Ring Purge function, and then releases a new token.

- *Lost token error* is detected as defined in Section 4.3.7.3 and causes the AM station to detect the absence of a token, execute its Ring Purge function, and release a new token.

- *Circulating priority token error* is detected as a token error defined in Section 4.3.7.3 and causes the AM station to destroy the circulating priority token, execute its Ring Purge function, and release a new token. Some implementations change the circulating priority token to a new token on the fly and do not execute the Ring Purge function, providing a faster recovery time for this error. This error can also be caused by the multiple AM error defined earlier under SE_T1 errors.

- *Circulating frame error* is detected as a token error defined in Section 4.3.7.3 and causes the AM station to execute its Ring Purge function (insures removal of the circulating frame) and release a new token. This error can also be caused by the multiple AM error defined earlier under SE_T1 errors.

These SE_T2 error conditions cause frames and tokens to be lost.

Type 3 Soft Errors (SE_T3)

Type 3 soft errors consist of two error conditions: lost monitor and frequency error. Both of these errors cause the Claim Token function to be executed until a new AM is selected and executes its Ring Purge function. These SE_T3 error conditions cause frames and tokens to be lost.

Type 4 Soft Errors (SE_T4)

The Type 4 soft error is caused when the Claim Token function cannot be resolved. If the Claim Token function is resolved, it is classified as a SE_T4 error. If this condition is caused by a hard error, then the Beacon Recovery process (see Section 4.6.2) is executed. These SE_T4 error conditions cause frames and tokens to be lost.

4.6.3.2   Soft Error Recovery Time Line

Figure 4.73 illustrates and explains the Soft Error Recovery time line caused when one of the four soft error types is detected.

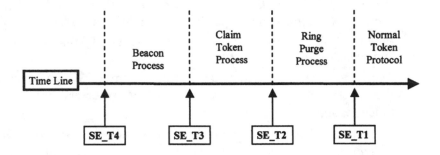

Where:

- **SE_T1** affects the 802.5 Token Ring Protocols, but no ring recovery functions are started.
- **SE_T3** is recovered by the execution of the Ring Purge Processes.
- **SE_T3** is recovered by the execution of the Claim Token and Ring Purge Processes.
- **SE_T4** is recovered by the execution of the Beacon, Claim Token and Ring Purge Processes.

Figure 4.73 Soft Error Recovery time line.

5

Dedicated Token Ring

5.1 Overview

When IEEE 802.5 completed the update that specified requirements for the operation of both 4- and 16-Mbit/s Token Ring on telephone wire, they formed a study group to look into the next Token Ring enhancement requiring standardization. Because the first step in improved Token Ring performance was to increase speed from 4 to 16 Mbit/s, most committee members assumed that another increase in speed was in order. However, after reviewing actual user requirements and technical feasibility, it became apparent that a more useful enhancement would be to deliver more of Token Ring's bandwidth to each of the stations attached to the LAN and utilize high-speed internal backplanes in hubs to bridge/switch the data between ports. The committee was turning from higher speed Token Ring to Token Ring switching. Later focus was on the need for full backward compatibility with Classic Token Ring and the requirement to support a full range of solutions that delivered the full range of performance options, each at an appropriate price/performance point. It became clear that the Token Ring switching solution provided a solution at the high end and that Classic Token Ring provided the basis for the performance range presently being addressed. Further definition of topologies and expansion of interconnect capabilities to basic Switched Token Ring Stations and Concentrators resulted in the creation of Dedicated Token Ring (DTR), the family of solutions with the capability to deliver the full range of performance options, each at an appropriate price/performance point.

5.1.1 Toward Higher Performance

The development of DTR began with a change in how LANs were viewed. During the early 1980s LANs were seen as a way to interconnect active devices such as workstations, servers, and main frame computers together. When required, active repeaters were placed in the cabling plant to extend drive distance. Bridges were used to move data from one LAN segment to another. Concentrators used to interconnect active stations were generally passive. Even the concentrator's relay circuits required to switch stations onto and off the ring were controlled remotely by that station's phantom current.

The development of the Token Ring standard for transmission over telephone wire forced the Standards development committee to focus on and define the use of active concentrators that retimed and amplified the signal each time it reached the wiring closet. Along with active concentrators came the introduction of intelligent hubs to the network. Within this framework the question, "How do we increase the performance of Token Ring?" brought a larger array of solutions than simply, "Increase its transmission speed."

The following requirements, developed by the committee to answer this question, led directly to the creation of DTR as the next major development for the IEEE 802.5 Standard.

- Solve the problems of backbone congestion, local ring congestion, and server congestion;
- Support high-performance applications such as multimedia;
- Allow graceful migration from Classic Token Ring operation;
- Minimize overall costs to the user by maintaining compatibility with existing hardware, application software, and drivers;
- Utilize the existing cabling plant, including Category 3 UTP telephone wire.

5.1.2 Developing a Solution

The obvious solution of increasing the Token Ring data rate had a number of drawbacks. Adapters operating at higher speeds could not be intermixed with old adapters. Operation on existing cabling, especially Category 3 UTP cabling, would be difficult or impossible without significantly increasing the cost of the adapters and concentrators. By developing a solution based on dedicated connections and bandwidth, and with the added capability of backward compatibility, all of the requirements could be met. By dedicating a full 16 Mbit/s to each end station, high-bandwidth multimedia applications could be supported. A high-speed backplane contained in a DTR concentrator could provide for significantly increased data flow between stations and servers. Frame formats and data rates were defined as the same for both Classic and DTR. Therefore, adapters designed for DTR could easily accommodate the requirements demanded by the Classic Token Ring protocol. In addition, DTR concentrators could be designed to allow attachment of Classic Token Ring stations, Classic concentrators, and the new DTR adapters. Therefore, graceful migration minimizing costs to the user were a natural fallout of the technology.

5.2 Architecture Enhancements

The committee that had developed station architectures over the past 10 years was now faced with developing an architecture that described the interaction between two dissimilar entities. If you have read the Standard, you will find that this "system" view is difficult to see.[1]

In Classic Token Ring the architectural model consists of one or more stations attached to a ring. The total bandwidth of the ring (4 or 16 Mbit/s) is shared amongst the attached stations. This is referred to as a *shared media architecture*. A token, which is required by the station to transmit data, is passed between stations on the ring. Network connectivity is accomplished by the station connecting to the ring via a TCU. The station controls this access by use of phantom signaling. When a phantom signal is not asserted by the station, the station is not connected to the network. In this state a repeat path is created that allows the station to run self-diagnostics on its transmit and receive hardware without disrupting the remainder of the network. This also serves as a way to test out the lobe attachment.[2] The Access protocol defined by the then-current Standard is supported in this environment. This Access protocol is referred to as the *Token Passing Access protocol* (TKP).

In DTR, there are two MAC entities directly attached to a lobe: a *dedicated media architecture*. The bandwidth available is dependent on the ring speed[3] and the Access protocol in use. One entity is the station, and the other is the C-Port. The C-Port supplies an interface to the rest of the network after the station and C-Port complete a Join process. In this environment a new Access protocol is introduced that does not require the use of a token to transmit, nor is it limited to a simplex mode of operation. This new access mode is referred to as the *Transmit Immediate Access protocol* (TXI). TKP and TXI Access protocols are supported in this environment.

When TKP is used, the C-Port provides the necessary interface to the station and appears to have the characteristics of a TCU plus a station. Phantom signaling is accepted and a return path is provided in order to support the

---

1. The Standard defines each end of the wire separately; the system view was discussed during the meetings, but was not included into the Standard.
2. Lobe Media test.
3. Either 4 or 16 Mbit/s.

station's wire fault circuitry. The bandwidth is shared between the C-Port and the station.

When TXI is used, the transmit and receive circuitry in both the station and the C-Port operate independently when transferring data. A token is not used for data transmission. Unlike Classic Token Ring where a frame transmitted by a station is returned to and stripped off the ring by the station, a frame sent to the C-Port from the station *does not return to the station.*[4] The bandwidth is not shared on this connection. Since both the C-Port and the station may transmit simultaneously, there is 32-Mbit/s bandwidth available on the lobe.

5.2.1 The DTR Components

Three components need to be introduced to discuss a basic DTR topology: the DTR station, the C-Port, and the DTR concentrator. The basic DTR system, with which the Standard concerns itself most, consists of a C-Port connected to a DTR station, as shown in Figure 5.1.[5]

5.2.1.1 The DTR Station

The DTR station consists of the MAC protocol layer, the PHY layer, the bridge, and management interfaces. In fact, the DTR station has all the capabilities of the Classic Token Ring station, with the addition of the MAC protocol that allows it to run in TXI Mode. TXI is not new to Token Ring; this mode of data transmission was simply restricted to the recovery process (i.e., Beaconing) and not used to transport user data. In this mode of operation the station does not require a token to transmit data. When user data is available, the station simply transmits the data.

5.2.1.2 The C-Port

This is a new MAC protocol entity introduced in the DTR Standard. Its role in the DTR universe is to control the attached station's access to the network. It does not behave like a station and its role is to support the station. Typical MAC layer management functions (CRS[6] and RPS[7]) are not recommended for use in managing the C-Port. In order to support Classic stations, it was

---

4. There are exceptions to every case. The MAC defined for the C-Port and station has no facility for returning (sometimes referred to as Loopback) a frame to its source. However, the Management Routing Interface (MRI) can be configured to forward a management MAC frame back to a station. See Section 5.4.8.1.
5. In Figure 5.1 the familiar terms of MAC and PHY are now preceded by either an "S" for station or "P" for C-Port. This is the notation that was adopted by the committee.
6. Configuration Report Server.
7. Ring parameter server.

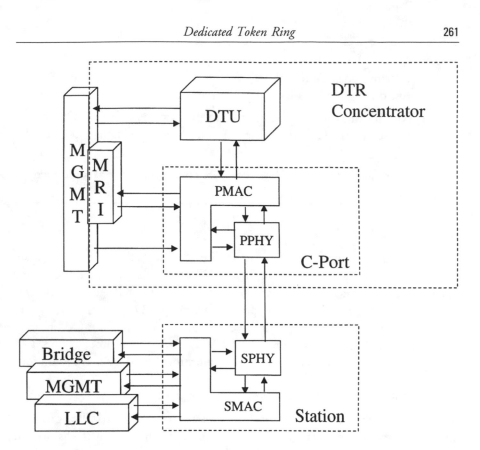

Figure 5.1 Architectural model of a DTR system.

decided that the C-Port has the facilities necessary to "fool" a Classic station to behave as if it were on a two-station ring.

5.2.1.3 C-Port Operational Modes

There are two operational modes for the C-Port: the C-Port mode and the station emulation mode.

C-Port Mode

In this operational mode, the C-Port is directly attached to a station. The station may be running either the TKP or TXI Access protocols. Additionally, the C-Port may be connected to another C-Port that is operating in the station emulation mode.

TKP Access Protocol Support

The C-Port provides facilities that allow a Classic station to operate as if it were connected to a two-station ring. These facilities take the form of:

- A recovered clock repeat path as required by the Classic station for lobe test during the Classic Join process and during Hard Error Recovery;

- Modifications to the Classic station protocol described in Chapter 4, as used by the C-Port, by the removal of phantom signaling, and lobe test;

- Modification to the Hard Error Recovery process, as described in Section 5.5.3.2.

Station Emulation Mode

In this operational mode, the C-Port is directly attached to either a Classic concentrator or a C-Port operating in the C-Port mode. When connected to a Classic concentrator, the C-Port may use only the TKP Access protocol, which is a modified version of the protocol used by a DTR station using the TKP Access protocol. When connected to another C-Port, the C-Port may use either the TXI Access protocol or the TKP Access protocol. The TXI Access protocol is a modified version of the protocol used by a DTR station.

5.2.1.4 The DTR Concentrator

The DTR concentrator was the committee's term for a "switch." The problem the committee had with using the term in the Standard is that it has no real meaning. All you had to do was look at the marketing hype of the time and you would see that anything was a switch![8] Initially the committee decided to define the DTR concentrator as a collection of C-Ports and existing MAC bridging entities. By January 1995 concerns for interoperability between different implementations of DTR concentrators—that is, control of loops in networks (there were others)—forced the committee to change its collective mind and produce an informative section to describe the DTR concentrator and the Data Transfer Unit (DTU) in detail.

The DTR concentrator consists of a collection of C-Ports that are interconnected via a DTU. Management function connectivity for the network, no longer supplied by a ring, is provided by the *management routing interface* (MRI). While box management is not defined by the Standard, its interfaces to the C-Port and the DTU are described in the Standard.

---

8. Everyone was trying to convince the customer that they had a revolutionary switch. Technically, there was nothing really new. A switch was built from upgraded multiport bridges and utilized crossport switching fabrics, time division multiplexed data bases, or large shared memory which was accessed by all ports. Switches either moved data using store-and-forward techniques or touted innovations such as cut through switches. This hype was no doubt confusing to the customer.

5.2.2 The DTR Network

The most obvious topology to use with DTR is direct attachment of DTR stations to the DTR concentrator, as shown in Figure 5.2. Here, each DTR station is attached to a DTR concentrator port, uses the TXI Access protocol, and has access to 32-Mbit/s bandwidth. Communication between the stations takes place via the concentrator's DTU, which may be implemented as a high-speed switching backplane. This configuration is both the most straightforward and the highest performance configuration for DTR. Moreover, it is the most expensive to implement. Each station must be DTR enabled, and the attachment to the ring must be via a DTR concentrator. All Classic concentrators and any existing adapters that are not DTR capable must be replaced. If maximum performance is not required, and if a dedicated 16-Mbit/s channel between station and concentrator is acceptable, then existing older generation Token Ring adapters do not have to be replaced. The DTR Standard requires all DTR concentrators to accept attachment of Classic Token Ring stations. Compared to Classic Token Ring attachment, Classic Token Ring stations connected to a DTR concentrator would have 10 to 100 times the bandwidth available for data transmission and reception.[9]

An attachment mechanism is required to provide for a more modest performance enhancement at a more modest cost since, in the first stages of DTR migration, it may not be necessary to increase the available station

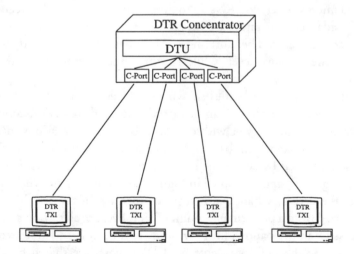

Figure 5.2 DTR direct attachment of stations to the concentrator.

9. This is true since a Classic station connected to a dedicated connection is not sharing the bandwidth of the ring with 10 to 100 other stations.

bandwidth by factors of 10 or 100. This incremental performance enhancement is provided by the microsegmentation of rings, as shown in Figure 5.3.

In this configuration the end stations are not connected directly to the DTR concentrator. Instead, they remain attached to Classic concentrators forming smaller rings. These rings attach to the LAN via the DTR concentrator, as shown in Figure 5.3. Each port of the DTR Concentrator, called a *C-Port*, attaches directly to a port of the Classic concentrator, with one C-Port to Classic concentrator port connection per microsegmented ring. Because the DTR concentrator backplane's bandwidth is much higher than 16 Mbit/s,[10] the resultant configuration provides much higher bandwidth than microsegmented rings bridged to a Token Ring backbone.

Upon close observation, it is seen that the C-Ports must operate in an entirely different mode than they do for the configuration shown in Figure 5.2. In Figure 5.3, a C-Port must be capable of operating in station emulation mode, which has the ability to source the phantom current necessary for insertion onto the Classic concentrator. Each Classic concentrator treats the attached C-Port as a single attached station on its local ring. The C-Port, in turn, acts in a fashion similar to a bridge, transmitting frames that must be forwarded to the backplane of the DTR concentrator and passing frames from that backplane to the local ring, as required. As the network shown in Figure 5.3 grows, additional stations can be added by increasing the station count of the microsegmented LANs that are operating below capacity or by creating additional microsegmented LANs up to the port capacity of the DTR concentrator. DTR adapters, operating in Classic station mode, can attach directly to a Classic concentrator. Therefore, new DTR adapters can be used to increase the size of the network, even while retaining the microsegmented network topology.

Combining the capabilities shown in Figures 5.2 and 5.3, high-performance stations such as servers can be directly attached to a DTR concentrator C-Port to provide that device with dedicated bandwidth. Stations that do not require the full Token Ring bandwidth can be configured in small rings, the size of each adjusted based on the bandwidth needs of the attaching stations. The resulting configuration shown in Figure 5.4 produces a network optimized to deliver the necessary bandwidth required for each of the attaching devices. It is optimized from a performance point of view because all station attachments have access to all the bandwidth they need. It is optimized from a cost point of view because the DTR concentrator C-Port costs are shared by all of the

---

10. Each manufacturer supplies backplanes of different capabilities. A nonblocking concentrator's backplane will supply sufficient bandwidth to support traffic between all ports running at the maximum media speed of 16 Mbit/s as defined in the 1995 Standard.

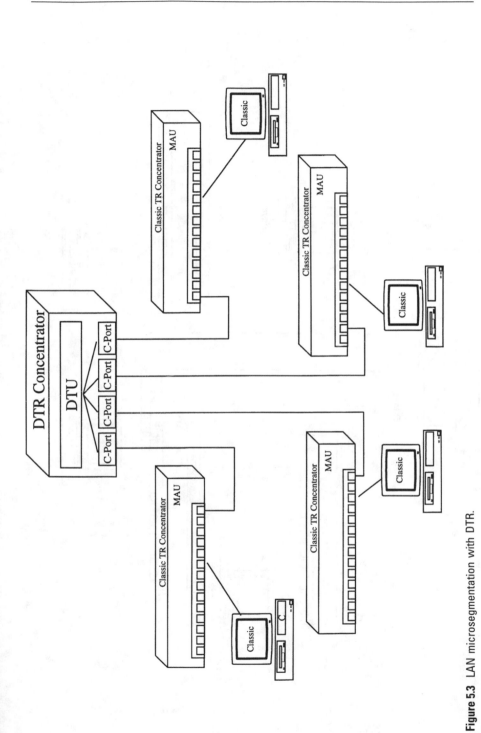

Figure 5.3 LAN microsegmentation with DTR.

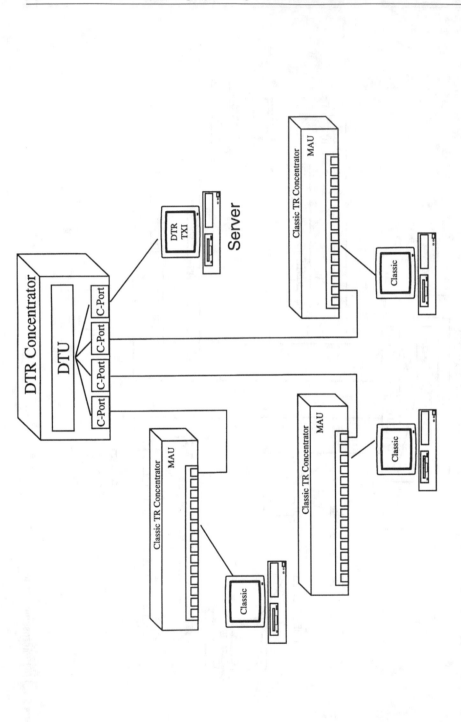

Figure 5.4 Bandwidth tailored to need.

stations on the microsegmented ring while generally retaining use of existing concentrators. Only the servers bear the full attachment cost of their C-Ports, and here that cost is justified because the servers require the entire bandwidth. Indeed, if they did not, then multiple servers could be grouped into their own small microLAN or grouped with the end stations they serve.

These configurations were for relatively small networks where only a single DTR concentrator is used. For larger networks DTR concentrators must be interconnected. Figure 5.5 shows one straightforward method of achieving this interconnection. When the C-Ports are connected to each other, they must either be manually configured with one in port mode and the other in station emulation mode, or they must go through a discovery process to determine which port operates in what mode. The DTR Standard provides an Autodetection protocol that can be used to allow both the autoconfiguration required here and attachment of the C-Ports to Classic stations, DTR stations, and Classic concentrators. Note that from a signaling and performance viewpoint, it does not matter which C-Port enters station emulation mode and which stays in port mode.

One final capability required of DTR concentrators is high-speed connectivity either to other DTR concentrators or to a higher speed backbone (e.g., 155-Mbit/s ATM). The DTR Standard provides for the capability to add one or more ports (other than 4- or 16-Mbit/s Token Ring), which it describes as a data transfer service. Figure 5.6 shows an ATM backbone network linked with DTR concentrators.

5.3 Migration From Classic Token Ring to DTR, Some Examples

5.3.1 Example 1: Collapsed Backbone Ring

Consider the multistory building shown in Figure 5.7, where each floor is served by a separate ring and all the servers are centrally managed. The floor rings connect to a hub via bridges and the servers are directly attached to the hub, which serves as the building's collapsed backbone. This building's LAN can be migrated to DTR by replacing the bridges and the hub by a DTR concentrator shown in Figure 5.8.

There will be two areas that see an increase in available bandwidth as a result of this change. First, and most obviously, the backbone will no longer be limited by 16 Mbit/s but by the backplane bandwidth of the DTR concentrator, which typically is in the hundreds of Megabit/second or Gigabit/second range. Second, the servers were initially directly attached to the collapsed backbone

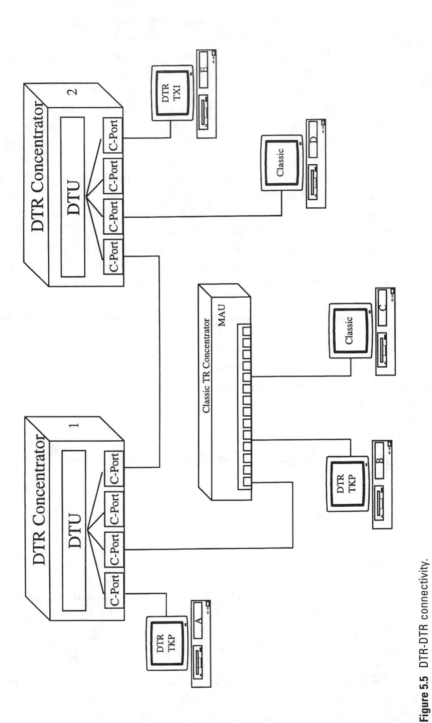

Figure 5.5 DTR-DTR connectivity.

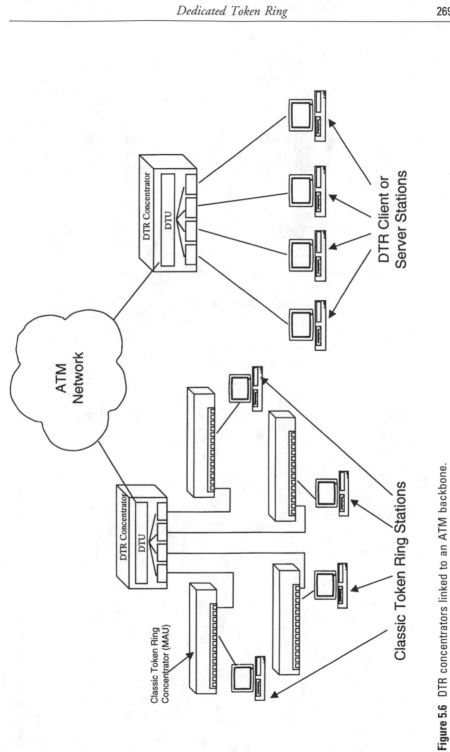

Figure 5.6 DTR concentrators linked to an ATM backbone.

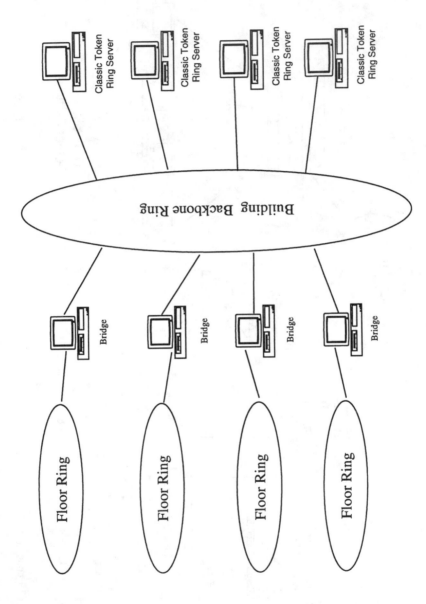

Figure 5.7 Multistory building with one ring per floor, each bridged to the backbone ring.

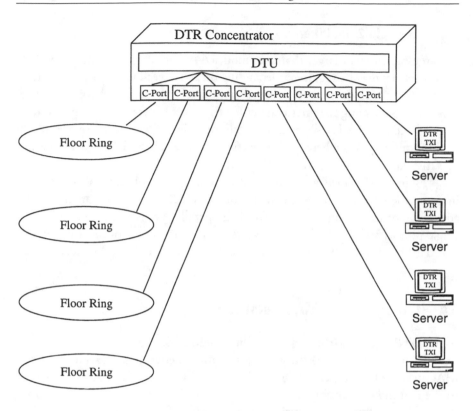

Figure 5.8 Switch replaces bridges and backbone ring.

for high performance. However, that high performance was limited by the access to the backbone that each server had. If all of the servers were relatively busy, and if we make the simplifying assumption that all of the traffic from the floor rings to the backbone was to access a server, then the available bandwidth for the server to receive and send data would be 16 Mbit/s divided by the number of servers on the ring, which in this case is 4. Therefore, the average bandwidth available to each server in this collapsed backbone example is 4 Mbit/s. As soon as the servers are attached to DTR concentrator C-Ports, the available bandwidth increases to 16 Mbit/s, assuming the servers have Classic Token Ring cards installed. If they have DTR-enabled Token Ring cards, then the total bandwidth available to each server to send and receive frames is 32 Mbit/s. In addition, the bridges shown in Figure 5.7 introduce significant delay to the network. Unlike bridges, DTR concentrators do not have the requirement of completely store and then forward, frames. The network in Figure 5.7 will show significant improvement in frame latency, which is important for applications such as real-time video conferencing.

5.3.2   Example 2: Multitiered LAN

For an establishment larger than that modeled in Section 5.3.1, multiple DTR concentrators may be required. Depending on the overall network connectivity needs, one topology that may be suitable is to connect each of the DTR concentrators serving the microsegmented rings to a central DTR concentrator, which becomes the LAN's central backbone. Figure 5.9 shows one example of this configuration where the connectivity to the central LAN backbone can be satisfied with a single 32-Mbit/s pipe.

Figure 5.10 shows an example of attaching multiple up-link C-Ports to increase the data pipe from 32 to 64 or even 128 Mbit/s. In order to take advantage of the multiple connections, the parallel paths must be treated as a single pipe by the attached concentrators. This function is not described in the Standard but is available on many commercially available DTR concentrators.

5.4   DTR-Enhanced MAC Facilities

In the DTR Standard, some of the facilities used to describe the MAC protocols have been renamed to identify the association between the facility and the MAC protocol entity, either the station or the C-Port. This caused a change to the naming conventions that were used in the 1995 Standard. In addition, new facility types, policy variables, and interface flags were introduced. The following subsections summarize these changes introduced by the DTR Standard.

5.4.1   Flags and Timers

The names of flags and timers are indicative of their application.

- The first letter indicates the type of facility, either "T" for timer or "F" for flag.
- The second letter indicates the MAC entity with which the facility is associated; "S" is used for station, while "P" is used for C-Port.
- The remaining letters are an abbreviation of the function name; option flags always end with "O."

For example, TSLMT is the timer used in the Station MAC for the LMT. FPASO is a flag, in this case an option flag, used in the C-Port MAC controlling the setting of the AC-bits in the repeat path supplied by the C-Port.

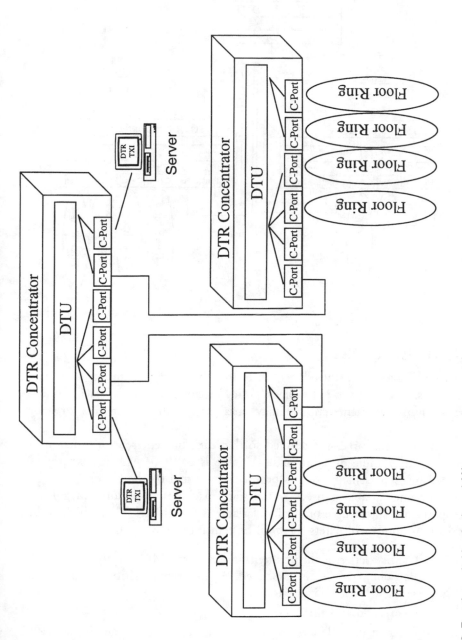

Figure 5.9 Two-tier switching for large LANs.

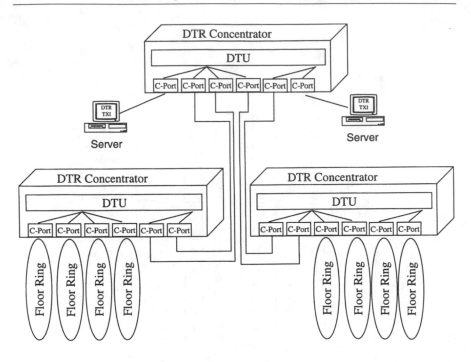

Figure 5.10 Multiple-channel connection.

5.4.2 Policy Variables

Policy variables were introduced into the Standard to simplify the state machines. In contrast to policy flags, policy variables are *multibit* facilities that are set by management action. Policy variables are either a bit mask or value type.

A bit mask variable is defined so that each bit represents a function or feature. An example of a bit mask variable is PPV(AP_MASK), where the defined bits represent the Access protocol supported by the C-Port.

A value variable is used by the state machines to perform magnitude comparisons. An example of a value variable is PPV(MAX_TX), which is the maximum number of octets that will be transmitted by the C-Port.

- The first letter of the acronym indicates the MAC entity with which the policy variable is associated. As for flags and timers, "S" is for station while "P" is for C-Port.
- The second and third letters *must* be "PV."
- The remaining letters consist of a pair of matched parenthesis "( )" enclosing an abbreviation of the function name.

For example, SPV(AP_MASK) is the Access Protocol Mask used in the station MAC.

5.4.3 Interface Flags

The C-Port's station emulation modes and its support of the TKP Access Protocol in port mode operate in a fashion that is very similar to the MAC protocols that were already written (TKP) or were being written for the DTR station's TXI AP support. In order to take advantage of these similarities, the following strategies were adopted.

- TXI station emulation mode was to be described within clause 9.2. Differences between actions taken by a C-Port in station emulation mode as compared to a DTR station would be added as transitions using interface flags. For example, there are transitions in the Station Operation Tables that deal with transfers from the station MAC to the bridge interface. Because there is no bridge interface in a C-Port, regardless of the operation mode, the interface flag FIPTXIS is used to deactivate the transitions that use the bridge interface (M_UNITDATA.indication) and activate the transitions that use the DTU interface (DTU_UNITDATA.indication).

- TKP station emulation mode was to be described within clause 9.5. Differences between actions taken by a C-Port in station emulation mode as compared to a DTR station would be added as transitions using interface flags. An example for this case is similar to that just stated, except the interface flag used is FIPTKPS.

- TKP port mode was described in clauses 9.3 and 9.4. Clause 9.3 contained the Join *finite state machine* (FSM), while clause 9.4 contained the remaining FSMs (Monitor, Transmit, Error Handling, Interface, and Miscellaneous Frame Handling). The remaining FSMs were modifications of the Classical Token Ring FSMs that are described in Clause 4. Unique transitions used the interface flags for two reasons: to allow communication between the Join FSM defined in clause 9.3 and to allow eventual integration of these transitions back into clause 4. An example of this is the conditions the C-Port uses to enter Beacon test during Hard Error Recovery. Since the C-Port does not drop the phantom signal and perform the LMT but instead supports the station in performing that action, the interface flag FIPTKPP is used to activate the transitions used to detect the station raising and dropping the phantom signal.

The interface flags were not intended to imply any preferred implementation, they were used as a method of communication between the FSMs defined in the various sections of clause 9.

- The first three letters indicate that this was an interface flag and was required to be "FIP."
- The next three letters indicate the Access protocol for which the flag was used; "TKP" and "TXI" were used.
- The seventh letter indicates the operational mode for which the flag was used; "P" indicated C-Port mode and "S" indicated station emulation mode.
- An optional suffix letter "E" was added for interface flags used to communicate error conditions.

An example is FIPTXIS. This flag is used in the Station Operation Tables by a C-Port operating in station emulation mode using the TXI Access protocol (clause 9.2). The only example of an error condition interface flag is FIPTKPPE. This interface flag is used by a C-Port operating in port mode using the TKP Access protocol to report an error, such as an attached station did not successfully complete its LMT during Hard Error Recovery to the Join FSM defined in the clause 9.3.

5.4.4 New Timers, Policy Flags, and Variables

5.4.4.1 Timers

Tables 5.1 and 5.2 describe the timers introduced by the DTR Standard.

Table 5.1
SMAC Timers

Timer Name	Abbreviation	Description
Station Insert Process Timer	TSIP	This timer is introduced as part of the Registration and Join process for the TXI Access protocol. This timer controls the rate at which Registration MAC frames are sent. A value of 200 ms is recommended by the Standard.
Station Initial Sequence Timer	TSIS	This timer is introduced as part of the Registration and Join process for the TXI Access protocol. This timer controls the duration of idle transmission that precedes the transmission of the first Registration Request MAC frame. This time is used by the C-Port to acquire phase lock to the station's clock. A value of 40 ms is recommended by the Standard.

Table 5.1 *(continued)*
SMAC Timers

Timer Name	Abbreviation	Description
Station Internal Test Timer	TSIT	This timer is introduced as part of the TXI Hard Error Recovery process. This timer is used to delay the station's dropping of the phantom signal in order to assure the detection of a possible wire fault condition. A value of 7 sec is recommended by the Standard.
Station Join Complete Timer	TSJC	This timer is introduced as part of the Join process. Once the station has completed its Lobe test, this timer is used to limit the amount of time the station will wait for an Insert Response MAC frame. If this timer expires before the receipt of the Insert Response MAC frame, the station closes and the Join process has failed. A value between 17.8 sec and 18 sec is required by the Standard.
Station LMT	TSLMT	This timer is introduced as part of the TXI Hard Error Recovery process. This timer is used to delay the start of the Station's LMT. A value of 10 sec is recommended by the Standard.
Station LMT Complete Timer	TSLMTC	This timer is introduced to help further define and control the requirements for the station's LMT. During the development of the DTR Standard, it was determined that limiting the duration of a station's LMT was an improvement to the Standard. This timer is started when the station begins its LMT. If the timer expires before the end of the test, the station fails LMT and will close. A value between 2.3 sec and 2.5 sec is required by the Standard.
Station LMT Delay Timer	TSLMTD	This timer is introduced to further define and control the requirements for the station's LMT In order to assure that the C-Port is allowed sufficient time to configure a repeat path for the station's LMT, this timer is used to delay the start of the test. A value between 200 ms and 250 ms is required by the Standard.
Station Queue Heart Beat Timer	TSQHB	This timer is introduced as part of the Heart Beat process. Each time the timer expires, the station will queue a Station Heart MAC frame for transmission. A value of 1 sec is recommended by the Standard.

Table 5.1 *(continued)*
SMAC Timers

Timer Name	Abbreviation	Description
Station Registration Request Timer	TSREQ	This timer is introduced as part of the Registration and Join process This timer is used to pace the rate at which the station transmits Registration MAC frames. A value of 40 ms is recommended by the Standard.
Station Receive Heart Beat Timer	TSRHB	This timer is introduced as part of the Heart Beat process This timer is reset each time a C-Port Heart MAC frame is received by the station. If this timer should expire, the station will start the TXI Hard Error Recovery process. A value of 5 sec is recommended by the Standard.
Station Registration Wait Timer	TSRHB	This timer is introduced as part of the Registration Query process. When a station MAC is responding to a Registration Query MAC frame by dropping the phantom signal, this timer is used to allow the C-Port sufficient time to configure itself for the reception of the station's Registration MAC frame. A minimum value of 200 ms is required by the Standard. Station implementation considerations may require that the value chosen be larger than the minimum; however, it should be noted that the smaller the value implemented, the sooner the station is allowed to enter the network.

Table 5.2
PMAC Timers

Timer Name	Abbreviation	Description
C-Port Break Lobe Test Timer	TPBLT	This timer is introduced to support the setting of the C-Port policy variable PPV(AP_MASK). When PPV(AP_MASK) indicates that the TKP Access mode is not supported, expiration of the timer is used to start disruption of a station's LMT, which is detected by the C-Port. This disruption assures that a station using the TKP Access protocol will not open as a single station ring when attached to a C-Port so configured. A value between 10 ms to 30 ms is required by the Standard.

Table 5.2 *(continued)*
PMAC Timers

Timer Name	Abbreviation	Description
C-Port Disrupt Lobe Test Timer	TPDLT	This timer is introduced to support the setting of the C-Port policy variable PPV(AP_MASK). When PPV(AP_MASK) indicates that the TKP Access mode is not supported, the C-Port transmits idles while this timer is running, thus disrupting the station's Lobe test. A value between 2.6 sec and 2.8 sec is required by the Standard. Note that the selection of this timer range exceeds the allowable duration of a station's LMT, assuring the failure of the LMT.
C-Port Insert Request Delay Timer	TPIRD	This timer is introduced to support clocking requirements during the Duplicate Address Check process. The station issues Insert MAC frames using its own clock and then reverts to a recovered clock after the transmission of the frame. The C-Port allows sufficient time for the station's internal circuitry to lock on to the C-Port's idle transmission before transmitting an Insert MAC frame. A value between 10 ms and 30 ms is required by the Standard.
C-Port Internal Test	TPIT	This timer is introduced as part of Hard Error Recovery process. While this timer is running, the C-Port transmits Beacon MAC frames. A value of 600 ms is recommended by the Standard.
C-Port LMT Failure	TPLMTF	This timer is introduced to help further define and control the requirements for the station's LMT. During Beacon test a method was required to determine when a station had failed the LMT, allowing the C-Port to close in a timely fashion. The duration of LMT was not defined for Classic Token Ring stations; however, it was determined by the committee that there were no current implementations whose LMT exceeded a duration of 15 sec. The value for this timer was selected so that implementations would interoperate with the definition of a C-Port supporting a Classic station. A value between 15 sec and 18 sec is required by the Standard.

Table 5.2 *(continued)*
PMAC Timers

Timer Name	Abbreviation	Description
C-Port LMT Running	TPLMTR	This timer is introduced to help further define and control the requirements for the station's LMT. During the development of the DTR Standard, it was determined that limiting the duration of a station's LMT was an improvement to the Standard. This timer is started when the C-Port begins its support of the station's LMT. If the timer expires before the C-Port receives an Insert Request MAC frame from the station, the C-Port determines that the station has failed LMT. A value at the lower end of the range between 6 sec to 7.4 sec is recommended by the Standard.
C-Port Queue Heart Beat Timer	TPQHB	This timer is introduced as part of the Heart Beat process. Each time the timer expires, the C-Port queues a C-Port Heart Beat MAC frame for transmission. A value of 1 sec is recommended by the Standard.
C-Port Receive Heart Beat Timer	TPRHB	This timer is introduced as part of the Heart Beat process. This timer is reset each time a Station Heart Beat MAC frame is received by the C-Port. If this timer should expire, the C-Port will start the TXI Hard Error process. A value of 5 sec is recommended by the Standard.
C-Port Registration Query Delay Timer	TPRQD	This timer is introduced as part of the Registration Query process. This timer is used to delay the end of the process and allow the station time to drop a phantom signal in response to the last Registration Query MAC frame and time for the C-Port to detect the event. A value between 20 ms and 200 ms is required by the Standard. The value selected by an implementation must account for the time required by the C-Port to detect the loss of the phantom signal.

5.4.4.2 Policy Flags and Variables

Tables 5.3 and 5.4 describe station and port policy flags and variables.

Table 5.3
Station Policy Flags and Variables

Policy Name	Abbreviation	Description
Station Open Option	FSOPO	This policy flag is introduced to control the Registration and Join process. When the station does not receive any Registration Response MAC frames in response to transmitted Registration Request MAC frames, this flag controls whether the station closes (1) or attempts to Join using the TKP Access protocol (0).
Station Registration Denied Option	FSRDO	This policy flag is introduced to control the Registration and Join process. When the station receives a Registration Response frame denying its request to Join using the TXI Access protocol, this flag controls whether the station closes (1) or attempts to join using the TKP Access protocol (0).
Station Registration Option	FSREGO	This policy flag is introduced to control the Join process. When a station is directed by management to open, this option flag controls whether the station opens using the Registration process (0) or Joins using the Classic Join process (1).
Station Registration Query Option	FSRQO	This policy flag is introduced to control the use of the Registration Query process by the station. A station will either ignore (1) or respond by dropping the phantom signal (0) when a Registration Query MAC frame is received.
Station Access Protocol Mask	SPV(AP_MASK)	This is a 2-octet mask used to indicate the supported Access protocols for this station. Only three values of the mask are supported at this time: X'0001' The station supports TKP Access protocol *only*. X'0002' The station supports the TXI Access protocol *only*. X'0003' The station supports *both* the TXI and TKP Access protocols. The protocol that the station executes is dependent on the results of the Join process.

Table 5.3 (continued)
Station Policy Flags and Variables

Policy Name	Abbreviation	Description
Station Individual Address Count	SPV(IAC)	This 2-octet field is used to indicate the number of individual addresses supported by this station. This policy variable has its roots in the IAC subvector. Unfortunately this subvector was not well defined in previous versions of the Standard, nor was the implication of a station with multiple individual addresses ever integrated into the definition of the MAC protocol. There appear to be two definitions in use by manufacturers: 1. The value represents a power of two. Hence a value of 0 indicates that the station has a single individual address and a 1 indicates 2 addresses, for example. 2. The value represents a numerical value of the number of stations, with the following encoding: • 0 indicates a single address and the station is not enabled for multiple individual addresses. • 1 indicates a single address and the station may be enabled for multiple individual addresses. • A value greater than 1 indicates the number of individual addresses supported. Due to the differences in definitions and the small use that this function has seen, it was decided to leave the issue of multiple individual addresses to a later standardization process and define this field to be X'0000' for a DTR station.
Station Maximum Transmitted Octet Count	SPV(MAX_TX)	This 2-octet value represents the maximum frame size supported by the station. The maximum value that can be assigned to this field is constrained by the media speed maximums supported on a Classic Ring. The maximum value allowed for 4 Mbit/s is 4550 and the maximum value allowed for 16 Mbit/s is 18200. Due to differing implementations, some implementations used timers while others used octet counters to enforce this policy variable; a station is compliant as long as it

Table 5.3 *(continued)*
Station Policy Flags and Variables

Policy Name	Abbreviation	Description
Station Phantom Signaling	SPV(PD)	does not transmit a frame longer than this value. A station may determine that a frame near this value is in violation as well.[11] This 2-octet field indicates the phantom signaling definition in use by this station. Currently there is only one in defined for use.[12] The definition for phantom signaling and wire fault detection is defined in ISO/IEC 8802-5:1995.

Table 5.4
C-Port Policy Flags and Variables

Policy Name	Abbreviation	Description
C-Port AC Repeat Path Option	FPACO	This option flag is used by the C-Port Operation Tables to allow for differences in the implementation of the repeat path supplied by the C-Port (see Section 5.4.8.2). It was found that in some implementations, the repeat path used by the C-Port[13] contained the address recognition circuitry that set the A-bits and C-bits in the Frame Status byte (1). Other implementations did not (0). This option flag allowed both types of repeat paths to be compliant. This was an acceptable solution since the LMT was being more stringently defined.
C-Port Abort Sequence Option	FPASO	This option provided the implementor with two methods for ending frames that were determined to be too long (exceeded the value

---

11. This is a very fine point and is only interesting if you are working on compliance tests. See the details in clause 10 of the Standard and how this policy variable is used for comparison within the station operation tables of clause 9.2.
12. There were other ideas on the table during the committee meetings. These ranged from 3v implementations to dropping phantom signaling from the interface. The committee was unable to reach a consensus in a reasonable amount of time, so this discussion is left for further standardization.
13. Keep in mind that one of the objectives of the DTR Standard was to allow reuse of existing MAC chip sets.

Table 5.4 *(continued)*
C-Port Policy Flags and Variables

Policy Name	Abbreviation	Description
		set by PPV(MAX_TX). The traditional method is to terminate the frame with an Abort Sequence (0). The newer method (1) permits terminating the frame with an invalid FCS and setting the *error detected bit* (E) in the Frame Status octet to 1. Both methods assure that the frame is not receivable by a station. However, the traditional method will render a (shared) ring connected to the C-Port inoperable for 10 ms. This is the time it takes for the AM to notice that there is no token and needs to clean up the ring and release one. The newer method does not have this problem; however, current chip sets are unable to provide this function since the previous standard allowed the E-bit to be transmitted as a 1 *only*.
C-Port Frame Control Option	FPFCO	This option flag introduced the concept of "cut through." Most of the implementations that were under development included a mode that was not a bridge traditional store and forward. The idea was to include a mode of operation that assured a low first octet in first octet out latency. This is done by forwarding the frame to its output destination port when enough of the frame has been received to allow this determination. This option flag is used to determine when the frame is indicated to the DTU for forwarding, either when the entire frame has been received (FR) or when the frame control octet has been received (FR_FC). Receipt of the Frame Control octet allows discrimination between an LLC or MAC frame, the latter is never forwarded to the DTU.
C-Port Operation Table Option	FPOTO	The C-Port can operate as a C-Port (1) or in station emulation mode (0). This option flag controls in which of these two modes the C-Port operates.
C-Port Access Protocol Mask	PPV(AP_MASK)	This is a 2-octet mask used to indicate the supported Access protocols for this C-Port. Only three values of the mask are supported at this time:

Table 5.4 *(continued)*
C-Port Policy Flags and Variables

Policy Name	Abbreviation	Description
		X'0001' The C-Port supports TKP Access protocol *only*. X'0002' The C-Port supports the TXI Access protocol *only*. X'0003' The C-Port supports *both* the TXI and TKP Access protocols. The C-Port uses this policy variable to determine if it will support the station's requested Access protocol during the Join process.
C-Port Maximum Transmitted Octet Count	PPV(MAX_TX)	This 2-octet value represents the maximum frame size supported by the C-Port. The maximum value that can be assigned to this field is constrained by the media speed maximums supported on a Classic Ring and the capabilities of the DTU implemented in the DTR concentrator that this C-Port is attached. The maximum value allowed for 4 Mbit/s is 4550 and the maximum value allowed for 16 Mbit/s is 18200. Due to differing implementations, some implementations used timers while others used octet counters to enforce this policy variable; a C-Port is compliant as long as it does not transmit a frame longer than this value. A C-Port may abort transmission of a frame if the frame size is *near* this value as well.
C-Port Phantom Signaling	PPV(PD)	This 2-octet field indicates the phantom signaling definition supported by this C-Port. Currently there is only one value defined. The definition for phantom signaling and wire fault detection is in ISO/IEC 8802-5:1995.

5.4.4.3 Interface Flags

Table 5.5 describes the interface flags used in the FSMs in clause 9 of the DTR Standard. These flags have a meaning only when set to a value of 1. When these flags are set to zero, no information about the state of the MAC or the MAC entity can be obtained from these flags.

Table 5.5
Interface Flags

Interface Flag Name	Abbreviation	Description
C-Port TXI Station Emulation	FIPTXIS	When set to 1, this flag is used to activate transitions in the station TXI FSM (clause 9.2) for use by a C-Port operating in station emulation mode and using the TXI Access protocol.
C-Port TKP Port Mode	FIPTKPP	When set to 1, this flag is used to activate transitions in the TKP Port FSM (clause 9.3) for use by the C-Port operation in the port mode and using the TKP Access protocol.
C-Port TKP Port Error	FIPTKPPE	When set to 1, this flag indicates that the Monitor FSM defined in clause 9.4 detected a LMT failure while in the Beacon test state. This interface flag is used to communicate this condition to the Join FSM defined in clause 9.3. The Join FSM of clause 9.3 closes the C-Port when FIPTKPPE is set to 1.
C-Port TKP Station Emulation	FIPTKPS	When set to 1, this flag is used to activate transitions in the TKP FSM defined in clause 9.5 for use by a C-Port operating in station emulation mode and using the TKP Access protocol.

5.4.5 Summary of "Illegal" MAC Frames

One of the concerns during the development of the Standard was to be sure that a station or C-Port using the TXI access mode would "know" if it had been moved into a shared media ring.[14] A station or C-Port in station emulation mode using the TXI Access protocol would break any ring into which it was plugged.

    To this end, a set of MAC frames that were not used in the TXI Access protocol were identified and should any of them be received while running the TXI Access protocol, the station would close. The MAC frames are Claim

---

14. There was much concern over unauthorized or unplanned changes in the wiring closet. However, once it was understood that most customers tended to carefully control wire closet changes, the committee decided that the additional complexity to the Standard was not warranted.

Token, Ring Purge, AMP (prior to the start of the Heart Beat process), and SMP (prior to the start of the Heart Beat process).

5.4.6 A New Vector Class

A concentrator port class (x'3') is added to the Standard to identify frames that are either destined to or originate from the C-Port MAC. The C-Port is not limited to receiving frames that have a destination class of a concentrator port, it also receives and transmits frames with a ring station *vector class* (VC). MAC frames with new *vector identifiers* (VI) will use this new class.

5.4.7 Summary of New MAC Frames

5.4.7.1 Vector Descriptions

Table 5.6 is a summary of the MAC frames that were added to support the new DTR protocols.

Table 5.6
New MAC Frames

Vector Name	VI	VC	Subvectors	
C-Port Heart Beat	X'05'	X'00'	X'02'	UNA
			X'0B'	Physical drop number
Station Heart Beat	X'06'	X'00'	X'02'	UNA
			X'0B'	Physical drop number
Registration Request	X'11'	X'30'	X'0C'	Phantom
			X'0E'	Access protocol request
			X'21'	Individual address count
Registration Response	X'12'	X'03'	X'0F'	Access protocol response
Insert Request	X'13'	X'30'		
Insert Response	X'14'	X'03'	X'0D'	DTR response code
Registration Query	X'15'	X'03'		

C-Port and Station Heart Beat

In Classic Token Ring, the presence of a token on the ring was an indication that the AM was functioning and that the ring was in an operational state. The Heart Beat frames, Station Heart Beat and C-Port Heart Beat, are used in the Heart Beat process (see Section 5.5.1.4) to provide a method of determining that the MAC entities at both ends of the lobe are functioning. The vector identifier values (Table 5.6) used for these frames are the same as the Classic AMP and SMP MAC frames.

Registration Request

The TXI Access protocol Join process includes a registration process (see Section 5.5.1.1) that allows the C-Port and the station to verify operational characteristics of each other. The station provides address and Access protocol information to the C-Port using this frame.

Registration Response

As part of the registration process (see Section 5.5.1.1), the Registration Response MAC frame is used by the C-Port to indicate if the information provided by the received Registration Request MAC frame allows the interoperation of the station and the C-Port.

Insert Request

This frame is used by the station at the completion of a TXI LMT to signal the C-Port of a successful test. After the transmission of this frame, the source of the clock on the lobe changes from the station to the C-Port.

Insert Response

The Insert Response MAC frame is used by the C-Port to indicate to the station if it may connect to the network through the DTR concentrator. The Insert Response MAC frame is sent only in response to an Insert Request MAC frame. Transmission of the frame is delayed by 10 ms to 30 ms to allow the station to change its clocking mode from crystal to recovered clock. A negative response forces the station to close.

Registration Query

This frame is used by the C-Port to start the Registration Query process (see Section 5.5.4). This frame informs a DTR station that is currently connected to the C-Port and, using the TKP Access mode, that the C-Port is capable of supporting the TXI Access protocol. A Classic station will discard this frame.

5.4.7.2   Subvector Descriptions

Access Protocol Request

The Access Protocol Request subvector, used in the Registration Request MAC frame, is sent by the station to request use of a specific access protocol. Because registration for the TKP Access protocol was removed late in the development of the Standard, the only valid value defined is for the TXI Access protocol (x'0002'). The C-Port compares this value (bit wise logical AND) to its Access Protocol mask (PPV(AP_MASK) to determine if it can support the requested protocol.

Access Protocol Response

The Access Protocol Response subvector, used in the Registration Response MAC frame, is sent by the C-Port in response to a Registration Request MAC frame. The results of the bit wise logical AND of the Access Protocol mask (PPV(AP_MASK) and the value in the Received Access Protocol Request are used to set the value of the Access Protocol Response. Two values are defined, TXI Access protocol (x'0002') accepted or access denied (x'0000').

DTR Response Code

The DTR Response Code subvector, used in the Insert Response MAC frame, is used by the C-Port to communicate to the station the results of the duplicate address check performed by the DTU. Two values are defined, insert request accepted (x'0000') and insert request denied (x'8020'). If a station receives an insert request denied response, the station closes.

5.4.8 C-Port Interfaces

Figure 5.11 illustrates the interfaces to the C-Port that are described in this section.

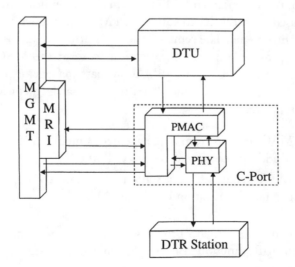

Figure 5.11 C-Port interfaces.

5.4.8.1   PMAC

There are three interfaces to the PMAC: the DTU and the MRI are explicitly defined in the Operation Tables; the management interface is implicit and is defined in clause 11.

Data Transfer Unit Interface

This interface is used by the C-Port to send and receive all non-MAC traffic. The DTU interface is defined for both store and forward transfers and for cut through transfers. duplicate address check requests and results are also sent across this interface.

The Management Routing Interface

Since the DTU does not provide services to forward MAC frames, a new service is required to allow the management connectivity model used in the base standard to continue to be used. The *Management Routing Interface* (MRI) provides a set of services that provide for the routing of management MAC frames to management servers, such as the CRS, REM,[15] or RPS.

A management MAC frame is a MAC frame where the VC octet takes on a nonzero value. A nonzero VC indicates that the frame is either coming from or going to a management function.

To complete the definition of the MRI, an explanation of how the forwarding of frames is determined is required. The forwarding of the MAC frames between C-Ports is defined within the MIB proposed by Annex K of the DTR Standard. The two services provided by the MRI to the PMAC are MRI_UNITDATA.request and MRI_UNITDATA.indication. The request primitive is used by the MRI to cause the PMAC to transmit a MAC frame onto the attached media. The semantics for this primitive are described in the Standard and provide sufficient information to form the MAC frame. The indication primitive is used by the PMAC to pass management MAC frames to the MRI. A MAC frame is considered to be a management MAC frame if it meets the following tests:

- The MAC frame is from a management function and is not addressed to any address recognized by the C-Port.
- The MAC frame is from a station MAC and the destination for the frame is neither a station or C-Port MAC.

Annex K of the Standard provides a definition of management objects that support the MRI function in a LAN. Specifically the dtrMRITable provides

15. Ring Error Monitor.

the necessary information to configure a DTR concentrator to properly forward management MAC frames between C-Ports.

Each row of the dtrMRITable contains three entries that are defined as follows:

- dtrMRICRFIndex: Identifies the CRF for which this MRI forwarding entry is intended. More on CRFs later.

- dtrMRIMgmtType: Identifies the *destination class* (DC) for this entry. The DC is the first four bits of the VC octet in the MAC frame. When the DC is zero and the Source Class (SC which are the last four bits of the VC) is not zero, the destination address of the MAC frame is used to access the filtering database of the CRF in order to determine how to forward the MAC frame.

- dtrMRIOutMask: Identifies the output ports to forward a MAC frame matching this entry.

When configuring the MRI for use on a LAN, the administrator may either direct the management traffic to the stations that contain the management functions or simply set the dtrMRIOutMask to broadcast the frame to all C-Ports.

MRI Example

One of the more common functions that a network management application supplies for Token Ring is a map of the ring. This map shows all the stations attached to a C-Port and the order in which the stations are connected on the ring. The method used to collect this information is well known. Starting with the C-Port MAC, each station on the ring is sent a Request Station Addresses MAC frame. The MAC protocol in the station (and the C-Port in station emulation mode), will respond to this MAC frame with a Report Station Addresses MAC frame. One of the subvectors of that MAC frame contains the UNA of the station. Using the UNA as the destination of the next Request Station Addresses MAC frame allows the network management application to "walk the ring" and build a database that represents the ring order of the stations attached to the ring.

Assume the topology shown in Figure 5.12. The topology shows a single CRF[16] connected to four rings. On C-Port 3[17] a network management station

---

16. You may need to jump ahead to Section 5.6, the DTR concentrator-Annex K, for the discussion of CRFs.

17. Note that the C-Port number is the same as the CRF port number.

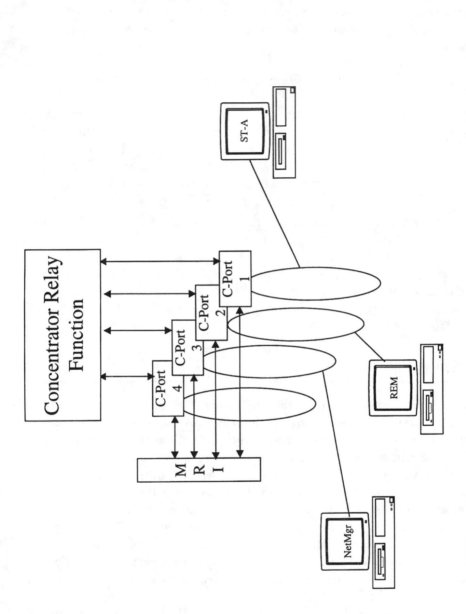

Figure 5.12 Topology for MRI forwarding example.

has been installed. This example will illustrate an exchange between the network management station and a station connected to C-Port 1.

In this example, the network administrator has created the entries indicated in Table 5.7.

The network administrator has configured the table so that all the RPS traffic (i.e., Request Initialization MAC frames) is discarded by the MRI. In addition, the destination for all CRS-directed frames is the same as for network management, which allows the network management application to detect when a change to the ring topologies has occurred (i.e., Report New AM or Report SUA Change). Error reports will be sent to the REM station that is connected to C-Port 2.

A partial list of the contents of the CRF's Forwarding Database is shown in Table 5.8.

The following sequence of events occurs when the network management application continues to "walk the ring" at ST-A.

1. The network management application sends a Request Station Addresses MAC frame to ST-A. The MAC frame contains the information: SA = NetMgr, DA = ST-A, VC = x'80', VI = x'0E'.

2. C-Port 3 receives the frame. Because the frame is not addressed to the C-Port MAC and the source class does not indicate a ring station (x'0'), the C-Port Operation Tables send the frame to the MRI.

Table 5.7
Entries Used by the Network Administrator in the dtrMRITable

dtrMRICRFIndex	dtrMRIMgmtType	dtrMRIOutMask '1234'
1	CRS (x'4')	'0010'b
1	RPS (x'5')	'0000'b
1	REM (x'6')	'0100'b
1	NM (x'8')	'0010'b

Table 5.8
Filtering Database for CRF (Partial)

MAC Address	Output Port	DRD	Output Port
ST-A	1		
NM	3		

3. The MRI examines the frame. Because the destination class is a ring station, the MRI accesses the forwarding database of the CRF. Examination of the database shows that ST-A is found at CRF port 1, where the frame is sent.

4. C-Port 1 PMAC receives a request from the MRI to transmit the frame. The frame is queued for transmission by the C-Port and eventually sent onto the media and received by ST-A.

5. ST-A responds to the Request Station Addresses MAC frame with a Report Station Addresses MAC frame containing the information: SA = ST-A, DA = NetMgr, VC = x'08', VI = x'22'. The SVI includes the UNA, enabled group and functional addresses, the individual address count, and physical drop.

6. C-Port 1 receives the Report Station Addresses MAC frame. Because the destination class is network management (x'8') and the source class is ring station (x'0'), the C-Port MAC sends the frame to the MRI.

7. The MRI receives the frame. Because the destination class is network management, the MRI examines its own table (dtrMRITable), locates the entry, and sends the frame to C-Port 3.

8. C-Port 3 PMAC receives a request from the MRI to transmit the frame. The frame is queued for transmission by the C-Port and eventually sent onto the media and received by NetMgr.

9. The sequence starts again at step 1, except the DA is now set to the value of the UNA found in the received Report Station Addresses MAC frame.

The "ring walk" ends when the UNA reported is the C-Port MAC address, where the ring walk started.

5.4.8.2 PPHY

The PPHY must provide all the functions required by a station implementation and an additional function that is required to implement the C-Port Operation Tables. During the Join process, the PPHY must provide a repeat path suitable for both Classic and DTR stations. Upon Join completion, it must be able to sink phantom current and complete the associated insert request. Upon insertion of the attached station, the PPHY is required to transmit with timing based on its own crystal rather than from the recovered clock derived from the signal transmitted by the attached station. For a full discussion of the PHY layer function, see Chapter 3. A more detailed illustration of the PPHY showing the PSC and PMC is shown in Figure 5.13.

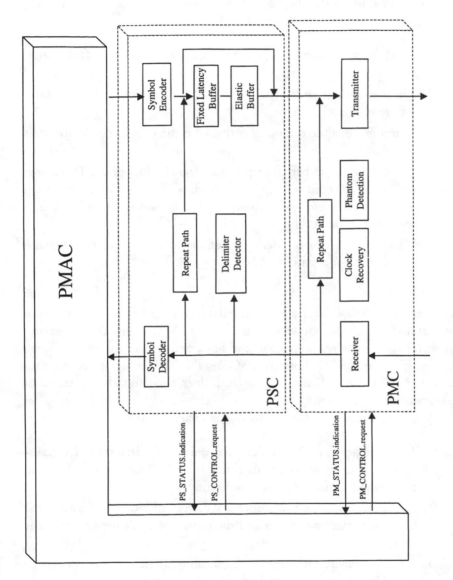

Figure 5.13 A detailed view of the PPHY showing the PSC and PMC boundaries and the permitted repeat paths.

The PMC in Port Mode

When the C-Port is operating in station emulation mode, the C-Port has all of the characteristics of a station as described in Chapter 3. When the C-Port is operating in port mode, the PMC has all of these characteristics and the following:

1. Origination of phantom signaling and detection of a wire fault condition are not supported.

2. The media interface connector's contact specifications for a concentrator are used.

3. Phantom signal detection, as defined for the Classic concentrator, is included.

4. The current return path for a phantom signal is supported. This allows the station to confirm the presence of the C-Port and supports station detection of an open-wire condition and certain short-circuit conditions in the lobe cabling.

5. Within 190 ms of the detection of the loss of a phantom signal a repeat path for a station's LMT is provided.

C-Port Repeat Path

Implementors of the C-Port are permitted to select the type of repeat path that is available to the station when supporting a LMT. Figure 5.13 illustrates two repeat paths, one repeat path in the PSC, and one in the PMC. When repeat paths are available in both the PSC and the PMC, only one repeat path is active at any time. A disabled repeat path does not affect the operation of the transmit path in any way.

A C-Port repeat path has the following characteristics:

1. A C-Port repeat path includes a transmitter, all circuitry in the transmission path between the data retiming (latching) mechanism and the media interface transmit connections.

2. A C-Port repeat path includes a receiver, all circuitry between the media interface connectors and the recovered clock used to latch the data.

3. A C-Port repeat path repeats all transmitted ones and zeros between the SD and the ED fields inclusive when repeating a frame.

4. A C-Port repeat path uses the recovered clock.

5. The average repeat path latency should not exceed 100 symbols. This is necessary to reduce latency introduced by the C-Port.

6. A C-Port repeat path may modify the A-bits and C-bits if the frame's destination address is recognized by the C-Port MAC. *This may have a serious impact on any LMT implementation that compares the received frame with the transmitted frame when determining an error.*

7. A C-Port repeat path may set the E-bit when a frame with an error is detected.

8. A C-Port repeat path may optionally correct a BURST-5 error to BURST 4.

9. A C-Port repeat path may optionally cause a polarity inversion of the data stream.

The characteristics in items 6 to 9 were added to allow reuse of existing MAC chip sets in the implementation of a C-Port.

5.4.9 DTR Station and C-Port Management

The station and C-Port MAC entities are similar in many ways, more so due to the addition of the C-Port's station emulation mode. The following subsections describe the management services and objects recommended by the Standard.

DTR management is defined as a set of services and objects. The services define the interaction between the MAC layer; the management layer; and, for a C-Port, the MRI. The objects define the information collected by management and are available to network management applications to aid in problem determination.

5.4.9.1 DTR Station Management Services

Figure 5.14 illustrates the management services defined for a DTR station.

The following services are defined for a DTR station.

MGT_ACTION is used for communication between the management entity and the SMAC. This service is used to open and close the MAC. Status for the requested action is returned to management. This is the only service that is required to be implemented by the SMAC.

The *MGT_ACTION.request* service is used to either open or close the station. Parameters that may be passed during a management-directed open action include the Access protocol used during the Registration process, the station's individual address, the media rate, and the value of the Access protocol mask.[18]

---

18. SPV(AP_MASK) value.

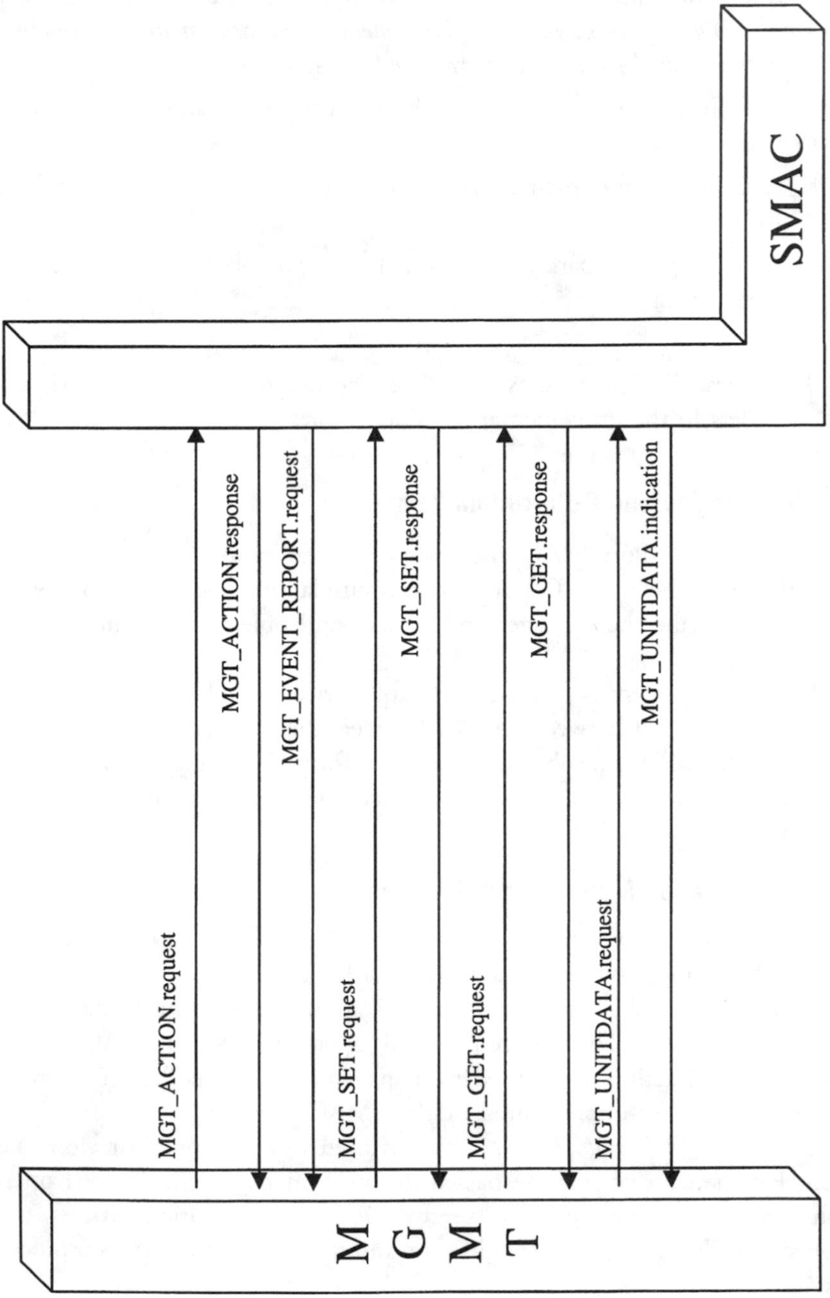

Figure 5.14 Set of management services defined for a DTR station.

The MGT_ACTION service is a confirmed service. As such, each request requires a response from the SMAC. The MGT_ACTION.response service returns the results of an open or close action taken by the SMAC. Table 5.9 illustrates the results of an open action requesting the TXI Access protocol.[19]

When a successful open action occurs, the SMAC may respond indicating the Access protocol used and the UNA.[20]

MGT_EVENT_REPORT is necessary in order for the MAC to send unsolicited event information. It used by the SMAC to inform management about unexpected events. The events may or may not require any action by network management but are available to diagnose problems with the network.

Table 5.10 illustrates the class of the status reported by the station. MGT_EVENT_REPORT contains sufficient information to indicate the cause, as detected by the SMAC, of the event.

MGT_UNITDATA is defined to allow management to send and receive management MAC frames.

To manipulate the objects defined for management the MGT_SET and MGT_GET services are defined. *MGT_SET* sets the value of an object in the information database. *MGT_GET* retrieves the value of an object from the information database.

Station management is expanded for a DTR station. In addition to RFC 1748, "IEEE 802.5 Token Ring MIB," additional objects unique to the DTR environment and unique to the TXI Access protocol are added. See Section 5.4.9.3 for a discussion of the objects provided.

5.4.9.2   C-Port Station Management Services

Figure 5.15 illustrates the management services defined for a C-Port. While the capability in the PMAC exists to manage a C-Port in the same fashion as a station, that is, use of management MAC frames, this practice is discouraged. Clause 11 of the Standard describes services and objects that should be implemented in order to manage a C-Port.

The C-Port MAC does not support an interface to management that permits the transmission or reception of management MAC frames. However, the transport of management MAC frames is supported through the DTR concentrator by use of the MRI and the MRI_UNITDATA service.

The following services are defined for a C-Port.

MGT_ACTION is used by the management entity to request that the PMAC either open or close. Status for the requested action is returned to management. This is the only service required to be implemented by the PMAC.

---

19. The results of a TKP Access protocol open are the same as described in Chapter 4.
20. Which would be the C-Port for a TXI connection.

Table 5.9
Results of an Open Action Requesting the Access Protocol

Open Result . . .	. . . Due to SMAC Action
Success	The SMAC completed the Join process and is now able to transmit data into the network
Standby Received	An SMP MAC frame is received, indicating that a violation of the TXI protocol check has occurred.
Beacon Received	A Beacon MAC frame is received before the station is allowed to participate in Hard Error Recovery
Claim Received	A Claim Token MAC frame is received, indicating that a violation of the TXI protocol check has occurred.
Protocol Not Supported	The requested Access protocol is rejected by the C-Port and management has not set up the option allowing the SMAC to attempt the TKP Access protocol.
Remove Received	A Remove Ring Station MAC frame is received, forcing the station to close unless the option permitting the station to ignore the frame is enabled.
Purge Received	A ring purge MAC frame is received, indicating that a violation of the TXI protocol check has occurred.
Internal Error	The station has detected an internal error and is unable to continue operation.
Station Error	The station has detected an error during a transmit operation and is unable to continue operation.
Registration Time-Out	Registration Request Time-Out, TSREQ = E. The station has issued up to six Registration Request MAC frames and has not received a response from a C-Port, forcing the station to close unless the option permitting the station to attempt the TKP Access protocol is enabled (FSOPO = 0).
Lobe Test Failure	The station's test of the lobe failed.
Lobe Test Time-Out	The station's test of the lobe failed due to excessive duration.
Heart Beat Received	A Heart Beat frame is detected either prior to the Station sending its first Insert Request MAC frame, or a Heart Beat frame has been received from a C-Port other than that identified during registration. In either case, something unexpected has occurred on the link and the station closes.
Duplicate Address Failure	The C-Port's Insert Response MAC frame has indicated to the station that the DTR reports a duplicate address.
Join Time-Out	The station has completed its LMT and has sent Insert Request MAC frames. The C-Port has failed to respond to the station within the time specified by the Standard (18 sec).
Heart Beat Time-Out	The station has completed its LMT and has sent Insert Request MAC frames. The C-Port has either not started or failed to continue its Heart Beat process.
Invalid Source Address	A MAC frame has been received at the station from a C-Port other than that which the station identified during registration. Something unexpected has occurred on the link and the station closes.

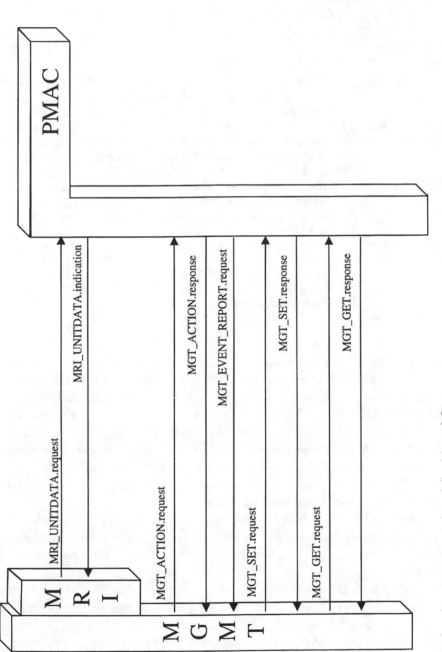

Figure 5.15 Set of management services defined for a C-Port.

Table 5.10
Class of the Status Reported by the Station

Station Status	. . . What Happened
Station Operational	This status is sent to management when the station either completes Hard Error Recovery due to the reception of an Insert Response MAC frame or is reporting a soft error.
Station Not Operational	This status is returned when the station has entered Hard Error Recovery due to Heart Beat Failure or receipt of a Beacon MAC frame. The station is no longer transmitting user data since the network is no longer available.
Station Failure	This status is returned when the station closes due to the receipt of a Remove Station MAC frame or the detection of a wire fault, internal error, or an error during the transmission of a frame.
Station Detected Protocol Error	This status is returned when the station closes due to the detection of a protocol error. A Protocol error occurs when either a restricted[21] TKP Access Protocol MAC frame or a MAC frame from a C-Port other than that which is identified during the Registration process.

The *MGT_ACTION.request* service is used to either open or close the C-Port. Parameters that may be passed during a management-directed open action include the Access protocol mask,[22] the C-Port's individual address, and the media rate.

The MGT_ACTION service is a confirmed service that returns the status of the MGT_ACTION.request with a MGT_ACTION.response. The status returned by the MGT_ACTION.response are indicated in Table 5.11.

When a successful open action occurs, the PMAC may return the media rate and the address of the attached station.

MGT_EVENT_REPORT is used by the PMAC to report an unexpected event to the management entity. The events may or may not require any action by network management but are available to diagnose problems with the network.

Table 5.12 illustrates the class of the status reported by the C-Port. The MGT_EVENT_REPORT contains sufficient information to indicate the cause, as detected by the PMAC, of the event.

MRI_UNITDATA is used to send (.indicate) to the MRI or notifies the receipt of a management MAC frame from (.request) the MRI. Management

---

21. Claim Token, Ring Purge, or SM Active.
22. PPV(AP_MASK) value.

Table 5.11
Status Returned by the MGT_ACTION

Open Result . . .	. . . Due to PMAC Action
Success	The C-Port has completed the Join process with the attached station. The C-Port is open and able to provide the attached station with connectivity to the remainder of the network.
Beacon Received	A Beacon MAC frame is received before the C-Port is allowed to participate in Hard Error Recovery. The C-Port is closed.
Claim Received	A Claim Token MAC frame is received, indicating that a violation of the TXI Access protocol has occurred. The C-Port is closed.
Purge Received	A Ring Purge MAC frame is received, indicating that a violation of the TXI Access protocol has occurred. The C-Port is closed.
Internal Error	The C-Port has detected an internal error and is unable to continue operation. The C-Port is closed.
C-Port Error	The C-Port has detected an error during a transmit operation and is unable to continue operation. The C-Port is closed.
AM Received	An AMP MAC frame is received, indicating that a violation of the TXI Access protocol has occurred. The C-Port is closed.
Invalid Source Address	A MAC frame has been received with a source address that does not match the address of the station with which the C-Port entered the Registration process. Something unexpected has occurred on the link, and the C-Port closes.
Heart Beat Time-Out	No Heart Beat, TPRHB = E. The C-Port is waiting for the DTU to post its response to the Duplicate Address test prior to completing the Join process and after the station has completed its LMT. During this time, the C-Port Heart Beat timer expires, indicating that a Heart Beat MAC frame has not been received from the station for over 5 sec. The C-Port closes.

MAC frames received by the PMAC from the media are sent to the MRI. Frames received from the MRI are transmitted on to the media.

There is no frame-passing interface between the PMAC and management. However, the MRI supports passing management MAC frames[23] to other C-Ports residing on the same DTR concentrator. This permits Classic management functions to continue to receive management MAC frames in a DTR environment.[24]

23. Valid MAC frames with a nonzero vector class and neither the source or destination class is 3.
24. This is not to say that unmodified Classic management applications, such as REM and CRS, can be used. Since the topology has changed significantly, these applications will have to change. However, access to the necessary MAC frames anywhere in a DTR network is supported by this architecture.

Table 5.12
Class of Status Reported by the C-Port

C-Port Event Status . . .	. . . What Happened
C-Port Operational	This status is sent to management when the C-Port completes Hard Error Recovery or when it is reporting that a soft error has been detected and is being reported using the MRI interface.
C-Port Not Operational	This status is sent to management when the C-Port enters Hard Error Recovery or detects an event that requires the C-Port to close. The C-Port closes when the attached station closes (loss of phantom), when a MAC frame directed to the PMAC with a source address other than the station's address is received, or when an address that is a duplicate of the C-Port's is detected.
C-Port Failure	This status is sent to management when the C-Port detects an internal error with which it is unable to continue to operate or an error during the transmission of a frame. The C-Port is closed.
Protocol Error Detected	The C-Port, while operating TXI Access protocol, received a restricted TKP Access protocol MAC frame. The C-Port is closed.

To manipulate the objects defined for management, MGT_SET and MGT_GET are defined. *MGT_SET* sets the value of an object in the information database. *MGT_GET* retrieves the value of an object from the information database.

Management objects defined for the C-Port include objects previously defined for Token Ring stations and new objects unique to the C-Port. Because there are many common objects used for both the C-Port and the DTR station, the Standard specifies a single MIB that is used by both.

5.4.9.3 DTR MAC MIB

The DTR MAC MIB is defined in clause 11 of the Standard and should be reviewed by anyone planning to implement a browser application or a C-Port. The following summarizes the major objects found in the MIB.

Protocol Table

This table provides information about the state of a MAC entity[25] that is using the TXI Access protocol. It contains information identifying the type of entity,

---

25. Either a C-Port or station.

station, or C-Port; the operational mode of the C-Port, either port mode or station emulation mode; functional address; UNA; vital product data; authorized function classes; the value of the error report timer; ring number; ring status; Join and monitor FSM states; and Hard Error Recovery information, such as the address of received Beacon frames, the Beacon type, and the last protocol event (see MGT_EVENT_REPORT).

Station Table

This table provides station characteristics information. It contains objects that identify the type of station, either a DTR Station or a C-Port in station emulation mode, the Access protocol in use, the Access protocol that will be requested during the Registration process, and the response from an attached C-Port during the last Registration event.

Many of the objects can be used to determine the current operational state of the station and to modify that operational state. These objects include the Access protocols supported by the station,[26] the maximum frame size supported by the station, the error reporting option, the use of the Registration process, the action taken by the station when a C-Port fails to respond to the station's Registration requests, action taken by the station when its Registration request is denied, action taken by the station on the receipt of a Remove Station MAC frame, the media speed at which the station operates, and support of the Registration Query protocol.

C-Port Table

The characteristics of a C-Port are available from the C-Port table. Information on the C-Port's current Access protocol in use is available.

Objects that are used to report the current state of the C-Port and can be used to modify that state are the Access protocols supported by the C-Port; the maximum frame size supported; supported phantom drive options;[27] the error reporting option; the media rate currently in use which can be modified for the next open of the C-Port; the operational mode of the C-Port, either port mode or station emulation; the option to set the AC-bits in the repeat path on frames with a DA recognized by the C-Port; the option to either abort a frame that exceeds the maximum frame length by use of an Abort Sequence or force a bad FCS and set the E-bit to one; setting of Beacon participation, either after Neighbor Notification or after Join completion; and the support for frame cut through.

---

26. This is the value of the SPV(AP_MASK).
27. This is the value of PPV(PD_MASK).

TXI Statistics

This table contains statistics collected for the TXI Access protocol. Statistics for the TKP Access protocol are maintained in the MIB defined by RFC 1748, "IEEE 802.5 Token Ring MIB." In addition to the counters that are defined for soft error reports, a counter for overlength frames is included.

Traps

Three traps are defined to indicate entry into Hard Error Recovery, events that cause the station to close due to internal errors or the receipt of a Remove Station MAC frame, and the use of a restricted TKP Access protocol MAC frame while operating in the TXI Access protocol.

5.5 DTR-Enhanced MAC Protocol

The DTR MAC protocol has many differences when compared with the Classic Token MAC protocols. There are two MAC entities working together in order to assure communication between the end user and the network, new processes to initiate station communication, new uses of old modes of transmit operation, and a few cautions to implementors of DTR stations that are starting their designs from a Classical base.

5.5.1 Join Process

The Join process defined by the committee is fairly robust. It does not require that the station have any prior knowledge of its connection to the network in order to support migration from a Classic Token Ring network to a mixed DTR and Classic network. A properly configured station uses the Registration and Join process to determine the highest performance protocol possible for the lobe to which it is connected.

5.5.1.1 DTR Registration and the Join Process

The timers and options in Table 5.13 control the DTR Registration process.

Once a station has performed all its internal testing and has received the command from management to open, the station must examine its policy flags to determine how to operate. Unlike a Classic station, a DTR station can open using either the TXI Access protocol or the TKP Access protocol. The Station Registration Option policy flag is set by management (as are all the policy flags and variables) prior to issuing an open command to the station.

Figure 5.16 is a high-level representation of a successful occurrence of the Join process for a DTR connection. The station starts the sequence by

Table 5.13

Timers, Policy Flags, and Variables for the DTR Registration Process

Station timers	
TSIS	Station Initial Sequence Timer (40 ms)
TSJC	Station Join Complete Timer (18 sec)
TSREQ	Station Registration Request Timer (40 ms)
Station policy flags and variables	
FSOPO	Station Open Option
FSRDO	Station Registration Denied Option
FSREGO	Station Registration Option
SPV(AP_MASK)	Station Access Protocol Mask
SPV(IAC)	Station Individual Address Count
C-Port policy flags and variables	
PPV(AP_MASK)	C-Port Access Protocol Mask
FPOTO	C-Port Operation Table Option

transmitting idles for about 40 ms, the duration is controlled by the Station Initial Sequence Timer. This is done to allow an attached C-Port's receiver electronics to synchronize with the station's clock. During this part of the process, the station supplies the clock for the connection while the C-Port recovers the clock from the data and idles received.

The DTR station then sends a Registration Request MAC frame to the attached C-Port. The Registration Request MAC frame (see Section 5.4.7.1) is used to request service from an attached C-Port. If a C-Port is present and configured properly, the C-Port responds with a Registration Response MAC frame (see Section 5.4.7.1). The Registration Response MAC frame indicates either acceptance or rejection of the station's request by the value of the Access Protocol Response subvector. When the Registration Response MAC frame indicates acceptance of the station's request, the C-Port has initiated the *Duplicate Address Check* (DAC) process (see Section 5.5.1.3) using the address information contained in the Registration Request MAC frame.

The standard assures the delivery of both the Registration Request MAC frame and the Registration Response MAC frame by including timers and special actions to detect when the frame appears to have been lost in transmission. The registration request may be sent up to six times, paced by the Station Registration Request Timer, if a Registration Response MAC frame is not received by the station. A registration response is allowed to be sent by the C-Port on each receipt of a registration request.

After responding to the Registration Request MAC frame, the C-Port configures its transmit machine to supply a repeat path that supports the station's LMT. The DTR Standard introduced guidelines for the LMT that

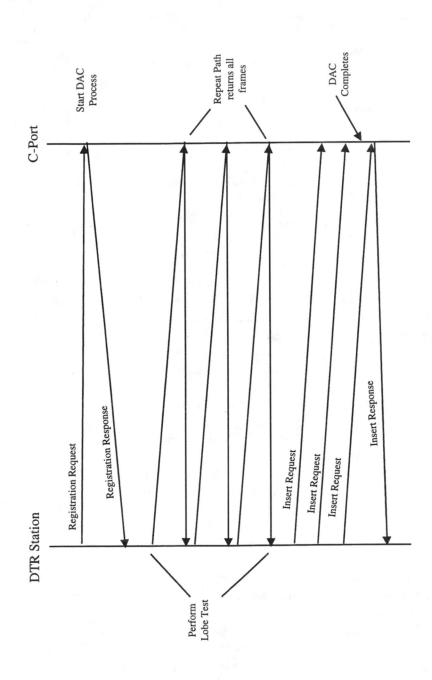

Figure 5.16 DTR Registration and Join process.

are discussed in Section 5.5.1.2. In addition to these guidelines, the Standard now enforces a maximum duration for Lobe test of 7.4 sec using the timer TPLMTR (see Section 5.4.1).

The station signals completion of its Lobe test by sending an Insert Request MAC frame (see Section 5.4.7.1) to the C-Port, and starts the Heart Beat process (see Section 5.5.1.4). Up to this point, the station was supplying the clock for the connection. After the transmission of the Insert Request MAC frame, the station switches to recovered clock and the C-Port supplies the clock for the lobe. On receipt of the first Insert Request MAC frame the C-Port starts its Heart Beat process if the C-Port is not going to immediately reject the station's request. If the DAC process is not completed when an Insert Request MAC frame is received, the C-Port does not respond. If the DAC process is completed, the C-Port responds to the Insert Request MAC frame with an Insert Response MAC frame. The DTR Response Code subvector carried in the Insert Response MAC frame indicates the status of the stations insert request. The C-Port can either accept the insert request, which allows the Join process to complete, or reject the request due to a DAC error.

The Standard assures delivery of the Insert Request and Insert Response MAC frames by including timers and special actions to detect when the frame appears to have been lost in transmission. Insert Request is paced by the station using the station Insert Process Timer. The station will continue to send Insert Request MAC frames until the Station Join Complete Timer expires or until an Insert Response MAC frame is received by the station. Insert response is allowed to be sent by the C-Port on each receipt of an insert request once the DAC process has completed.

The reception of an Insert Response MAC frame by the station signals the end of the Join process. When the insert response indicates that the station's request is accepted, the station and the C-Port may start to transmit user traffic.

Failed Registration

A failed registration occurs when there is no response to the station's Registration Request MAC frame or the Registration Request is denied, or when the station's Insert Request is rejected by the C-Port.

No Response

The most obvious failure from registration is when the station sends out six registration request MAC frames and there is no response. The action that the station takes next is controlled by the setting of the Station Open Option Policy flag and the Station Access Protocol mask. The policy flag allows the station to attempt to open using the TKP Access protocol. If this option is available, the station access protocol mask is set by management to permit use of the TKP Access protocol. When the option is not available, the station closes.

Registration Rejection

The Station Registration Denied Option policy flag and the Station Access Protocol mask allows the station to try opening using the TKP Access protocol. If the option is available, the Station Access Protocol mask is set by management to permit use of the TKP Access protocol. When this option is not available, the station closes.[28]

What Happens to the Station If Its Insert Request Is Denied?

In this case, there are no options for the station; it must close.

5.5.1.2   LMT

The timers in Table 5.14 control the LMT process.

The LMT was not well defined for the TKP Access protocol in the 1995 Token Ring Standard. For the TXI Access protocol it became necessary to add a few restrictions.

1. The transmission of Beacon, Claim Token, Ring Purge, AMP, and SMP MAC frames is not allowed. Within the TXI Access protocol detection of these frames is used to determine if the station has been

Table 5.14
Station and C-Port Timers for LMT

Station timers	
TSLMTC	Station LMT Complete Timer (2.4 sec)
TSLMTD	Station LMT Delay Timer (200 ms)
C-Port timers	
TPLMTR	C-Port LMT Running Timer (6 sec)

---

28. For migration purposes, it is recommended that this option be enabled.

moved to a shared media. Use of these MAC frames when the TXI Access protocol is active is discouraged.

2. The LMT may not exceed 2.3 sec in duration. The duration of LMT became an issue in several customer accounts. This information was shared during committee discussions and a limit on the duration was proposed and accepted.

3. Delay is added to the protocol that permits the C-Port the time necessary to configure a repeat path. This was accomplished by requiring the station to wait for the expiration of the Station LMT Delay timer before proceeding with LMT.

4. BER testing was defined as providing a valid operating range for the lobe with an attached C-Port and station when BER is less than or equal to 10^{-7} and an invalid operating range when BER is greater than or equal to 10^{-9} (with a probability greater than or equal to 99%).

Further, guidance was provided on the construction of LMTs that met these criteria. Specifically, given a LMT consisting of the transmission of frames that are between 256- and 1024-bits long, with a total of one million bits transmitted, and setting the pass/fail criteria as two or more frames with errors, the lobes that passed this test had a BER better than 10^{-7} with a probability greater than or equal to 99%. Lobes that failed the test had a BER worse than 10^{-5} with a probability greater than or equal to 99%.

5.5.1.3 Duplicate Address Check Process

Figure 5.17 is a high-level representation of the DAC process. The DAC process is a cooperative effort between two architectural constructs, the C-Port MAC and the DTU, to ensure that the address of an attached station is not a duplicate of any other station attached to another C-Port.

The DAC process starts when the C-Port receives the station's Registration Request MAC frame and concludes when the C-Port transmits an Insert Response MAC frame to the station.

When the C-Port receives the Registration Request MAC frame from the station, it captures the address of the station from the SA field of the MAC frame and reads the optional *individual address count* (IAC) subvector value. These fields are passed to the DTU interface to allow the DTU to determine if there is a duplicate of this address and address range attached to another C-Port in the DTR concentrator.

The DTU returns to the C-Port either a positive or negative response. This response is placed in the DTR Response Code subvector of the Insert

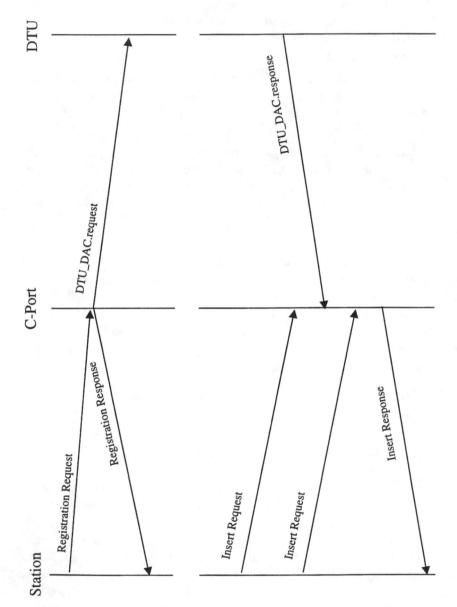

Figure 5.17 DAC process.

Response MAC frame, which is sent to the station upon receipt of the next Insert Request MAC frame.

A positive response code in the Insert Response MAC frame allows the station to complete the Join process. A negative response code forces the station to close.

When the station transmits its Insert Request MAC frame, the station is using its own crystal to generate the clock. At the end of the Lobe test this is necessary since the C-Port is not supplying the clock for the link; it is on recovered clock in order to supply a repeat path for the station's LMT. During subsequent transmissions of the Insert Request MAC frame, the station goes back on to its own clock since there is no assurance that the C-Port received the first insert request and may not currently be providing clock for the link. Because the station is changing its clock source every 200 ms (controlled by the station's insert process timer), there are windows of time during which the station will be unable to receive a transmitted frame from the C-Port. For this reason, the C-Port is only permitted to transmit its insert response MAC frame immediately after the receipt of the station's insert request. "Immediately" is controlled by the expiration of the C-Port's Insert Request Delay Timer (20 ms). This delay allows the station's receiver electronics to settle out after the clock change and allows it to receive the frame from the C-Port.

5.5.1.4 Heart Beat Process

The Heart Beat process is used by the TXI Access protocol to ensure that the MACs at each end of the lobe, the station and the C-Port, are operating normally. Failure of this process is the only event that will cause the TXI Hard Error Recovery process to be initiated.

The timers indicated in Table 5.15 control the Heart Beat process.

The Heart Beat process consists of two independent activities: Heart Beat transmission, and Heart Beat reception and checking. The Heart Beat process is an exchange of the Station Heart Beat MAC frame and the C-Port Heart Beat MAC frame (see Section 5.4.7.1).

Table 5.15
Timers That Control the Heart Beat Process

Station Timers	
TSQHB	Station Queue Heart Beat Timer (1 sec)
TSRHB	Station Receive Heart Beat Timer (5 sec)
C-Port Timers	
TPQHB	C-Port Queue Heart Beat Timer (1 sec)
TPRHB	C-Port Receive Heart Beat Timer (5 sec)

As illustrated in Figure 5.18, the station starts its Heart Beat process at the end of its LMT. When the station queues the Insert Request MAC frame for transmission, the station activates the Station Queue Heart Beat timer. When this timer expires a Heart Beat MAC frame is queued by the station for transmission and the station restarts the timer to continue the process. When the C-Port receives the first Insert Request MAC frame and is not going to immediately reject the request, the C-Port starts its own Queue Heart Beat timer to start transmission of the C-Port heart beat MAC frame. The station and the C-Port use the Queue Heart Beat timers to pace the rate (1 sec) at which a Heart Beat Frame is transmitted.

The reception and checking activity monitors the Heart Beat process. The station and the C-Port expect to receive a Heart Beat MAC frame at least once every 5 sec; this is controlled by the Receive Heat Beat timers. If this event fails to occur, the Heart Beat process is considered to have failed,[29] and the TXI Hard Error Recovery process is started.

5.5.1.5   C-Port Join Process Support for a Classic Station

The C-Port is specified to support both the TXI Access protocol and the TKP Access protocol. It must be capable of supporting the operation of either a DTR station or a Classic Token Ring station.

The C-Port supports a Classic station by providing an interface that appears to be a Classic TCU connection. Once a phantom signal is asserted, the ring appears to the station as a two-station ring. For a DTR connection, special support is added since the station's assumption of a dedicated lobe that can be switched out of the ring for a LMT is not true. This affected the development of the Join and Beacon process of the C-Port TKP Access protocol.

The Classic station proceeds through the Join process as normal. During the station's LMT, the C-Port provides the necessary repeat path. When the station asserts a phantom signal, the C-Port responds by starting the Claim Token process. Figure 5.19 is an example where the C-Port wins the Claim Token process, proceeds to become the AM, initiates the Ring Purge process, and starts the Neighbor Notification process. These actions allow the station to continue the classic Join process through completion.

5.5.1.6   A TXI Only C-Port

The Standard allows for a management configuration where the C-Port does not support a Classic station or a DTR station using the TKP Access protocol. Since a Classic station does not attempt to register, nor is there a method for

---

29. Heart Beat failure.

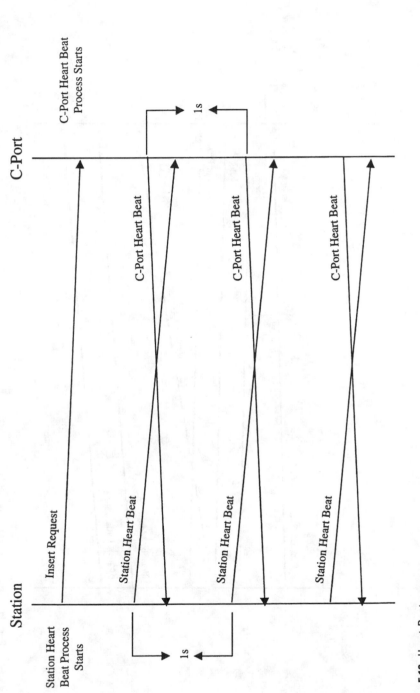

Figure 5.18 Heart Beat process.

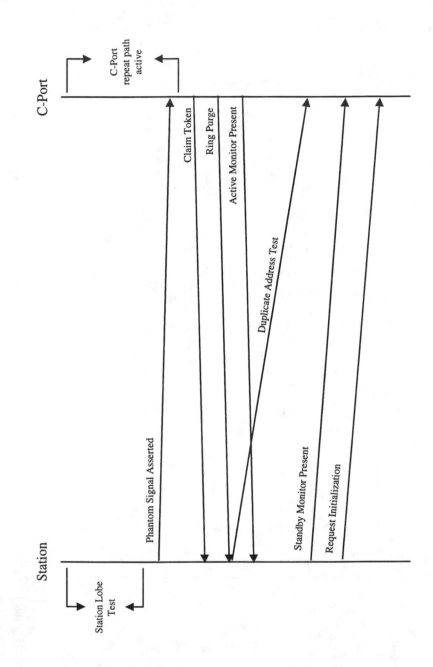

Figure 5.19 C-Port Join process support for a Classic station.

a DTR station to register for the TKP Access protocol, the Standard committee chose to cause the station's LMT to fail when this condition occurred.

A C-Port configured to support the TXI Access protocol will deny access to a Classic station by causing the station's LMT to fail. This is accomplished by the C-Port monitoring the lobe for a token or frame while in the registration state. If either event is detected, the C-Port will configure the transmit FSM to transmit idles for a period of at least 2.6 sec after the receipt of the last token or frame. This is accomplished in the Join FSM using the timers *C-Port Break Lobe Test* (TPBLT) and *C-Port Disrupt Lobe Test* (TPDLT). TBPLT is used to delay the disruption of the repeat path by 10 ms to 30 ms. This was done to accommodate the implementation of an auto detection mechanism that is described in Annex T of the Standard. The C-Port breaks the repeat path for at least 2.6 sec, which is long enough to cause the station's LMT to fail.

Configuration of a C-Port so that it supports only the TXI protocol cannot be recommended since it does not provide for a smooth migration of Classic concentrators to switches throughout an enterprise. A Classic station attaching to a C-Port configured in this fashion will close with an LMT failure and will not have any understanding that the problem is due to the C-Port's configuration. This causes problems for network management and decreases the value of Token Ring in general.

5.5.2 Transmit FSM

The concepts behind the Transmit FSM used by the TXI Access protocol are not new to Token Ring. The method of transmitting a frame without the use of a token is part of the Beacon process defined for Classic Token Ring.

The transmit FSM typically performs one of the following functions:

- Transmits frame traffic from either the MAC or LLC queue;
- Transmits idles;
- Supplies a repeat path (C-Port only).

Figure 5.20 illustrates the conceptual model of the TXI Access protocol's transmit FSM. Two queues are defined: one for MAC traffic required to establish and maintain the lobe and a queue used for LLC traffic and any management traffic.

The MAC queue is used by the MAC protocol to transmit all frames necessary to establish and maintain the lobe between the station and the C-Port. The frames placed in the MAC queue are transmitted before any frames in

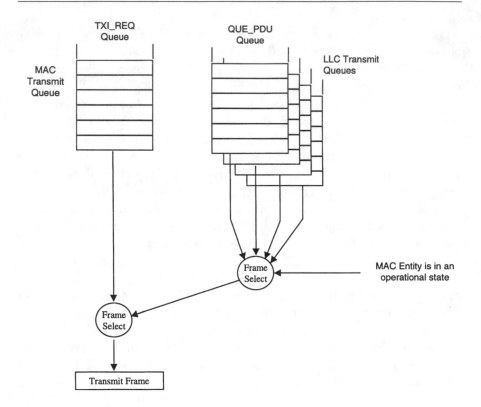

Figure 5.20 Transmit FSM queue model.

the LLC queue. Frames placed in the MAC queue are transmitted regardless of the operational state of the MAC.

The LLC queue is used to transmit user data and management MAC frames. The transmission of frames from the queue is not enabled until the MAC entity is in an operational state and all the frames from the MAC queue have been transmitted. A MAC entity becomes operational once Join is completed and remains in that state until it either closes or enters Hard Error Recovery. In Figure 5.20 the LLC queue is shown with multiple queues, which is neither described or prohibited by the Standard. Many existing implementations of C-Ports maintain two queues to manage priority traffic. Typically, priority levels 0 to 3 are mapped into a low-priority queue while priority levels 4 to 6 are mapped into a high-priority queue. Priority level 7 is reserved for the management MAC traffic that is sent to this queue. Designs with multiple priority queues are recommended since these designs will be able to provide control over both the jitter and latency of transmitted frames. Annex I of the 1995 Standard puts forth recommendations on how to best use the priority mechanisms provided by Token Ring. The user priority field found in the FC byte of the frame header is used to identify the priority queue to which a frame is sent. Using the user priority field instead of the AC byte's priority field is important since the user priority travels the entire path, from source to destination, while the access priority is useful only on the current ring.[30]

Once the station and the C-Port have completed the Join process, the transmit FSM is either transmitting frames or transmitting idles. There is no "quiet" time on the link. The C-Port is supplying the clocking for the lobe.

When the station requires support for its LMT, the transmit FSM in the C-Port supplies a recovered clock repeat path. The C-Port repeat path is described in Section 5.4.8.2.

5.5.2.1 Clocking for the Transmit FSM

The clocking used to transmit frames, illustrated in Figure 5.21, is dependent on the Join state of the MAC entity as well as the identity of the entity.

For the station, prior to the DAC state, the station's crystal is used. Once the Join process has completed, the station uses the recovered clock. During the DAC state, the station will use the station's crystal clock to transmit the insert request MAC frames.[31]

For the C-Port, prior to the DAC state, the recovered clock is used to transmit frames. The C-Port uses the recovered clock while it is supplying a repeat path for a station's LMT. Once the C-Port has completed the Join process, the C-Port's crystal is used to transmit frames.

5.5.3 Hard Error Recovery

The results of Hard Error Recovery are used by network management and network line support to isolate faults that exist in a DTR connection. The fault domain of a DTR connection is illustrated in Figure 5.22 and consists of the C-Port, the attached station, and the components between them.

5.5.3.1 TXI Hard Error Recovery

The TXI Hard Error Recovery process consists of five steps.

1. Heart Beat failure (see Section 5.5.1.4) is detected.

---

30. And is not useful at all in the TXI Access protocol.

31. Heart Beat MAC frames may be transmitted on either crystal or recovered clock. The clock used is dependent on when the Heart Beat frame is placed on the queue and the current clock in use. Recall that transmission of an Insert Request MAC frame requires the use of the crystal clock and the protocol returns to recovered clock once the frame is sent. Note that clocks will not be changed in the middle of a frame transmission.

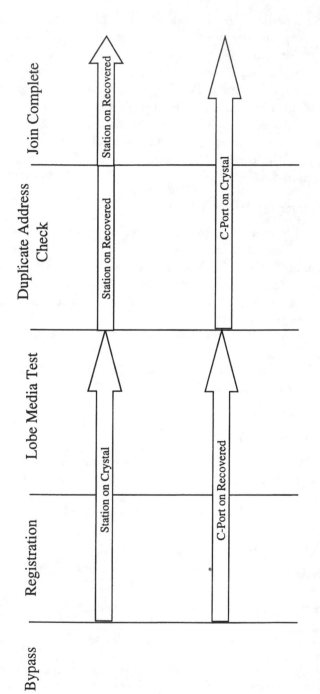

Figure 5.21 Clocking for the transmit FSM during different FSM states.

Fault Domain

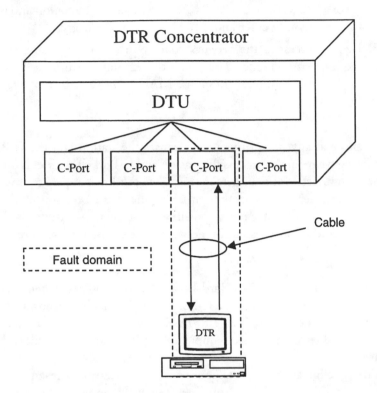

Figure 5.22 Fault domain for a DTR connection.

2. Start transmission of Beacon frames. The Beacon frame has two valid values for the Beacon type field, which indicates either Heart Beat failure or signal loss. Signal loss is reported as an additional indication to help in isolating the error. A MAC entity, either station or C-Port receiving a Beacon frame from its attached MAC (upstream neighbor) entity, will enter the Beacon process in a passive mode and will not transmit Beacon frames.

3. Submit sufficient time for the detection of a Wire Fault condition. Wire Fault is not immediately disabled when the TXI Hard Error Recovery process starts. Because a failure may have occurred due to a cable being cut or disconnected, Wire Fault isolates the fault to the components between the station and the C-Port.

4. Test the components between the station and the C-Port by performing an LMT. Unlike the TKP Hard Error Recovery process, this is always

performed during TXI Hard Error recovery. This removes the possibility of a lobe with an unacceptable BER, causing the station and C-Port to continuously enter and exit Hard Error Recovery.

5. Request access to the network again by the station issuing an insert request MAC frame. The C-Port does not perform a DAC and responds with an Insert Response MAC frame, which completes Hard Error Recovery.

A failure at any one of these steps causes the station and the C-Port to close. Noting where in the process the station and C-Port closed yields information that allows network management and network line support to isolate the fault. However, depending on the nature of the fault, isolation of the fault within the fault domain may not be possible using the results from Hard Error Recovery, and additional testing by network support may be required.

An Example of a C-Port Failure

Figure 5.23 illustrates the TXI Hard Error Recovery process when there is a failure in the C-Port, such that the C-Port MAC is unable to transmit frames. In this example, the C-Port is able to receive frames and configure its repeat path. Timer values used in this example are typical and recommended by the Standard.

The timeline at the top of Figure 5.23 indicates major events that can be observed on the media. The timelines below this indicate the duration of timers defined in the Standard that control the duration or the start of some events.

As described in the Heart Beat process (see Section 5.5.1.4), 5 sec will elapse before the station detects a heart beat failure, which starts the Hard Error Recovery Process for the station. The station sends out Beacon frames paced at 20 ms for about 7 sec (TSIT). This allows the detection of a possible wire fault condition. The reception of the Beacon frame by the C-Port causes the C-Port to enter the Hard Error Recovery process in a passive mode; the C-Port waits for the station to drop phantom signaling, which indicates the start of the lobe test.

Both the station and the C-Port use internal timers[32] to detect when a lobe test has exceeded the maximum allowed duration (see Section 5.5.1.2). In this example, the station passes its LMT and requests access to the network by issuing Insert Request MAC frames. This action is the same as that taken during the Registration process. The MAC frames are sent every 20 ms for

---

32. TSLMTC and TPLMTR. See Tables 5.1 and 5.2.

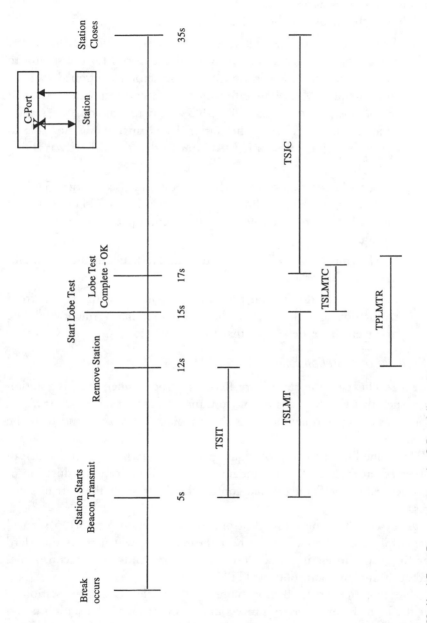

Figure 5.23 Hard Error Recovery example, C-Port failure.

up to 18 sec. At the end of the 18 sec, the station determines that the C-Port MAC is not responsive and closes. Once the station has dropped its phantom signaling, the C-Port also closes.

Observe the following about this example.

1. The C-Port completed Hard Error Recovery when the first insert request was received and is unaware of the reason for the station to close. From the C-Port's viewpoint, the station continued to send Insert Request MAC frames to which the C-Port responded. After 18 sec of this, the expiration of TSJC within the station, the station appeared to have closed. The station failed since it was not able to observe the Insert Response MAC frames that the C-Port was trying to send.

2. Because the location of the C-Port repeat path (see Section 5.4.8.2) is in either the PSC or PMC, a failure within the C-Port may not cause a failure of LMT since a working repeat path is available to the station.

3. Hard Error Recovery failed due to a failure between the PMC and the PMAC interface.

4. As a result of the Beacon process, a correlation of the results from the station and the C-Port allows network management and line management to determine that the failure occurred in the C-Port.

An Example of a Station Failure

Figure 5.24 illustrates the Hard Error Recovery process when there is a station failure, such that the station is unable to transmit frames to the C-Port. As in the previous example, timer values used are typical and recommended by the Standard.

The time line at the top of Figure 5.24 indicates major events that can be observed on the media. The time lines below this indicate the duration of timers defined in the Standard that control the duration or the start of some events.

As described in the Heart Beat process (see Section 5.5.1.4), 5 sec will elapse before the C-Port detects a heart beat failure, which starts the Hard Error Recovery process in the C-Port. The C-Port sends out Beacon frames paced at 20 ms for about 600 ms (TPIT).

The reception of the Beacon frame by the station causes the station to enter the Hard Error Recovery process in a passive mode. The station delays dropping phantom signaling for about 7 sec (TSIT). This allows the detection of a possible wire fault condition. When the station drops phantom signaling, the C-Port prepares for the start of the station's lobe test.

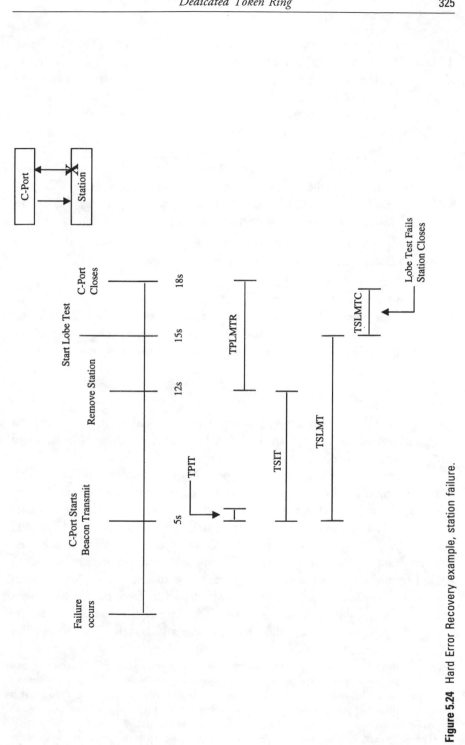

Figure 5.24 Hard Error Recovery example, station failure.

Both the station and the C-Port use internal timers (TSLMTC and TPLMTR, respectively) to detect when a lobe test has exceeded the maximum allowed duration (see Section 5.5.1.2). The C-Port uses its timer to determine that the station's lobe test has failed. The C-Port then closes. The station closes as soon as it fails its lobe test.

Observe the following about this example.

1. The C-Port started Hard Error Recovery due to heart beat failure and would have indicated loss of signal in the Beacon type field. Because the station did not issue an Insert Request MAC frame, the C-Port knows only that the station failed its lobe test.

2. The station started the Hard Error Recovery due to the reception of a Beacon frame. This indicates that the C-Port was unable to receive frames from the station.

3. A correlation of the results from the station and the C-Port does *not* allow network management and line management to isolate the failure any further than the transmit side of the station. The actual failure may be in C-Port, the attached station, and the components between them.

5.5.3.2 Support for a Classic Station

The Classic Hard Error Recovery process used by the C-Port in this configuration is modified to interoperate with a Classic station performing its Hard Error Recovery process.

As previously discussed, the C-Port does not perform an LMT. This requires a modification in the way the C-Port operates during Hard Error Recovery when it detects an error or receives a Beacon frame. Unlike a Classic station, the C-Port does not rely on timers to determine when an LMT should start; instead the C-Port waits for the station to indicate entry into its test state by dropping the phantom signal. Because the station expects to be disconnected from the network and have a private lobe to test, the C-Port provides a recovered clock repeat path. In a necessary departure from the Classic Token Ring standard,[33] there is a time limit of 15 sec imposed by the C-Port on the duration of the LMT. If a station fails to raise a phantom signal within this

---

33. This allows the C-Port to be able to determine if the application of a phantom signal indicates the return from Beacon test or if a new station open using the TKP Access protocol is occurring. This also keeps the C-Port from hanging in this state due to a station that is closed and is not going to reopen. A C-Port hung in this condition would only be able to report to management that it was not operational due to entrance into Hard Error Recovery. The addition of the time-out allows the C-Port to close and report this status to management.

constraint, the C-Port assumes that the station has failed its LMT and the C-Port closes.

5.5.4 Registration Query Protocol

The Registration Query protocol was included in the Standard late in the development cycle. During interoperability testing conducted at the University of New Hampshire, it was noted that under some typical conditions DTR stations and C-Ports failed to join using the TXI Access protocol as was expected. This usually occurred when the C-Port was not "ready" for the station to start the Registration process and missed the transmission of the Registration Request MAC frames. To eliminate this situation, the Registration Query protocol was added to the Join process of the C-Port, and the DTR station MAC was modified[34] to recognize and respond to the Registration Query MAC frame (see Section 5.4.7.1).

Figure 5.25 is a high-level representation of the Registration Query protocol. The station completes and fails the DTR Registration process (see Section 5.5.1.1) and is configured to attempt to join using the TKP Access protocol (see Section 5.5.1.5). The station fails DTR registration because the C-Port was unable to respond to the registration request MAC frames that were issued by the station.

Once the station completes its LMT, it will assert a phantom signal. Any time after this occurrence the C-Port MAC may become active and respond to the phantom signal. As shown in Figure 5.19, the C-Port's response to a phantom signal is to start the Claim Token process to elect a new AM on the lobe. Claim Token is followed by Ring Purge and the start of Neighbor Notification. The example in Figure 5.25 shows the case where the C-Port wins the Claim Token process and becomes the AM.

The end of the first Neighbor Notification cycle starts the Registration Query process when the C-Port transmits a Registration Query MAC frame. Transmission of this frame is augmented by an assured delivery process. The frame is sent out multiple times, controlled by a pacing timer and a counter. The frame is sent up to 5 times at 20-ms intervals.

The station responds to the Registration Query MAC frame by dropping the phantom signal and preparing to start the DTR Registration process. The station issues a Registration Request MAC frame no sooner than 200 ms after it has dropped a phantom signal. When the C-Port detects the loss of phantom

---

34. The Registration Query protocol is optional for the DTR station. Classic stations and DTR stations that do not implement this protocol will interoperate with a C-Port that uses this protocol since the registration query MAC frame will be ignored by the station MAC.

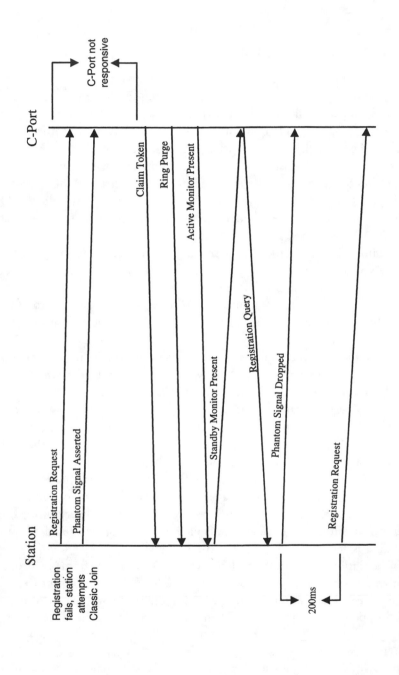

Figure 5.25 Registration Query protocol.

signaling, it prepares for the start of the DTR Registration process by establishing a repeat path and returning to recovered clock.

If the station fails to respond to the Registration Query MAC frames, the C-Port continues to the Join complete state and continues to use the TKP Access protocol.

5.5.5 Other Enhancements

5.5.5.1 Support for Frame Cut Through

To reduce the latency through implementations of DTR concentrators, there was significant interest to support the ability to start forwarding a frame from the source port to the destination port before the entire frame was received by the source port. Many implementations are able to start transmitting the first bytes of a frame onto the destination media before the last bytes of the frame have been received by the source port.

Support for cut through appears in the Operation Table as transitions that indicate the length of the current frame is unknown (FR_LTH = UNK) or in transitions using FR_FC, a frame received through the FC field. The transitions with unknown frame lengths are used to cut through frames from the DTU interface to the PMAC. Transitions using the FR_FC event are used to cut through frames from the media to the DTU interface. In addition, transitions are added to remove frames from the transmit queue or to force a frame transmission to be terminated when a cut through frame is determined to be in error due either to overlength or invalid FCS as detected by the source port.

Two options are included in the Standard that affect the implementation of cut through. FPASO controls the method used by the transmit FSM when terminating the transmission of a frame that is determined to be too long. The port policy variable PPV(MAX_TX) defines the maximum length of a frame that will be transmitted by the C-Port. The upper limit of this variable is determined by the media speed.[35]

FPASO permits the implementation to either terminate the transmission of a frame by using an Abort Sequence or by truncating the frame, generating bad CRC for the frame and setting the' E-bit in the ED. The first method, supported by many existing chip sets, requires that the ring recover after the Abort Sequence is sent since a token[36] is not released. The second method does not require any ring recovery since a token is released. Further, since the

---

35. At 4 Mbit/s the upper limit is 4,550 octets. At 16 Mbit/s the upper limit is 18,200 bytes. In general, most implementations do not support these maximums.

36. This comment is valid only when the C-Port is using the TKP Access protocol.

E-bit is set, the downstream neighbor of the C-Port will not count (line error) the terminated frame.

Cut through is not supported for DTR stations but is supported for C-Ports in both station emulation and port mode.

5.5.5.2 Explicit Loopback

All MAC service layers are required to support a method to return frames from the LLC layer, whose destination address is that of the transmitting station's. In Classic Token Ring this was supported by the nature of the ring; whatever was transmitted by a station would be returned to the station, and if the destination address was recognized, the station MAC would *indicate* the frame to the LLC.

DTR removed the ring, so what goes out *never comes back.*[37] The SMAC specification explicitly returns any frame from either the LLC or bridge interface, whose destination address is recognized by the station to the LLC layer as well as transmitting the frame onto the media. This has some very important impacts on the implementation of device drivers, adapters, and MAC bridges. Many Token Ring implementations have relied on the ring to return this class of frames, so no provision was made to test a frame's destination address on the "transmit channel" of an adapter. Implementors of DTR stations using Classic Token Ring station as a base need to ensure that this function is included in their design.

5.6 The DTR Concentrator—Annex K

Figure 5.26 illustrates the architectural model of a DTR concentrator. As previously stated, the Standard originally addressed only the station and C-Port MAC. The interfaces external to these MACs were to be defined, but the entities on the other side of these interfaces were left for further standardization. That all changed in November 1994, when a presentation concerning potential interoperablity problems was made. This presentation made it clear to many members of the committee that some statement needed to be made concerning issues such as spanning tree and frame forwarding. By May 1995, Annex K was added to the Standard as an informative annex that described a recommended implementation.

There are two major implementations of a DTR concentrator available on the market today. The main difference is the forwarding function defined

---

37. There is always an exception. If the attached C-Port's MRI has been configured properly, a management MAC frame will be returned to a transmitting station. However, this is an action taken by the MRI and not by either the SMAC or PMAC.

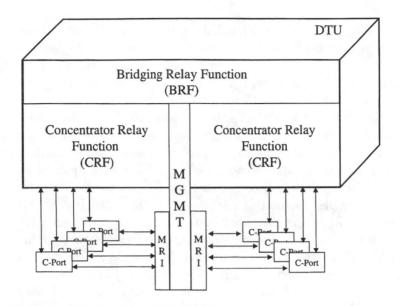

Figure 5.26 DTR concentrator functional organization.

in the DTU. In one implementation, the DTU is simply a source-routed bridge as originally defined in the IBM Token Ring architecture. The other implementation follows the recommendation provided for in Annex K. The argument for which implementation to select is complex. However, the architecture defined within Annex K is required when considering the development of a network where many connections are dedicated. An architecture that depends on a classical source-routed bridge limits the size the network can achieve since each dedicated connection must be assigned its own ring number. Additionally, a source-routed bridge solution will typically further limit the size of the network since many bridge implementations available limit the number of "hops" in the route information field to 7 (the Standard allows for 14). A network comprised of DTR concentrators using a SR DTU and classical source route bridges will find this a difficult limitation.

Figure 5.26 exposes the details of the DTU, which is comprised of a single, optional, *bridging relay function* (BRF) and one or more *concentrator relay functions* (CRF).

The BRF is defined to be an SRT Bridge as described in ISO/IEC 10038:1993 for MAC bridges and source routing transparent bridges. Ports of the BRF may be either external or internal to the DTU. Internal ports are connected to the concentrator relay functions as shown in the figure. External ports are not shown. As defined by the ISO standard, the BRF forwards frames based on either destination MAC addresses or by the contents of the RIF, if present. The presence of the RIF is determined by examining the RI-bit, which is the high-order bit in the SA field of the MAC frame. A "1" indicates that a RIF is present. The MAC Bridging Standard requires that when the RI-bit is set to 1, the RIF must be used to make forwarding decisions. When the RI-bit is set to 0, then the DA must be used.

The CRF is a newly defined construct. Unlike an SRT bridge, this forwarding entity will forward a frame based on either the MAC address or information contained in the RIF (if present), regardless of the setting of the RI-bit. The decision as to which field to use is dependent upon where in the path, from source to destination, a CRF receives the frame.

For example, assume that a frame is sent from station A on Ring 1 to station B on Ring 6 as illustrated in Figure 5.27. A CRF that receives the frame on Ring 1 (the source ring) forwards the frame based on information contained in the RIF. A CRF that receives the frame in the middle of the path at Ring B, where the CRF has no ports on either the source or the destination ring, also forwards the frame based on information in the RIF. Finally, when a CRF that has a port or ports on the destination ring receives the frame, the forwarding decision is based on the destination MAC address.

5.6.1   CRF Service Characteristics

When placed within a LAN, certain service characteristics are expected from a CRF. Adding this new functional entity should not disrupt the operation of an existing bridged LAN. The service characteristics of a CRF differ little from those of a MAC layer bridge, as defined in ISO/IEC 10038.

The CRF should be designed so that it does not significantly increase the number of frames lost in the network. Typically, frames are lost due to congestion within a CRF. This can be avoided through proper network design by providing for adequate bandwidth.

Within a priority level, frames transmitted from a source to a destination should be forwarded by the CRF in the same order as they are received. This allows frames of higher priority to be serviced ahead of frames of lower priority between stations. It is recommended that a CRF examines the user priority[38] of the frames it forwards and to service frames of higher priority ahead of those with lower priority. While the number of priority queues is not suggested in the Standard, most implementations support two queues and map priorities 0 to 3 to a low-priority queue and 4 to 7 to a high-priority queue.

The operation of a CRF should not of itself cause duplication of frames to occur within the network. Duplication of frames within a network is usually

---

38. The user priority is found in the FC byte of the MAC header.

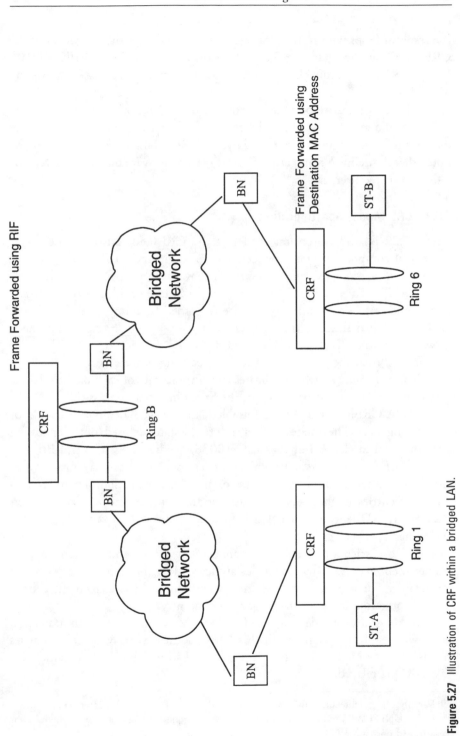

Figure 5.27 Illustration of CRF within a bridged LAN.

the result of loops within the network. This is avoided in the definition of the CRF by using the Spanning Tree protocol, as defined in ISO/IEC 10038. Using the same Spanning Tree protocol assures interoperability with existing MAC bridges.

The CRF must respect the maximum frame length limitations imposed by the media attached to the C-Ports. For Token Ring this means that the maximum frame lengths are governed by the attached media speed. It should be noted that the definition and use of the C-Port's policy variable PPV(MAX_TX) satisfies this requirement.

5.6.2   CRF Model and Operation

Figure 5.28 is an architectural model of the CRF and illustrates the major functional components and processes that make up the CRF.

Frame reception examines frames that have been presented[39] to the *CRF Port* (CRFP) from either a C-Port or from an internal bridge port. All MAC frames and frames with a FCS error are discarded by frame reception. If the CRFP state, as defined by the port state database, is Forwarding, the frame is forwarded to the Relay process and the Learning process. If the CRFP state is Learning, then the frame is sent to the Learning process only.

The *Port State Database* indicates the current state of each of the CRFPs. These states relate to the operation of the Spanning Tree process that the CRF supports. A CRFP may be in the Disabled, Listening, Blocking, Learning, or Forwarding State. These states are identical to the states defined for the Spanning Tree protocol in clause 4 of ISO/IEC 10038, Media Access Control Bridges.

The *Filtering Database* contains entries used by the Relay process and created by either the Learning process or management action. There are two classes of entries in the database: MAC address entries and *destination route descriptor* (DRD) entries. Each class is further subdivided into two types: static and dynamic.

Static entries are created by management action and are removed by management action. Dynamic entries are learned by the Learning process and are retained for a period of time that is configured by management action. This is sometimes referred to as *aging out* an entry.

MAC Address entries consist of a 48-bit MAC address and an indication of to which CRFP a frame with this MAC address is relayed. DRD entries consist of a 16-bit DRD and an indication of to which CRFP a frame with this DRD is relayed.

---

39. Within the Architecture, this presentation takes the form of the interface primitives DTU_UNITDATA.indicate or M_UNITDATA.request, depending on whether the attached port was a C-Port or an internal bridge port.

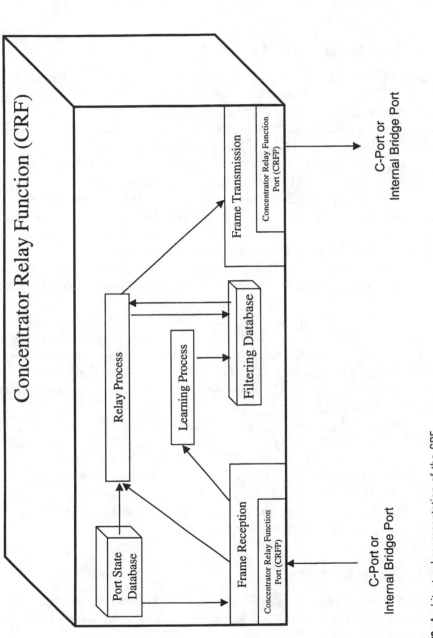

Figure 5.28 Architectural representation of the CRF.

Frame transmission simply presents[40] a frame forwarded from the Relay process to the C-Port or internal bridge port attached to the CRFP.

5.6.3   Route Information Field

The Token Ring frame format was introduced in Chapter 4. Referencing Figure 5.29, the RI-bit is the high-order bit of the SA. When the RI-bit is set to 1, the frame contains a RIF. The RIF consists of the *route control* (RC) field and the *route descriptor* (RD) entries. There can be up to 14 RDs contained in a RIF, although most bridge implementations support only 7.

The route control field consists of the route type, the length, the direction, and the largest frame fields (the "r" indicates a reserved field).

The route type indicates if the frame is an *All Routes Explorer* (ARE), a *Spanning Tree Explorer* (STE), or a *specifically routed frame* (SRF). The route type is recognized by SRT MAC bridges and is used to control the processing performed on the frame. Explorer frames are used to discover routes from one end station to another. The SRT bridge, after a series of checks, will insert a RD entry into an explorer frame before forwarding it. A SRF is simply forwarded by an SRT MAC bridge when the conditions for forwarding are met.

The LTH field indicates the total length of the RIF. The *direction* (D) bit is used by SRT bridges to determine the direction (left to right or right to left) to traverse the RIF when making forwarding decisions. The *largest frame* field (LF) is used by SRT MAC bridges to indicate the largest frame size in octets size supported by the bridge. This value may only be modified downward by a bridge.

The RD field consists of a series of RDs, which are 16-bit entries made up of a 12-bit LAN (or ring) ID and a 4-bit bridge ID. Within a bridged network the LAN ID is unique, while the bridge ID must be unique only between two LAN IDs. In an SRF, this series of RDs documents the path the frame is to take to reach the DA from the SA. In either type of Explorer frame, the series of RDs documents the path that the frame took to reach the point in the network where the frame has been observed.

5.6.4   The Relay Process

The Relay process examines frames received at a port of the CRF and determines if the frame is be relayed or discarded. If the frame is to be relayed, the Relay

---

40. This presentation takes the form of the interface primitives DTU_INUTDATA.request or M_UNITDATA.indicate, depending on whether the attached port is a C-Port or an internal bridge port.

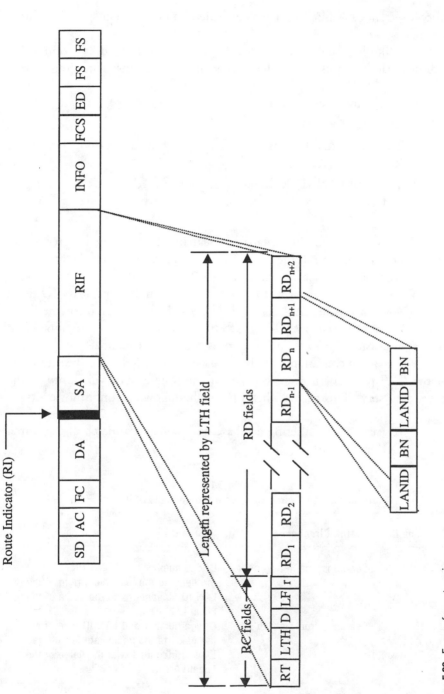

Figure 5.29 Frame format review.

process searches the filtering database to determine to which port the frame is sent.

The filtering database is searched for a match with the destination MAC address of the received frame when either of the following conditions exists:

- The frame does not contain a RIF;
- The RIF does not contain any RDs;
- The local LAN ID[41] is found in the rightmost RD position in the RIF and D = 0;
- The local LAN ID is found in the leftmost RD position in the RIF and D = 1.

Otherwise, the filtering database is searched by forming a search DRD from information found in the RIF of the received frame when the local LAN ID is known and a match is found in the RIF field.

The search DRD is formed from the bridge number and LAN ID fields that are adjacent to the LAN ID field that matched the search for the Local LAN ID. The *direction bit* (D-bit) affects the way the RIF is examined and the way the search DRD is constructed as shown in Table 5.16.

Under some conditions, the Relay process will send a received frame to all the CRF ports, with the exception of the source port, which are in the forwarding state. This occurs when one of the following conditions exists:

- A search of the filtering database does not find a match of the search DRD;

Table 5.16
Formation of the Search DRD From the Contents of the RIF Field

Direction Bit	Search Direction	Search DRD
D = 0	Search is from left to right	Bridge number and LAN ID taken from the fields to the immediate right of the LAN ID identified with the value of the Local LAN ID.
D = 1	Search is from right to left	Bridge number and LAN ID taken from the fields to the immediate left of the LAN ID identified with the value of the Local LAN ID.

41. The ring number where the CRF is located.

- The frame is a valid ARE or SRE frame (RT = 1xx);
- The frame is a valid specifically routed frame and the DA is either a group or broadcast address.

Under some conditions, the Relay process will discard a received frame. This occurs when one the following conditions exists:

- A frame's RIF is invalid;
- A frame is a valid specifically routed frame and the CRF's local LAN ID is unknown;
- A search of the frame's RIF does not find a match with the CRF's Local LAN ID.

To see how the Relay process works, let's consider the simple network topology shown in Figure 5.30. This network could be part of a larger Bridged Network. Shown are two workgroups connected via CRFs and LAN bridges to a server pool. The two paths the clients could use to reach the servers (S1 through S4) are installed to ensure continued availability of the servers to the clients in case there is a failure in either of the bridges or of CRF-1 or CRF-2. Note that the Spanning Tree will disable one of the four paths between the bridges (BA and BB) and the workgroup CRFs (CRF-3, and CRF-4). For this example, assume that the operation of the *Spanning Tree protocol* (STP) has placed the bridge port of BA connected to Ring 4 (that would be the dashed line connected to CRF-4) into the blocking state. The servers have dual-port Token Ring adapter cards installed that are connected to Rings 1 and 2. Further note that all ports of a CRF are always on the same ring. Hence, all ports of CRF-1 are on Ring 1, all ports on CRF-2 are on Ring 2 and so on. In the figure, the port identifiers are indicated within the smaller block at the point of connection.

Based on the topology and Spanning Tree assumptions, the filtering databases outlined in Table 5.17 will have been created by the Learning process.

There could have been many more entries in the filtering databases, but they are not shown in order to focus on the forwarding mechanisms in the following examples.

Example 1

Suppose that C1 wanted to request a file from Server S3. C1 has already discovered a path to S3 and places the following in the frame that contains the file transfer request

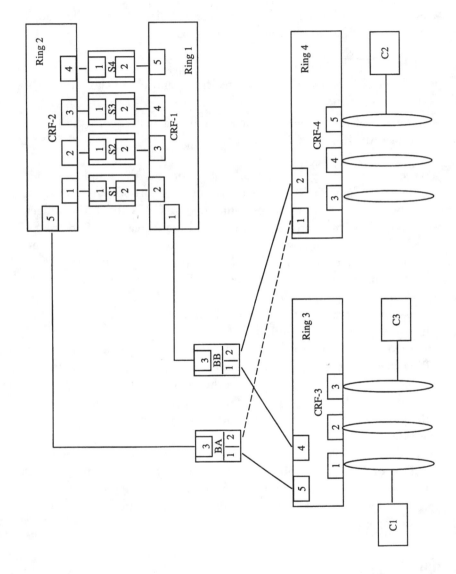

Figure 5.30 Example network topology.

Table 5.17
Filtering Databases Created by the Learning Process

Filtering Database for CRF-1			
MAC Address	**Output Port**	**DRD**	**Output Port**
S1-2	2	[BB, 3]	1
S2-2	3	[BB, 4]	1
S3-2	4		
S4-2	5		
Filtering Database for CRF-2			
MAC Address	**Output Port**	**DRD**	**Output Port**
S1-1	1	[BA, 3]	5
S2-1	2		
S3-1	3		
S4-1	4		
Filtering Database for CRF-3			
MAC Address	**Output Port**	**DRD**	**Output Port**
C1	1	[BB, 1]	4
C3	3	[BB, 4]	4
		[BA, 2]	5
Filtering Database for CRF-4			
MAC Address	**Output Port**	**DRD**	**Output Port**
C2	5	[BB, 1]	2
		[BB, 3]	2

DA = S3, SA = C1

RIF = { RT = '0xx'; LTH = 6; D = 0; RD_1 = [3, BB]; RD_2 = [1 , 0]}

The RIF documents a path from Ring 3 through Bridge BB to Ring 1. Note that the last bridge ID entered in a RIF is always 0. CRF-3 receives the frame on port 1 where the frame's destination MAC address and the RIF field is examined. The RIF field is searched from left to right (the D-bit is set to 0) for a LAN ID that matches the Local LAN ID. Because the LAN ID is located in the RIF and is not contained in the last RD field, the Relay process constructs a search DRD from the bridge number of RD_1 and the LAN ID of RD_2, yielding a value of [BB, 1]. An examination of the filtering database for CRF-3 shows that the output port to which this frame is to be relayed is port 4.

Bridge BB receives the frame on port 1 and relays the frame to port 3 after matching the bridge number and LAN IDs in the RIF field.

CRF-1 receives the frame on port 1 and examines the frame's destination MAC address and RIF. Examination of the RIF shows a match of the local LAN ID in the last RD, which requires the Relay process to search the filtering database for a destination MAC address entry of S3. Searching the database shows that the frame is relayed to port 4 where the server receives the frame.

Example 2

Server S4 is responding to a request from client C2. The RIF received from C2 is

$$RT = \text{'0xx'}; LTH = 6; D = 0; RD_1 = [4, BB]; RD_2 = [1, 0]$$

To send a frame back to C2, the server can use this cached RIF with the D-bit set to 1. The frame sent from S4 contains

DA = C2, SA = S4

RIF = {RT = '0xx'; LTH = 6; D = 1; RD_1 = [4, BB]; RD_2 = [1 , 0]}

S4 starts the frame on its way to C2 by transmitting on its port 2. The frame is received by CRF-1 at port 5 where the destination MAC address and the RIF are examined. CRF-1's local LAN ID is found in the RIF in RD_2. Unlike the previous example, D = 1, which indicates that the RIF field is traversed from right to left; hence, RD_2 is the first RD in the RIF instead of the last. The Relay process constructs a search DRD from the bridge number and LAN ID of RD1 yielding a value of [BB, 4]. Examination of CRF-1's filtering database shows that the output port the frame is relayed to port 1.

Bridge BB receives the frame on port 3 and forwards the frame to port 2 after examination of the RIF.

CRF-4 receives the frame on port 2. An examination of the RIF and the MAC destination address shows that the local LAN ID, 4, is found in RD_1, which is the last RD since D = 1. The Relay process uses the DA in the received frame to examine the filtering database. C2 is found in the filtering database and shows that the frame is to be relayed to port 5, where C2 receives the frame.

Example 3

To complete our set of examples, consider some peer-to-peer communication between clients C1 and C3. As can be seen from the figure, there is no need for a RIF field in any frames transmitted between the client stations since there are no bridges in the path. A frame from C1 to C3 would enter CRF-3 at port 1. The Relay process examines the frame and finds the RI-bit is 0, so only the DA found in the frame is used to search the filtering database. An examination of CRF3's filtering database shows that the output port the frame is relayed to is port 3.

5.6.5 The Learning Process

The Learning process builds the filtering database tables that were used in the previous section's examples. The process observes frames transmitted in the network to determine:

- The MAC addresses of stations that are directly attached to ports of the CRF;
- The bridge numbers of the bridges that are directly attached to ports of the CRF;
- The identifiers of the LANs to which these bridges are connected.

This information is used to build the MAC address and DRD entries of the CRF's filtering database. The standard contains a table in Annex K that defines conditions and the actions taken by the Learning process. These conditions and actions are summarized as follows:

- The frame is discarded by the Learning process when the RIF is illegal or the local LAN ID cannot be located in the RIF.
- A MAC address entry is added to the filtering database when the RIF does not contain any RD fields or when the local LAN ID is found in the first[42] RD. The MAC address added to the filtering database is the SA found in the examined frame. The output port is the port where the frame was observed.
- A DRD is added to the filtering database when an explorer frame is observed that contains RD fields or when a specifically routed frame is observed and the local LAN ID is found in an RD that is not the first in the RIF. The DRD is formed from the information found in the RIF and will be described in the examples that follow. The output port is the port where the frame was observed.

---

42. As previously stated, the D-bit determines the direction in which the RIF is examined. When D = 0, the first RD is the one in the leftmost position in the RIF; when D = 1, the first RD is the one in the rightmost position in the RIF.

5.6.5.1   An Example of the Learning Process Operation

Turning to the network illustrated in Figure 5.30, we start this example with all of the filtering databases cleared, the bridges configured, and the local LAN IDs known by all the CRFs. Client stations C1, C2, and C3 join the network and immediately start looking for the servers. In this example the clients send out a frame with an LLC payload of either a TEST or XID. A station receiving either frame responds by returning the frame to the sender.

The LLC TEST or XID frame is sent in the following formats:

- Sent as a SRF without any RD fields in the RIF, which is a form of a local frame that will not be forwarded by an SRT bridge.

- Sent as either an ARE or STE. Depending on the protocol in use, the receiving station can return either an SRF by flipping the D-bit to 1, an ARE, or an STE. In some cases, multiple responses may be received by the sending station, which selects and caches one of the routes returned in the response frames for future use.

- Sent without any RIF field (RI = 0). This is used last because the previous tests would have found paths through either SR or SRT bridges. If there is a path through a transparent bridge, then this frame will be responded to.

Using the first format, clients C1, C2, and C3 send out an SRF frame without any RD fields to the MAC address servers S1, S2, and S3, respectively.

- The frame from C1 will enter CRF-3 at port 1. The Learning process examines the frame and determines that the frame is local since there are no RD fields in the RIF. The Learning process creates a MAC address entry in the filtering database, indicating that C1 is found at port 1. The Relay process examines the frame. Because the RIF does not contain any RD fields, the destination MAC address is used to search the filtering database. Because S1 is not in the database, the search will not find a match and the frame will be sent out of all the ports (with the exception of the source port) of CRF-3. Because the frame does not have any RDs in the RIF, bridges BA and BB will not forward the frame. Stations on the rings of ports 2 and 3 will see the frame but will not receive or respond to the frame because the destination address will not match their address.

- The frame from C3 will enter CRF-3 at port 3, while the frame from C2 will enter CRF-4 at port 5. In exactly the same fashion as stated previously, the Learning process will create a MAC Address entry in the filtering database. In CRF-3, a MAC address entry will be created, indicating that C3 is found at port 3. In CRF-4, a MAC address entry will be created, indicating that C2 is found at port 5. In both CRFs the frames are forwarded to all ports but will not be responded to or forwarded by the bridges for the reasons given previously.

By the end of the first part of the discovery process the filtering databases in CRFs 3 and 4 have been updated as shown in Table 5.18.

After a management-defined period, in which the client stations have not received a response from the servers, the stations move on to the second step in the discovery process, sending out an STE frame to the servers.

- The frame from C1 enters CRF-3 at port 1. The Learning process examines the frame and determines that the frame is local because there are no RD fields in the RIF. Because there is already an entry in the filtering database, indicating that C1 is found at port 1, there is no further action taken by the Learning process. The Relay process examines the frame. Since the frame is an STE, the frame is sent out of all the ports (with the exception of the source port) of CRF-3. After adding the appropriate RD field required for a STE frame, bridges BA and BB forward the frame.
  - Stations on the rings of ports 2 and 3 will see the frame but will not receive or respond to the frame because the destination address will not match their address.
  - The frame sent out of bridge BA at port 3 contains a RIF field with the following information: RT = '1xx'; LTH = 6; D = 0; RD1 = [3, BA]; RD2 = [2 , 0].
  - The frame sent out of bridge BB at port 2 contains a RIF field with the following information: RT = '1xx'; LTH = 6; D = 0; RD1 = [3, BB]; RD2 = [4 , 0].

Table 5.18
Updated Filtering Databases in CRF-3 and CRF-4

MAC Address	Output Port	DRD	Output Port
Filtering database for CRF-3			
C1	1		
C3	3		
Filtering database for CRF-4			
C2	5		

- The frame sent out of bridge BB at port 3 contains a RIF field with the following information: RT = '1xx'; LTH = 6; D = 0; RD1 = [3, BB]; RD2 = [1 , 0].

- The frame leaving bridge BA at port 3 will be received by CRF-2 on port 5. The Learning process examines the frame and determines that the frame is an STE and contains RD fields. The DRD is formed from the bridge ID and LAN ID found in RD_1. A DRD entry is created in CRF-2 that indicates [BA, 3] is found at port 5. The Relay process examines the frame. Because the frame is an STE,[43] CRF-2 forwards the frame to all of the ports and is received by S1 on port 1.

- The frame leaving bridge BB at port 2 will be received by CRF-4 on port 2. The Learning process examines the frame and determines that the frame is an STE and contains RD fields. The DRD is formed from the bridge ID and LAN ID found in RD_1. A DRD entry is created in CRF-4 that indicates [BB, 3] is found at port 2. The Relay process examines the frame. Because the frame is an STE, CRF-4 forwards the frame to all of the ports. Stations on the rings of ports 3, 4, and 5 will see the frame but will not receive or respond to the frame because the DA will not match their address.

- The frame leaving bridge BB at port 3 will be received by CRF-1 on port 1. The Learning process examines the frame and determines that the frame is an STE and contains RD fields. The DRD is formed from the bridge ID and LAN ID found in RD_1. A DRD entry is created in CRF-1 that indicates [BB, 3] is found at port 1. The Relay process examines the frame. Because the frame is an STE, CRF-1 forwards the frame to all of the ports and is received by S1 on port 2.

- The frame sent from C3 follows the same path as the frame from C1 and will not cause any changes to be made in any of the filtering databases.

- The frame from C2 enters CRF-4 at port 5. The Learning process examines the frame and determines that the frame is local since there are no RD fields in the RIF. Because there is already an entry in the filtering database, indicating that C2 is found at port 5, there is no further action taken by the Learning process. The Relay process examines the frame. Because the frame is an STE, the frame is sent out all

---

43. Recall that any explorer frame, either STE or ARE, is forwarded to all ports of a CRF that are in the forwarding state as determined by the Spanning Tree protocol.

ports (with the exception of the source port) of CRF-4. Because the frame is a STE, bridge BB forwards the frame after adding the appropriate RD field. Port 2 on bridge BA is in the blocking state and does not forward the frame.

- The frame sent out of bridge BB at port 3 contains a RIF field with the following information: RT = '1xx'; LTH = 6; D = 0; RD_1 = [4, BB]; RD2 = [1 , 0].

- The frame leaving bridge BB at port 3 is received by CRF-1 on port 1. The Learning process examines the frame and determines that the frame is an STE and contains RD fields. The DRD is formed from the bridge ID and LAN ID found in RD_1. A DRD entry is created in CRF-1 that indicates [BB, 4] is found at port 1. The Relay process examines the frame. Because the frame is an STE, CRF-1 forwards the frame to all of the ports and is received by S2 on port 3.

The filtering databases now contain the information indicated in Table 5.19.

At this point in the example, the servers S1, S2, and S3 have received frames from C1, C2, and C3, respectively. The servers will respond by forming SRFs using the RIF fields found in the received frames. The frames indicated in Table 5.20 will be sent.

The Learning process in CRF-1 and CRF-2 will examine these frames as they enter the CRF. Since the local LAN ID is found in the first RD field, MAC address entries are created in the filtering databases for MAC addresses S1-1 and S3-1 in CRF-2 and for MAC addresses S1-2, S2-2, and S3-2 in CRF-1.

Table 5.19
Updated Filtering Databases After Client Discovery Frames

MAC Address	Output Port	DRD	Output Port
Filtering database for CRF-1			
		[BB, 3]	1
		[BB, 4]	1
Filtering database for CRF-2			
		[BA, 3}	5
Filtering database for CRF-3			
C1	1		
C3	3		
Filtering database for CRF-4			
C2	5		
		[BB, 3]	2

Table 5.20
Learning Process Example; Server Response to Client Discovery Frames

From	RIF	Server Port	Enters CRF at Port
S1-1	RT = '0xx'; LTH = 6; D = 1; RD_1 = [3, BA]; RD_2 = [2, 0];	1	CRF-2 @ Port 1
S1-2	RT = '0xx'; LTH = 6; D = 1; RD_1 = [3, BB]; RD_2 = [1, 0];	2	CRF-1 @ Port 2
S2-2	RT = '0xx'; LTH = 6; D = 1; RD_1 = [4, BB]; RD_2 = [1, 0];	2	CRF 1 @ Port 3
S3-1	RT = '0xx'; LTH = 6; D = 1; RD_1 = [3, BA]; RD_2 = [2, 0];	1	CRF-2 @ Port 3
S3-2	RT = '0xx'; LTH = 6; D = 1; RD_1 = [3, BB]; RD_2 = [1, 0];	2	CRF-1 @ Port 4

The Relay process in CRF-1 examines the frames for the local LAN ID, which is found in RD_2. Because D = 1, the search DRDs formed are [BB, 3] and [BB, 4], both found in the filtering database associated with port 1. In the same fashion, the Relay process in CRF-2 examines the frames it has received and forms the search DRD of [BA, 3], which is found in the filtering database associated with port 5.

After being forwarded by bridges BA and BB, the frames illustrated in Table 5.21 will be received at CRF-3 and CRF-4.

The Learning process in CRF-3 examines the frames and finds the local LAN ID in RD_1, which is the last RD since D = 1, and creates DRD entries [BB, 1] at port 4 and [BA, 2] at port 5. The Relay process searches the filtering database using destination MAC addresses found in the received frames and relays them to ports 1 (for the frames from S1) and port 3 (for the frames from S3).

Table 5.21
Server Response Frames Received at CRF-3 and CRF-4

From	RIF	At CRF Port
Frames received at CRF-3		
S1	RT = '0xx'; LTH = 6; D = 1; RD_1 = [3, BA]; RD_2 = [2, 0];	5
S1	RT = '0xx'; LTH = 6; D = 1; RD_1 = [3, BB]; RD_2 = [1, 0];	4
S3	RT = '0xx'; LTH = 6; D = 1; RD_1 = [3, BA]; RD_2 = [2, 0];	5
S3	RT = '0xx'; LTH = 6; D = 1; RD_1 = [3, BB]; RD_2 = [1, 0];	4
Frames received at CRF-4		
S2	RT = '0xx'; LTH = 6; D = 1; RD_1 = [4, BB]; RD_2 = [1, 0];	2

Similarly, CRF-4 creates a DRD entry [BB, 1] at port 2 and relays the frame to port 5. Finally, the filtering databases contain the frames indicated in Table 5.22.

5.6.6 Duplicate Address Check Process

The Duplicate Check process is a collaborative effort between the C-Port MAC and the DTU. The Standard defines the actions of the PMAC, namely, to collect the address of the attached station, to pass the information to the DTU, and using the result from the DTU to either accept or deny the insert request of the station.

The Standard does not define the actions that must be taken by the DTU. The DTU is defined as a collection of forwarding constructs, CRFs and a BRF. These constructs contain forwarding tables used to forward frames to C-Ports attached to the DTU. These tables need to be consistent; they must identify a unique exit port for an individual MAC address. If two stations with the same MAC address were to attach to the DTU, the tables would continuously be updated as the DTU learned the "new" location of the station. It would be unlikely that any frame forwarding to these stations would be successful.

When the DTU receives the address information from an attached PMAC, the DTU must examine all the forwarding tables in use. The examination must be done in such a fashion that takes into account possible asynchronous and simultaneous actions taken by all attached PMACs that are in the process of requesting duplicate address checks. One method of doing this is to have

Table 5.22

Updated Filtering Databases at the End of Example 3

MAC Address	Output Port	DRD	Output Port
Filtering database for CRF-1			
S1-2	2	[BB, 3]	1
S2-2	3	[BB, 4]	1
S3-2	4		
Filtering database for CRF-2			
S1-1	1	[BA, 3]	5
S3-1	3		
Filtering database for CRF-3			
C1	1	[BB, 1]	4
C3	3	[BA, 2]	5
Filtering database for CRF-4			
C2	5	[BB, 1]	2
		[BB, 3]	2

a single code thread within the DTU that performs the address checking. As part of the address check function, the forwarding tables are updated with the MAC address and the output port. This second step eliminates the window between the time a C-Port makes the duplicate address check request and when the first LLC frame transmitted from the station causes the update to the DTU's forwarding tables. Further, these entries are not allowed to age out but instead are removed only when the C-Port closes.

Finally, because the station will wait only 18 sec from the first Insert Request MAC frame for an Insert Response MAC frame, the preceding procedure must be completed within that time.

6

Onward to High-Speed Token Ring

With the introduction of DTR, significant increases in channel capacity became available for Token Ring networks. High-performance servers could be directly attached to switched ports, establishing two dedicated 16-Mbit/s channels for the server traffic. Although 16-Mbit/s channels generally had more bandwidth than was required for end station attachment, switching helped these attaching stations by allowing for ring microsegmentation. But perhaps the greatest benefits brought by the introduction of switching were the increased bandwidth it provided in both switch-to-server and switch-to-switch connections and the use of the switching fabric as a very high bandwidth collapsed backbone with a capacity of hundreds of megabits/second. With the availability of Token Ring switching, a 1996 survey of Token Ring customers showed that they were not generally interested in any further increases in data rate because switching would provide all the required increases in bandwidth for the foreseeable future. Even at the time of this writing, Token Ring networks are still in the early stages of exploiting the benefits of DTR for the additional network capacity it can provide. Then what sparked the demand and the intense push to develop 100-Mbit/s and Gigabit *High-Speed Token Ring* (HSTR)?

In late-1996 and early-1997 there was an avalanche of trade journal articles on Fast Ethernet, Gigabit Ethernet, and on Token Ring's lack of comparable data rates. Token Ring's deficiency in this area coupled with its higher cost for end stations, hubs, and switches prompted network managers to rethink their future Token Ring requirements. They began to focus more sharply on the sea change occurring in network computing. The old network model was client server, where most of the information required by and transmitted from the end stations remained on the local LAN segment. Today the networking model, as shown in Figure 6.1, is quickly moving to

351

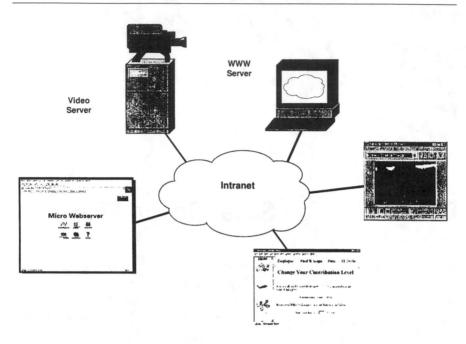

Figure 6.1 The new networking model.

an inter- and intranetcentric one, where much of the information flowing both into and from the attaching stations must traverse the network's backbone. This change pushed the demand for backbone bandwidth far more than client-server computing ever could. Further, the demand for streaming video and other bandwidth-intensive applications were beginning to emerge. In response to this new demand, network managers started placing the requirement on their LANs to have appropriate capacity enhancements either built in or planned for. They saw the lack of a high-speed follow-on for Token Ring in 1997 as a strong negative factor forcing them to begin evaluating and planning for replacement technologies for those LANs. In addition, negative Token Ring press from pro-Ethernet journalists and analysts placed more pressure on the Token Ring network manager to evaluate alternate technologies.

In the spring of 1997 there was an abundance of negative press, and the expectation was that Token Ring users would migrate to Fast Ethernet as the only practical frame-based high-speed LAN connection. Under this backdrop, Kevin Tolly, a well-known industry analyst, in conjunction with Network World, sponsored an HSTR round table for manufacturers. There he asked for commitment for an HSTR development effort. The vendors were noncommittal at that meeting but agreed to look at user demand and come back in August with an answer. Customer visits that followed that first round table confirmed significant Token Ring user interest in speed enhancements, although just one year earlier the interest level was low.

6.1 The Forces Driving Creation of High-Speed Token Ring

The research and customer surveys carried out between the June and August Round Table meetings did more than alert us to customer demand. It provided us with a much clearer understanding of why Token Ring was still looked at as the preferred LAN choice for our customer set. Token Ring was seen as a superior technology for handling *System Network Architecture* (SNA)-based traffic. In addition, it had superior network management and reliability. These attributes were critical to our customer set, which is heavily populated with the worlds' largest corporations and government agencies. Further, these users employ their Token Ring networks to carry their establishments' mission-critical applications. As a group, they tend to be conservative in their business decisions, and certainly in any decisions to migrate their mission-critical LAN applications. They foresaw many potential problems migrating their networks away from Token Ring, a technology that continues to serve their needs admirably well. They were concerned that Ethernet's inability to handle source routing would add a significant burden to effectively configuring their networks. Additionally, many were taking advantage of Token Ring's large frame sizes (commonly using at least 4KB-long frames and sometimes using 18.2KB frames compared with Ethernet's maximum frame size of 1518 bytes). In addition to these obvious differences between the technologies, there was the underlying uncertainty of not knowing what other differences would prevent their smoothly running applications from performing properly if the network was overhauled and replaced with Ethernet. In addition to gaining a better understanding of the loyalty of our Token Ring users, and despite the many stated claims that Token Ring was dying, Token Ring adapter sales continued to grow, as seen in Figure 6.2,[1] based on a February 1997 IDC study.

Armed with this positive news, many manufacturers began to warm up to the idea of a higher speed Token Ring offering.

6.1.1 Technology Choices for High Speed Token Ring

Once we decided to develop HSTR we needed to decide on a design approach and its capabilities. The first decision to be made was determining if we would

---

1. A recently released 1998 IDC study shows continued growth with 1997 adapter sales in the neighborhood of 4.5 million.

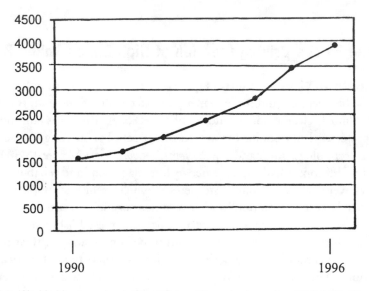

Worldwide Token Ring Adapter Sales, 1990 - 1996 (Thousands)

Figure 6.2 Worldwide annual volume of Token Ring adapter sales, 1990–1996. (*Source:* DataQuest 2/97 Market Intelligence.)

have a native Token Ring solution or an encapsulated one. Choosing an encapsulated solution would allow HSTR to be introduced without developing any new hardware. Token Ring frames would be encapsulated within a high-speed Ethernet frame and carried over existing and future Ethernet links, including 100-Mbit/s and Gigabit Ethernet. This solution would benefit those manufacturers that already announced encapsulated solutions and handicap those that did not have such a solution. The potentially divisive decision was made within the *High-Speed Token Ring Alliance* (HSTRA) based on customer input. A number of companies had received input from their customers indicating that they were concerned about any nonnative solution. As straightforward as encapsulation appeared, there were obvious limitations to it, and possibly more subtle ones that would be learned painfully while attempting to migrate the network. Two obvious limitations were Ethernet's lack of support for large frames, which are specifically allowed and widely used within Token Ring, and by its lack of support for source routing, an important bridging capability, especially for SNA-based networks.

With the decision made to have a native Token Ring solution, we next took up the issue of determining the best data rate. There was discussion of going directly to gigabit technologies as well as several data rates in the

100- to 200-Mbit/s range. The most likely early candidates considered were 100 Mbit/s, 128 Mbit/s, and 155 Mbit/s. The chief advantage of 155 Mbit/s was its 50% speed improvement over 100-Mbit/s Ethernet. It could use the existing ATM PHY layer providing established wiring rules and an established front-end. Disadvantages seen were the association of 155 Mbit/s with ATM and its relatively poor press, and with the relatively higher cost of ATM components compared with either Token Ring or Ethernet.

The next choice was 128 Mbit/s. This data rate could be achieved by taking 100-Mbit/s Ethernet front-ends and running them 28% faster. From a marketing point of view there was still a 28% advantage in speed over 100-Mbit/s Ethernet. It had the advantage of using relatively low-cost Ethernet PHY layer chips as HSTR front-ends. The disadvantages of this approach were the increased risk involved with running the Ethernet front-ends 28% faster and the new requirement to develop wiring rules (drive distance limitations) for the higher transmission speed.

The last choice was taking a copy-cat approach, setting the speed at 100 Mbit/s. This choice provided us with the ability to use Ethernet front end technology directly and Ethernet cabling distances, which were already established. This choice was initially greeted with much skepticism and disdain. It wasn't interesting to have a "me too" type solution. However, on closer reflection we saw that a 100-Mbit/s solution indeed met user requirements for a significantly higher speed capability. In addition, directly using the Ethernet front-ends would lead to a lowest cost solution and to earliest introduction of products. We also realized that the market for HSTR was with existing Token Ring users who need a migration path to higher performance. They required speeds at least comparable to fast Ethernet to provide the same speed benefits enjoyed by Ethernet users. Speed advantages over 100-Mbit/s Ethernet would have been nice but were not as critical as cost of product, minimization of risk, and time to market.

6.2 The 100-Mbit/s Token Ring Standard

Once a decision was reached to have a native 100-Mbit/s solution, we needed to determine its attributes. 100-Mbit/s Ethernet calls for both a shared LAN solution and a switched LAN solution. For Token Ring, a switched-only solution would be simpler and cheaper to design and develop and would lead to earlier available products. However, would it meet the needs of the marketplace? In examining this question we decided that the principle reason for wanting a shared HSTR solution was to allow for the relatively cheap connection of desktop devices to concentrators. If switching ports were priced low enough, there should be no driving reason to have a shared solution at all. Because the adapter costs would be lower with the switched-only offering, the balance was tilted in favor of not having a shared offering. A final argument for a switched-only offering was that the early requirements for HSTR were for backbone and server connections, where shared solutions were uninteresting. We took this input and hedged our answer. First, we said, we should develop a switched only solution. With the associated lower costs, and by implication, lower selling price, we might be able to totally avoid the requirement of offering a shared solution. However, in case we were wrong in our projections, our design approach should allow us to easily and quickly develop a shared solution in response to strong market pressures to do so. As a design point we took the MAC design from 16-Mbit/s DTR and coupled that with the PHY design for 100-Mbit/s Ethernet.

With data rate and signaling decided, we turned our attention to the media over which HSTR would operate. The obvious solution was to support the same media as 100BaseT, Category 5 cabling for the copper solution, and multimode fiber for a fiber solution. The fiber solution appeared acceptable. However, the copper solution had a significant hole in meeting the needs of our Token Ring users. Unlike Ethernet users, many of the Token Ring user set had installed the IBM Cabling System's 150Ω shielded twisted-pair cabling in their establishments in the late-1980s and early-1990s. This is excellent cabling whose transmission pairs have better electrical transmission properties than Category 5 cabling. The 100BaseT solution does not support it as a transmission medium because few Ethernet users had that cabling installed. We believed that it was a requirement to offer 150Ω support for HSTR.

Now that we had a technical approach, we needed to determine what it would take to implement it, both as a standard and as product.

6.3 Beginnings of HSTR: Timing Is Everything

As we embarked on a 100-Mbit/s Token Ring design and standard, we faced tremendous market pressures. Those same customers that had told us just a year ago that they foresaw no near-term requirements for HSTR were now saying that there was a crucial need to produce a product early. Even if their networks did not require the higher speed technology, they had to know that it was out there if they were to maintain Token Ring as their strategic LAN solution. In August 1997, we projected that we needed to have products available sometime in 1998. Further, our customers were accustomed to Standards-based products, so we would need to have the technical specifications of the Standard established within a year in order to base product designs on those specifications and deliver them in 1998. The technical people at the meeting who were experienced in both product design and standards development said that this was an unattainable schedule if we followed our usual standards development procedures. It would take unusual focus on our mission and cooperation among the companies developing both the standards and the products for the Token Ring marketplace if we were to have any chance of achieving such an aggressive goal. To achieve this focus, we agreed that the formation of an alliance was needed and that this alliance must have as one of its primary goals ensuring that all the Token Ring vendors keep the development of the HSTR Standard a top priority. HSTRA would facilitate issue resolution and serve as a forum for keeping all the players in agreement on any issues that could derail our aggressive schedule.

With the decision made to form HSTRA, Kevin Tolly threw out an additional challenge: We should be ready with technology demonstrations by the Spring 1998 NetWorld + Interop show in Las Vegas. This gave us until the beginning of May 1998. That left us just nine months from making a decision to develop HSTR to having operational hardware. Although the vendors swallowed hard, we all realized that for HSTR to succeed, we must show the world that we are serious and prove it to them with product demonstrations, so in the end we all agreed. As a result, we left the August 1997 High-Speed Token Ring Round Table agreeing to the following key checkpoints:

- Have a first pass straw-man standard for high-speed (100-Mbit/s) Token Ring available by 11/10/97;
- Commit to multivendor product demonstrations by spring (NetWorld + Interop May '98);
- Begin product introductions by the second half of 1998;
- Have all the technical work on the standard completed by July 1998 for a 100-Mbit/s copper solution—100 Mbit/s on fiber and gigabit efforts were to begin in parallel with the primary effort but could be completed later.

The following day, the IEEE 802.5 Token Ring Working Group held its interim meeting with most of the same cast. We continued where the round table discussions had stopped and established the following standards development schedule:

- Straw-Man available   10 November 1997
- Straw-Man circulated for comment   21 November 1997
- Straw-Man 2.0 circulated for comment   28 January 1998

- Committee ballot to approve Standard   27 March 1998
- Confirmation ballot following ballot resolution   June 1998
- Forward to LMSC and ISO for balloting   July 1998

As this final chapter is being written (June 1998), the standard development is still on schedule. The first HSTR product demonstrations took place
at NetWorld + Interop in the spring of 1998 and the standard is moving
forward on schedule.

6.4 Migration to High-Speed Token Ring

Given the availability of HSTR, it is still necessary to plan a smooth migration
from 4- and 16-Mbit/s shared technology, to 16/4 switching, and finally to
the deployment of HSTR with minimal disruption of the network and virtually
no disruption of service. The following migration plan, first published by IBM
[1] provides a general overview of how a typical user might achieve these
objectives.

6.4.1 The Starting Point

Consider the migration path from a small, Classic Token Ring network consisting of a number of concentrators interconnected in a single wiring closet
on a single ring as shown in Figure 6.3. Over a short interval, such as the time
it takes for a large file transfer, the average bandwidth available to a station is
the ring data rate (4 or 16 Mbit/s) divided by the average number of stations
demanding bandwidth during that interval. Consider a typical large ring of

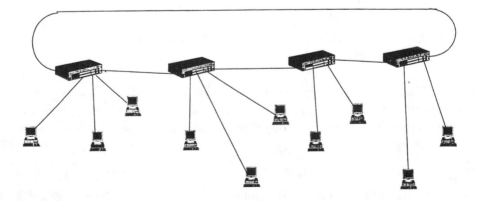

Figure 6.3 Classic Token Ring: single-ring configuration.

100 stations operating at 16 Mbit/s with 10 users simultaneously accessing the network over the duration of a large file transfer. For this example, the average available bandwidth per station (and specifically, for the station trying to receive that file) is 1.6 Mbit/s. Because this average bandwidth exceeds the requirements of most of today's legacy applications, no steps need be taken now to segment this network. However, plans should be made to reevaluate the adequacy of the present topology when new applications and hardware, such as bandwidth-intensive applications and high-speed servers, are brought online.

6.4.2 Exploit Switching and Workgroup Segmentation

6.4.2.1 Backbone Switching

A popular configuration for large Token Ring networks has always been to interconnect local rings to each other via a backbone ring. Typically, local rings have from 50 to 150 stations each. The lower number is based principally on topological considerations, such as how many stations are cabled to a specific wiring closet. This network design is used even when the total number of stations on the LAN are fewer than the maximum supportable on a single ring. A major advantage of this backbone topology, with bridges attaching the local rings, is a significant increase in network bandwidth capacity, especially if there is substantial traffic that is limited to each of the local rings. The potential drawbacks to this configuration are based on the latency associated with the bridges, which will delay packet transport across the network compared to a single-ring configuration. Evaluation of this configuration must be based on the expected traffic flow and traffic patterns. Consider the network shown in Figure 6.4. As a first cut, assume that the 80/20 rule applies and that 80% of the traffic is confined to the local rings while 20% of the traffic goes across rings. Further, assume that half of the traffic going across rings is of a broadcast nature and goes to every ring, while the remainder of the bridged traffic goes to a single station. Then, for each local ring, assuming the traffic load on each local ring = L, we have

$$80\% \text{ of traffic confined to 1 ring} = 0.8 \text{ L}$$
$$10\% \text{ of traffic broadcast to 8 rings} = 0.1 \text{ L}$$
$$10\% \text{ of traffic confined to 2 rings} = 0.1 \text{ L}$$

Then

$$\text{Backbone load} = 8 \times [(0.1 \text{ L} \div 8) + (0.1 \text{ L} \div 2)] = 0.5 \text{ L}$$

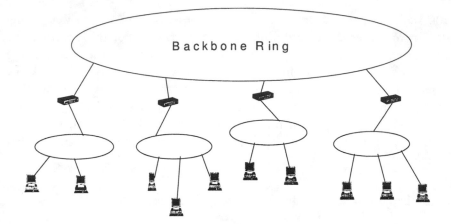

Figure 6.4 Classic Token Ring: bridged backbone configuration.

The backbone traffic load will be approximately half that of each of the local rings.

With the desire to centrally manage servers and the increased use of corporate intranets and internets, the 80/20 rule is quickly losing viability as a reasonable estimator. Instead, a much higher percentage of the network traffic is or will soon be flowing onto the backbone. Therefore, when we recompute the loading based on new traffic patterns, we may find that this configuration creates a bottleneck, the backbone ring.

With DTR, a switch is used to interconnect these rings as shown in Figure 6.5. Used this way, all existing shared Token Ring stations and concentrators are retained. The switch serves as a high-speed collapsed backbone for data transfer between the individual rings. This configuration provides much greater backbone capacity than the bridged configuration shown in Figure 6.4 and delivers

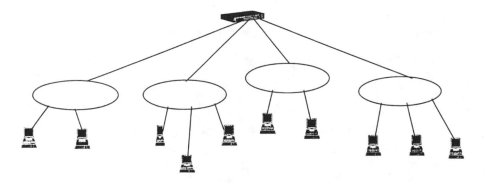

Figure 6.5 Replacement of bridges and backbone ring with a switch.

data with lower latency and at lower cost. Therefore, beginning the migration process of a bridge-laden backbone, the first step will be to replace the backbone and the bridges with a switch as shown in Figure 6.5.

With a large network, it is likely that key resources such as high-speed servers will need to be directly attached to the switch. As attachment increases, switch selection may include consideration of port density because as soon as port capacity for the first switch is exceeded, multiple, interconnected switches will be required.

Dedicated 16 Mbit/s for servers and other key resources make a big difference to the users of these devices. But eventually even 16 Mbit/s may not be enough to satisfy some users. At that time, full-duplex Token Ring can be used to provide an additional boost in bandwidth.

6.4.2.2 Workgroup Segmentation

As the network grows, and especially as the balance between client/server and peer-to-peer traffic changes, it may become necessary to split up heavily populated rings into multiple segments, each with fewer users. The main drivers for ring segmentation are power users with local servers. By subdividing the local rings, with power users and their local servers on the same segment, the available bandwidth for the power users is increased while off-loading that traffic from the rest of the network. Additionally, servers used by a broader range of end stations are directly attached to switch ports, gaining access to the full Token Ring bandwidth. The resulting performance improvement for all users is achieved at the cost of a few ports in the backbone switch. Figure 6.6 shows the network with servers both added to the local rings and directly attached to the switch.

When segmenting a workgroup ring with local servers, a workgroup switch cascaded from the backbone switch is usually the most cost-effective approach, allowing the ring to be split into many segments and minimizing the ports consumed on the backbone switch. However, if the workgroup does not have local servers it may not need much segmentation. Thus, depending

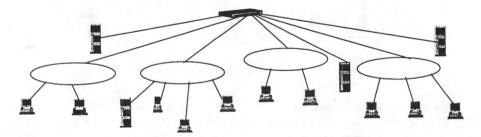

Figure 6.6 Network with servers added to local rings and directly attached to the switch.

on the extent of peer-to-peer traffic and workgroup server deployment in the network, some of this demand is best met with additional backbone ports and some with workgroup switches.

6.4.3 Building a Multiswitch Backbone

One obvious way to build a multiswitch backbone is to use full-duplex Token Ring as the interswitch connection. After all, this is a Token Ring environment, and Token Ring in the backbone eliminates any requirement for frame reformatting. One key advantage of this network architecture for small networks is that it is possible to build an entirely cut-through switching architecture from backbone to desktop. With cut-through switching, frames are forwarded while they are still being received rather than waiting for completion of the arrival of the frame. This protocol has the obvious advantage of greatly decreasing the delay through the switch. It is a potential problem in Ethernet, where collisions could corrupt the frame and thereby cause a significant portion of the network's bandwidth to be cluttered by corrupted transmissions being forwarded throughout the network. However, Token Ring has no such problems. With very high probability, any frame that starts being received will be received error-free. Virtual LANs are used with multiswitched LANs as a means of segmenting network traffic.[2] With a small number of switches, it is appropriate to attach them directly to each other, as shown in Figure 6.7.

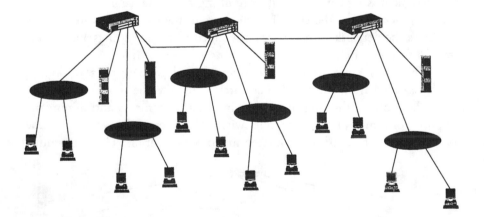

Figure 6.7 Multiple switches directly attached.

---

2. A virtual LAN is a network made up of a logical group of end stations that are physically attached to different ports of a switch or to ports on different switches. Using network management software, end stations in such a logical group can be managed with increased flexibility.

The configuration in Figure 6.7 provides full-duplex, 16-Mbit/s connectivity between the switches. Although this configuration is adequate for small networks, it is not as extendible as a two-tiered network design, where each of the switches is connected directly to a primary switch serving as a collapsed backbone for the entire network. The two-tiered configuration, as seen in Figure 6.8, provides full-duplex links to the primary switch, which provides 32-Mbit/s total bandwidth communication between each of the second-tier switches and the primary switch.

The two-tiered switch topology provides both very low end-to-end latency and a very high-speed backbone without changing any of the end-station adapters or any of the connections between stations and the concentrators or switches that are supporting them. The one change that may be considered is moving some of the high-speed server attachments directly to the backbone switch.

With ever-increasing demands for network capacity, however, even this configuration is not a final step; ultimately, higher speed connections will be required in the high-speed backbone and to the end stations. In addition, for many business establishments, having the high-speed backbone confined to a single electronic device, or even to a single building, is not a viable option. The next step in the migration path is likely to be in the backbone design.

6.4.4 Native 100-Mbit/s Token Ring

For many Token Ring users, an appropriate strategic plan is to migrate their networks to ATM to capitalize on its bandwidth, quality of service, and network

Figure 6.8 Two-tiered switch topology providing a high-speed collapsed backbone.

management capabilities. But this path is not for everyone. Many Token Ring users with high-speed requirements don't require or want ATM-enabled applications. These users prefer a high-speed option within the Token Ring technology family for both direct station attachment and backbone connectivity.

6.4.4.1   Upgrade the Network Infrastructure With 100-Mbit/s Token Ring

When upgrading the network infrastructure, key shared resources, such as backbone links and server attachment, should be migrated to 100 Mbit/s first. As an example, when 100-Mbit/s Token Ring is incorporated into Figure 6.7 or Figure 6.8, a full 100 Mbit/s of nonblocking bandwidth is available on the switch-to-switch and switch-to-server links.

A network incorporating 100-Mbit/s connections is depicted in Figure 6.9. For this configuration, each of the local LAN segments has a high-speed uplink to the backbone switch and to locally attached servers. Such a configuration provides vastly increased backbone and server bandwidth.

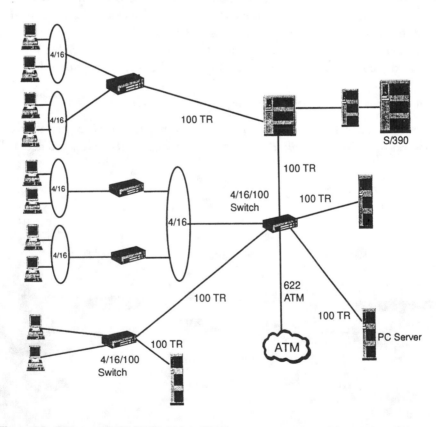

Figure 6.9 High-speed 100-Mbit/s Token Ring.

Additionally, with HSTR capability in the switches, high-end workstations can be directly attached as needed.

6.4.4.2 Exploit 100 Mbit/s at the End Stations

Higher bandwidth to the end stations can be achieved in two ways.

As discussed earlier, the first method is to attach end stations directly to switches. The bottom-tier concentrators can be phased out, replacing them with switches that contain high-speed uplinks. Even with old-generation Token Ring adapters, stations thus attached get an instantaneous bandwidth increase to 16 Mbit/s for their traffic requirements. If they have today's full-duplex-enabled Token Ring adapters, that bandwidth doubles to 32 Mbit/s total for both transmission and reception. Most end stations will be adequately served by 16-Mbit/s dedicated bandwidth and can remain at present data speeds, thereby preserving and extending their useful life.

The second method is to increase the attachment speed to 100 Mbit/s. As end-station bandwidth requirements continue to grow, 100 Mbit/s-Token Ring ports may be selectively enabled. This will provide another order of magnitude increase in available bandwidth for those stations requiring that performance level.

6.4.5 Putting It All Together

Switched collapsed backbones, desktop switching, and LAN segmentation are widely recognized as the ways to maintain high performance at the desktop while maximizing available bandwidth. Beyond LAN switching, a scaleable high-speed backbone is needed that will enable migration to the next generation of high-speed networking. For the Token Ring user choosing to stay with Token Ring technology, a typical migration strategy comprises two primary phases:

1. Exploit switching and workgroup segmentation:
   a. Collapse the upper-tier backbone into a backbone switch and provide dedicated ports for key resources.
   b. Replace the midtier rings with switches, directly attaching servers to the ports as appropriate.
   c. Microsegment heavily used workgroup rings, thereby increasing the port utilization of midtier switches.

2. Move to high-speed 100-Mbit/s Token Ring:
   a. Install client and server adapters and run them in 4/16-Mbit/s mode until ready. As required, increase the attachment data rate to 100 Mbit/s, exploiting that latent capability.

   b. Upgrade infrastructure, continue to exploit switching and work-group segmentation by incorporating 100-Mbit/s backbone links, and provide dedicated ports for key resources.
   c. Selectively upgrade power users to direct-attached connections to 100-Mbit/s HSTR. Phase out concentrators as more bandwidth is required to the end-stations, replacing them with 16/4 switches with 100-Mbit/s high-speed up-links for the bulk of end-user station attachments.

If we look at the evolution of a large network it's migration from a 16/4-Mbit/s network to one utilizing 100-Mbit/s Token Ring is likely to follow the progression shown in Figures 6.10(a–c). Starting with a Classic Shared Media Token Ring Network, as shown in Figure 6.10(a):

1. Exploit switching and workgroup segmentation as shown in Figure 6.10(b).
   a. Collapse the upper tier backbone;
   b. Replace the midtier rings with switches;
   c. Microsegment heavily used workgroup rings.
2. Migrate to HSTR, as shown in Figure 6.10(c).
   a. "Future proof" client workstations by installing 4/16/100-Mbit/s adapters;
   b. Upgrade infrastructure migrate key shared resources (such as backbone links and servers) to high-speed connections;
   c. Selectively turn on 100-Mbit/s clients and switch-to-switch links as required.

6.5 The Road Ahead

As we look down the road for Token Ring there are further areas that need development and refining to support anticipated applications over the next several years. The IEEE 802.5 Token Ring Working Group has approved project authorization to develop Gigabit Token Ring. It is anticipated to use the same MAC architecture as that used for 100-Mbit/s Token Ring and the PHY used by Gigabit Ethernet (which comes from fiber channel). That Standard should be completed by the second half of 1999. However, there is an economic battle going on that is more important to the future of Token Ring than ultra high speed development. Despite the superior architectural design of Token Ring, Ethernet has captured most of the LAN market with its significantly lower prices. Token Ring vendors will have to aggressively price adapters,

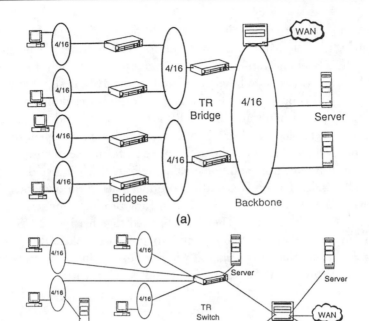

(a)

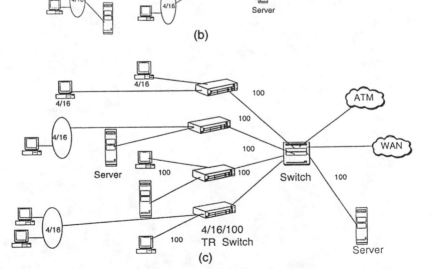

Figure 6.10 (a) Shared Token Ring and deployment of (b) switching and (c) 100-Mbit/s technology.

switches, and concentrators if they are to hold onto their loyal customer base. Fortunately, the 100-Mbit/s design point for Token Ring is based on the Ethernet PHY layer and can use PHY hardware specifically built for Ethernet. The MAC layer is also simplified compared to the 16/4 design because it is switched only. It is up to the manufacturers to exploit this design and deliver competitively priced hardware. Token Ring users understand the benefits of their chosen technology and are generally pleased with the performance it delivers, but will continue to evaluate the price performance to ensure that the premium paid is worth the additional value delivered. They give their answer in each LAN sale. Today Token Ring product sales form a market that is worth over one billion dollars a year. Token Ring product suppliers are proud and protective of this market. They have banded together in the HSTRA[3] to continue to develop the technology and to promote Token Ring's present benefits. If they do their job well, you will see Token Ring as a strategic LAN offering, especially in the largest business and government enterprises of the world, for years to come.

References

[1]   Love, R. D., and J. Lynch, "Token Ring Migration," IBM White Paper, Nov. 6, 1997; http:/www.networking.ibm.com/tra/tramigwp.pdf.

---

3. Visit their website at www.hstra.com.

Token-Ring-Local-Area Networks and Their Performance

(The following is an excerpt from the Werner Bux paper originally published in the Proceedings of the IEEE, *Vol 77, No. 2, February 1989. The entire latter portion of his paper dealing with Token Ring performance has been included. Also included is his comprehensive bibliography of early papers on Token Ring.)*

C. Performance

We start our discussion of the IEEE 802.5 token-ring performance by mentioning some fundamental results for the delay-throughput characteristic obtained through simulation and analysis (where applicable).

A fundamental performance characteristic of any LAN medium-access protocol is its sensitivity to transmission speed and distance. Fig. 8 illustrates the mean frame-transfer delay as a function of the information throughput for 4 Mbps IEEE 802.5 token rings with 1 and 5 km cable lengths. Transfer delay is the time from the generation of a frame in the source station until its reception at the destination station. It is assumed that all 100 stations generate the same amount of traffic and that frames are generated according to Poisson processes. Only one frame per access opportunity can be transmitted. It can be seen that increasing the ring length from 1 to 5 km has virtually no impact on the delay-throughput characteristic. Under the same assumptions, but for a transmission rate of 16 Mbps, Fig. 9 exhibits the same performance measures as Fig. 8.

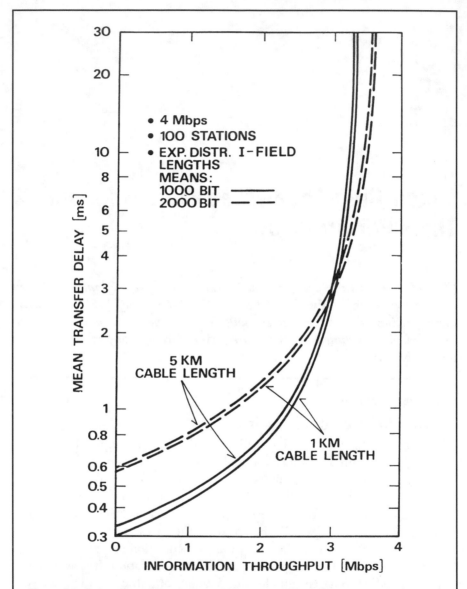

Figure A.8 Delay-throughput characteristic of the IEEE 802.5 protocol. 4 Mbps transmission rate. Symmetrical traffic pattern. From [98], © by Springer-Verlag Berlin Heidelberg 1985.

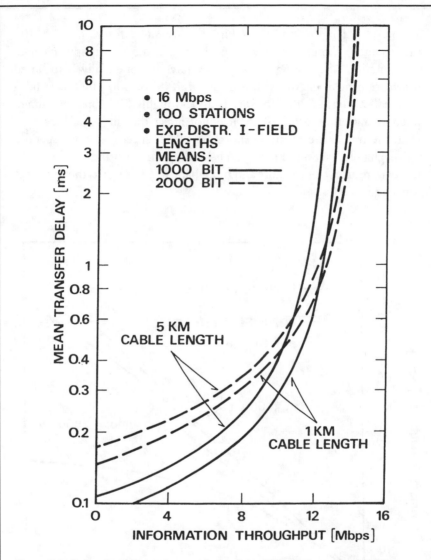

Figure A.9 Delay-throughput characteristic of the IEEE 802.5 protocol. 16 Mbps transmission rate. Symmetrical traffic pattern. From [98], © by Springer-Verlag Berlin Heidelberg 1985.

Increasing the cable length from 1 to 5 km leads to slightly more noticeable differences here.

In these examples, the single-token rule had little impact on the ring performance. This changes when transmission speeds become higher and/or cable lengths longer. In Fig. 10, we show for rings with 100 stations and 5 km cable length, how the maximum throughput changes as a function of the ring transmission rate. Exponentially distributed information-field lengths with a mean of 1000 or 4000 bits have been assumed. We observe that for speeds up to roughly 20 Mbps, the single-token operation of IEEE 802.5 does not affect the ring efficiency even when the frames are relatively short. At higher speeds, stations spend a non-negligible fraction of their token-holding time waiting for the frame header to return. Hence, more and more bandwidth is lost as the transmission rate increases.

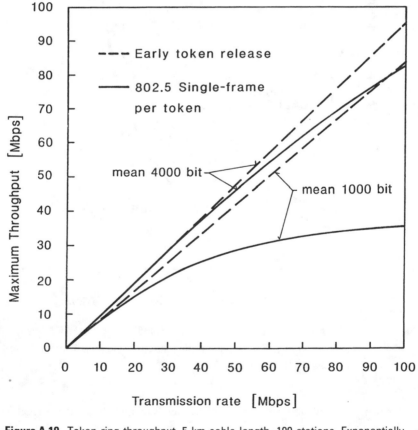

Figure A.10 Token-ring throughput. 5 km cable length. 100 stations. Exponentially distributed information field lengths. From [99], © by Springer-Verlag Berlin Heidelberg 1987.

There are two ways of overcoming this problem:

1. Allow stations to transmit more than one frame per token. Before issuing a new token, a transmitting station waits only until the header of its *first* packet has returned. This is sufficient for the priority-reservation scheme to work. The IEEE 802.5 protocol supports this kind of operation; it should be noted, however, that this alleviates the problem only if (i) stations have queued up a sufficient number of frames to fill the ring and (ii) their ring adapters are able to transmit more than one frame per token.

2. Release the token immediately after a frame has been completely transmitted. Without further protocol changes, this would, under certain circumstances, disable the priority-reservation mechanism described, but may nevertheless be an acceptable operation in cases where efficiency is of key and responsiveness of the priority operation of lesser importance. Another possibility of releasing the token early is not to use the priority-reservation mechanism but to use a timed-token protocol [66]. The latter is the approach taken by the FDDI Standards Committee [13], as will be described in Section V.

For comparison, we have included the efficiency results of the early token release protocol in Fig. 10. As expected, this protocol is not sensitive to the speed/distance product.

Overall efficiency of an access protocol is the most basic performance property; a further important criterion is the quality of service given to individual stations, especially in the case of unbalanced traffic situations. This service can differ significantly, depending on the rule defining the time a station is allowed to transmit per access opportunity. The IEEE 802.5 Standard specifies the use of a token-holding timer which limits the time a station is allowed to transmit continuously. To demonstrate the impact of this timer, we subsequently consider two extreme cases, a very short timeout, such that stations can only transmit one frame per token (Fig. 11), and a very long timeout, such that stations can always completely empty their transmit queues at each transmission opportunity (Fig. 12). In both examples, messages are generated according to Poisson processes; the message lengths are distributed according to a hyperexponential distribution with a coefficient of variation equal to 2. In cases where a message is longer than the maximum information-field length

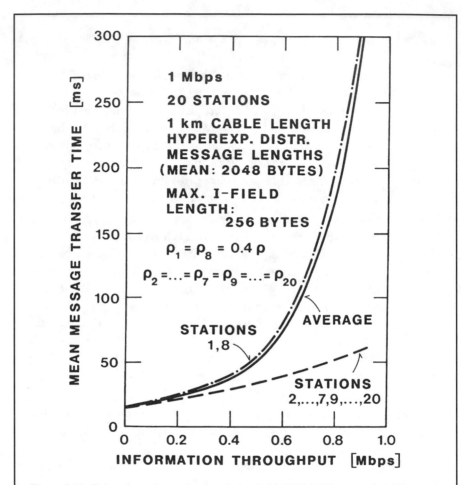

Figure A.11 Delay-throughput characteristic of the IEEE 802.5 protocol. 1 Mbps transmission rate. Asymmetrical traffic pattern. Short token-holding time-out. From [98].

of a frame (256 bytes), the message is segmented. In both examples, the traffic pattern assumed is very unbalanced: two of the 20 stations (Nos. 1 and 8) each generate 40% of the total traffic; each of the other 18 stations generates only 1.1% of the total traffic.

For the single-frame-per-token operation, Fig. 11 shows the mean transfer delay of the messages (not frames) as a function of the total information throughput. The delay of the messages transmitted by the heavy-traffic stations 1 and 8 increases steeply as the offered traffic approaches the ring capacity, while the delay experienced by the light-

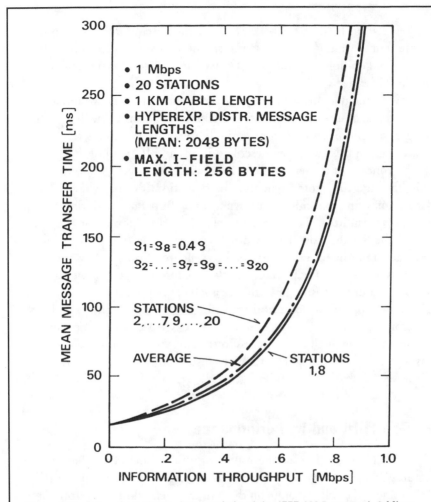

Figure A.12 Delay-throughput characteristic of the IEEE 802.5 protocol. 1 Mbps transmission rate. Asymmetrical traffic pattern. Long token-holding time-out. From [98], © by Springer-Verlag Berlin Heidelberg 1985.

traffic stations remains rather small even at very high utilizations. In this sense, the token-passing protocol combined with a single-frame-per-token operation provides fair access to all users.

From Fig. 12, it can be seen that the relationship of the delay experienced by light and heavy users is reversed when the token-holding time is long. The mean message-transfer delay of light-traffic stations is higher than that of heavy-traffic stations. This is due to the fact that messages generated at a heavy-traffic station have a relatively good chance of their station holding the token and, in this case, of being transmitted before frames waiting in other stations. These two examples demonstrate that the token-holding timer can be used to control the station-specific quality of service.

We conclude this section by briefly commenting on two interesting applications of the IEEE 802.5 priority mechanism. The first is the integration of voice and data on a single token ring. Various techniques to accomplish this have been described in [8] and [64], all of them based on the concept of a "Synchronous Bandwidth Manager" responsible for switching the ring from asynchronous operation, during which normal data frames are transmitted, to synchronous operation, during which voice is transmitted. To guarantee delay and throughput for the voice traffic, the Synchronous Bandwidth Manager makes use of the priority-reservation mechanism to interrupt the asynchronous data flow at regular intervals. Details and refinements of this technique together with a thorough and comprehensive performance analysis are given in [67]. A second important application of priorities is in the area of multi-ring LANs interconnected through bridges. As will be described later in Section VI, giving bridges higher priority access to the rings than normal user stations has significant advantages both with respect to overall efficiency and fairness.

V. The FDDI and Its Performance

As will become clear from the following discussion, FDDI has certain features that result in rather different protocol behavior from the IEEE 802.5 rings. We will elaborate on these in the subsection on performance. For a better understanding of that discussion, we will begin by describing the FDDI frame format and its timed-token protocol [13].

A. Frame Format

The general structure of the FDDI frames is the same as for the IEEE 802 LANs, see Fig. 13, with starting delimiter, frame control field, destination and source addresses, a variable-length information field, frame check sequence, ending delimiter, and frame status field. An obvious difference from the 802.5 frame format is the presence of a preamble preceding every transmitted frame (and token). The preamble is transmitted by the source station as a minimum of 16 idle symbols. Subsequent

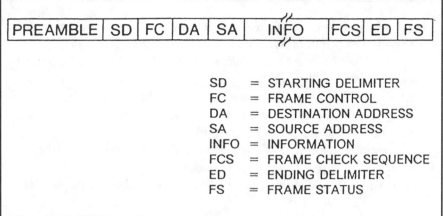

Figure A.13 FDDI frame format

repeating stations may change the length of the idle pattern consistent with the clocking requirements. As pointed out in Section II, FDDI employs a distributed clocking scheme which requires the insertion or removal of idle symbols between frames, depending on the relative clock rates of the ring segments. Thus, repeating stations may see a variable-length preamble which is shorter or longer than the one originally transmitted.

An FDDI *token* consists of a preamble, starting delimiter, frame control field, and ending delimiter. Tokens and the various frame types are distinguished by different values of the frame control field.

B. MAC Protocol

FDDI employs a token protocol which allows the transmitting station to pass the token immediately after the end of frame transmission. This is in contrast to the IEEE 802.5 protocol, which requires the transmitting station to delay issuing a new token until the header of the transmitted frame has returned, see Section IV.

The FDDI priority mechanism is designed to provide different classes of service that enable the network to simultaneously support traffic with different transmission requirements. The highest priority "synchronous" class guarantees bandwidth and transmission delay, and is therefore suitable for applications such as voice. All remaining traffic in FDDI belongs to the "asynchronous" class, which is subdivided into eight priority levels. The asynchronous priority class also has a restricted token mode which can be used to temporarily reserve all bandwidth not used for synchronous transmission for a specific asynchronous dialog.

FDDI employs a "timed-token" protocol in which the length of time the token may be held for transmitting frames of a given class (i.e., the token-holding time) depends on the time between successive arrivals of the token at the transmitting station (i.e., the token-rotation time). The FDDI protocol uses a number of timers and variables at each station to determine these times. We will assume that all timers are initialized with zero, and expire when they have counted up to their target time, which represents a slightly different description of the timer operations than in the FDDI standard [13].

As part of the ring-initialization process, all stations negotiate a Target Token-Rotation Time (TTRT). The protocol guarantees that the long-term average token-rotation time does not exceed the TTRT and that the maximum token-rotation time does not exceed 2 * TTRT [68], [69]. So stations with strict transmission-delay requirements must request a TTRT equal to one-half of the maximum acceptable delay. At the end of the negotiation, the shortest TTRT requested becomes the operative TTRT and is used to set the variable T_Opr identically in each station.

A Token-Rotation Timer (TRT) is used in each station to measure the time between successive arrivals of the token at that station. Normally, the TRT is reset each time the token is received in order to time the next token rotation. The TRT will expire if it counts up to T_Opr before the token has arrived back at the station. When the TRT expires, Late_Ct, a counter which is initially zero, is incremented and the TRT is reset to zero and continues timing. When the token arrives late at a station (Late_Ct = 1) the TRT is not reset, rather it is allowed to continue timing, thus accumulating the lateness of the current token rotation into the next token-rotation time. The result of accumulating lateness is that following a token rotation that exceeds T_Opr by time A, asynchronous transmission will be restricted until this lateness has been compensated for by token rotations less than T_Opr by a total of time A. This ensures that the average token-rotation time is at most T_Opr. If Late_Ct ever exceeds 1, error recovery is initiated [70]. Late_Ct is reset to zero each time the token is received.

A Token-Holding Timer (THT) is used by each station to control the amount of time the token is held for transmitting asynchronous frames. The THT is loaded with the current value of the TRT when the token is received on time (Late_Ct = 0) at a station. When the THT reaches the token-holding time threshold for a particular priority level, the token may no longer be used for transmitting frames of that level. Transmissions already in progress when the THT expires are completed. T_Pri(i) (i = 1 to 8) defines the token-holding time threshold for asynchronous priority level i. The convention adopted is that the priority increases from 1 to 8. A larger threshold value allows more time to elapse from the THT before the token must be passed. Therefore, the associated priority level has a greater transmission window, and consequently higher priority, than priority levels with smaller token-holding time thresholds. The maximum threshold value for a priority level is T_Opr.

C. Performance

We start this section with a brief review of the literature on FDDI performance. In [71], Ulm uses a simple approximation to examine the performance of a timed-token protocol similar to FDDI. Results are presented to demonstrate the effects of the network configuration and the protocol parameters, on the overall maximum ring utilization. Sevcik and Johnson prove two fundamental properties of the protocol regarding the average, and maximum time the token will take to circulate around the ring [68]. They also present an equation that characterizes some of the effects of synchronous traffic on the performance of the asynchronous class. A detailed simulation model is used by Dykeman and Bux to investigate the throughput and delay performance of the FDDI MAC protocol for a range of network configurations and protocol parameters [72]. In [73], the same authors describe a procedure to calculate estimates for the throughput of each asynchronous priority level and also provide a method that can be used to tune the FDDI parameters so that given performance objectives for the various priority levels are achieved. Goyal and Dias use deterministic techniques and a simulation model to compare FDDI with the IEEE 802.5 token ring, in the presence of a variety of traffic types [74]. A quantitative investigation of packetized voice on FDDI is provided by Frontini and Watson [75]. The FDDI protocol also incorporates reliability mechanisms to provide fault detection, isolation, monitoring, and configuration functions. These mechanisms are discussed by Johnson [70]. A variant of the timed-token protocol is being used in the IEEE 802.4 Token Bus standard [76], hence analyses of this protocol, e.g. [77]–[79], may be applied to FDDI, if appropriately modified.

In the following discussion of the FDDI performance, we follow [72] and [73], and first examine the effects of ring latency (r_l), target token-rotation time (T_Opr), and the number of actively transmitting stations (N) on the maximum total throughput (γ_{max}) obtainable on an FDDI token ring. We assume that each actively transmitting station always has frames queued for transmission, and that all transmissions beyond expiration of the token-holding timer, owing to frame transmissions in progress, are of equal length. Frame transmission times are assumed to be of constant length F.

As shown in [80], the following expression for the maximum total throughput holds for an FDDI token ring with only one asynchronous priority level:

$$\gamma_{max} = \frac{(N*tot\_tx\_time + N^2*tx\_window)*v}{(N*tot\_tx\_time + N^2*tx\_window + (N^2 + 2*N + 1)*r\_l)} \tag{7}$$

where

$$v = \text{the transmission rate}$$

$$tot\_tx\_time = CEILING(tx\_window/F)*F$$

$$tx\_window = T\text{-}Opr - r\_l$$

The limit of (7), as the number of active stations N goes to infinity, is in agreement with the simple formula given in [71] for the FDDI ring utilization:

$$\lim_{n \to \infty} \gamma_{max} = \frac{tx\_window*v}{tx\_window + r\_l} = \frac{T\_Opr - r\_l}{T\_Opr} *v \tag{8}$$

Fig. 14 illustrates the effects of ring latency (r_l) on the total maximum throughput. We consider a system with ten active stations and a frame size of 1.6 kbytes. The ring latency is varied from 0.011 msec, representing a ring with ten connected stations and a 1 km fiber, to 1.62 msec, which represents a ring with 1000 connected stations (990 of which are not transmitting) and a 200 km fiber. We can observe that with a target token-rotation time of 10 millisec, the maximum total throughput remains high, in this case over 82 Mbps, even when the ring latency is very large (1000 stations and 200 kilometer fiber length are FDDI maximum

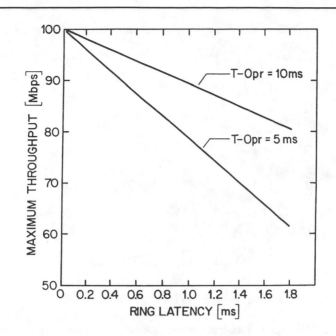

Figure A.14 FDDI maximum throughput versus ring latency for T_Opr = 5 and 10 ms (10 active stations). From [72], © 1987 Information Gatekeepers, Inc., Boston, MA.

values). When the target token-rotation time is reduced to 5 msec, the maximum total throughput corresponding to the largest ring latency drops to 65 Mbps. We can see that the ratio of ring latency and target token-rotation time has a significant effect on the maximum total throughput. As a rule of thumb, T_Opr should be at least five times the ring latency to maintain throughput levels above 80 Mbps.

In the remainder of this section, we focus on various aspects of the FDDI *priority* mechanism. First, we consider an FDDI ring in which there are eight stations, each attempting to transmit frames at one of the eight asynchronous priority levels. The asynchronous-restricted and synchronous ring-access classes are not used in this example. The arrival rate of frames to be transmitted is identical at each station. In total, there are 11 stations connected to the ring, three of which are idle. Fig. 15 shows the throughput plotted against the arrival rate for each of the eight priority levels, along with the overall ring throughput. At low arrival rates, all classes of traffic receive some bandwidth since all token-holding time thresholds are greater than the total ring latency. However, as the

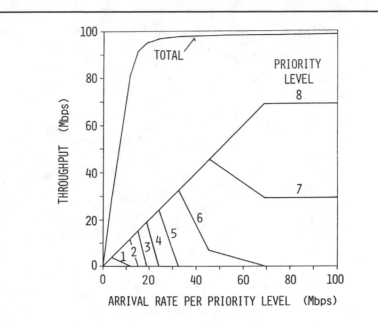

Figure A.15 FDDI throughput versus arrival rate per priority level. (Token-holding time thresholds: 100, 76.5, 56.2, 39, 25, 14, 6.2, and 1.5 msec; ring latency: 1.0236 sec). From [73].

arrival rate of frames increases, the throughput of lower-priority traffic begins to decrease, and is eventually reduced to zero. This characteristic of the asynchronous priority class is potentially very useful. Low-priority traffic (e.g., large file transfers) can be assigned to a priority level for which the throughput is reduced to zero when the total offered load exceeds a certain value. In this way, low-priority traffic is automatically deferred during busy periods, until the load on the ring has been reduced. Different levels of low-priority traffic can be defined to be deferred at different loads. Also, multiple levels of higher-priority traffic can be defined to give preferred service to some traffic, while still guaranteeing a certain throughput for the competing traffic even when the load offered to the ring exceeds its capacity.

We consider an additional scenario to demonstrate the effects of the token-holding time thresholds and the number of transmitting stations on the throughput received by each priority level. In this example, two asynchronous priority levels (1 and 2) are used. Fig. 16 shows the ratio of the guaranteed throughputs for the two priority levels as a function of the ratio of the token-holding time thresholds, when the number of

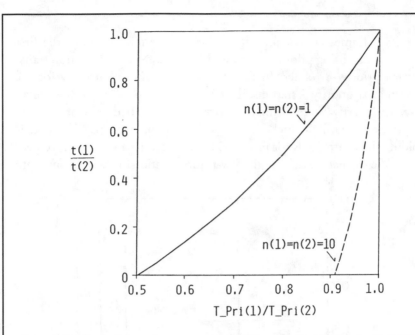

Figure A.16 Throughput ratio versus token-holding time threshold ratio. (T_Pri(2): 5 msec; frame length: 1 kbyte; ring latency: 63 μsec.) From [73].

stations transmitting frames of each priority is 1 and 10, respectively. When T_Pri(1) is too small, the lower priority level has no guaranteed throughput, and when T_Pri(1) = T_Pri(2) the priority levels are indistinguishable, thus the guaranteed throughputs are equal. The relative guaranteed throughputs become more sensitive to the token-holding time thresholds as the ratio approaches unity and as the number of transmitting stations increases. This demonstrates an inherent problem of the FDDI priority mechanism, namely that the relative performance of the priority levels is very sensitive to the number of active stations. In environments where the number of active stations on an FDDI ring varies dynamically, it will be very difficult to tune the protocol parameters to reliably meet given performance requirements.

The final result presented in this section addresses the issue of fairness in the FDDI protocol. If a protocol is completely fair, the transmission delay experienced by a frame will depend on its priority not on the location of the transmitting station. We consider a system in which ten stations are connected to a 10 km ring. One of these stations transmits priority-8 frames, while the other nine stations transmit priority-7 frames. We shall assume that the arrival rates at each of the stations transmitting

priority-7 frames are equal. "First priority-7 station" refers to the first
station transmitting priority-7 frames downstream from the station transmitting priority-8 frames. In Fig. 17, we show the mean delay for the
first and last priority-7 stations, the priority-8 station, and for the overall
ring, as the arrival rate of high-priority traffic is varied. The arrival rate
of priority-7 traffic at each station is fixed at 6.33 Mbps. The token-
holding time thresholds for priorities 8 and 7 are 10 and 2 msec, respectively. We see that the delay at the low-priority station furthest downstream

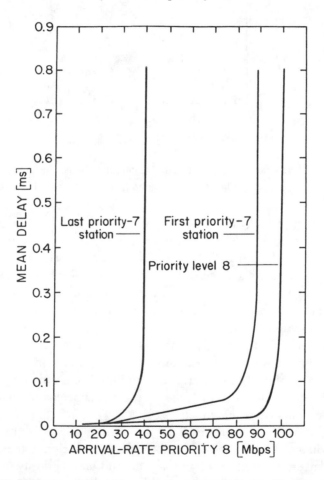

Figure A.17 FDDI mean delay versus arrival rate of priority-8 traffic. (One priority-8
station; nine priority-7 stations with a constant arrival rate of 6.33
Mbps; I_Pri(8)&rbl.= 10 msec; T_Pri(7) = 2 msec). From [72]. © 1987
Information Gatekeepers, Inc., Boston, MA.

from the high-priority station increases drastically, long before the first priority-7 station is affected. The explanation for this unfairness is as follows. When a high-priority station does not have a full queue of frames to transmit, the unused portion of its bandwidth will be available for use by the lower-priority stations on the ring. This unused bandwidth will circulate sequentially among the stations transmitting lower-priority traffic, beginning with the *first* low-priority station following the high-priority station, until it is "reclaimed" for use by the high-priority station. Each time the high-priority bandwidth is reclaimed and re-relinquished, it begins circulating with the first station following the high-priority station. When the offered load is low, there will frequently be unused bandwidth, but it will circulate among all low-priority stations, and therefore no unfairness will result. But as the offered load increases, the high-priority station will relinquish portions of its bandwidth for shorter durations only, thereby favoring the stations immediately following it on the ring.

VI. Interconnected Token Rings

Local-area networks must be capable of interconnecting a large number of stations over maximum distances of several kilometers. Whenever the limitations of a single-ring network are reached with respect to the maximum number of attachments or maximum distance, means for interconnection become necessary. The functions needed for token-ring interconnection are provided in specific nodes called bridges. These functions need to be simple in order to achieve the high throughput values required for the interconnection of high-speed subnetworks at reasonable cost. Simplicity of bridges can be achieved through a connectionless interworking of the subnetworks. This implies that bridges do not perform any complex flow or error control; such functions are only provided end-to-end, i.e., between the communicating stations. The interface of the bridge to a ring is the same as that of any other station, except that it must recognize and copy frames destined for one of the other rings within the network. A bridge must have the capability to temporarily buffer several frames awaiting transfer to the next ring. Received frames which cannot be buffered upon their arrival at a bridge will be lost. The bridge does not take any action when such a frame loss occurs; the responsibility for detecting and recovering from lost frames lies with the communicating end stations.

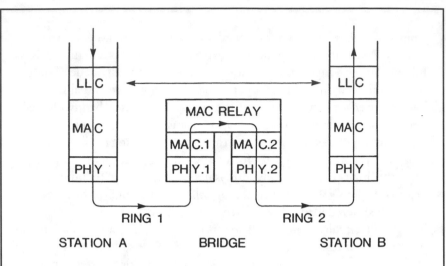

Figure A.18 Bridge architecture model.

Fig. 18 shows the architecture model of a bridge performing a MAC-layer relay function. The routing function of such a bridge is confined to the data-link layer. Such a bridge is currently under consideration for inclusion into the IEEE 802.5 standard [81]. The specific routing technique suggested for multi-token-ring networks is called source routing and will be described in the following subsection. It should be mentioned here that IEEE 802 is also pursuing the standardization of another bridging technique called "Transparent" or "Spanning Tree" bridging. For details on this technique and its relationship to source routing, the reader is referred to [82] and [83].

A. Source Routing

The source-routing technique is based on the concept that the station which originates a frame designates the route the frame will travel from source to destination [84]–[89]. This information is supplied by imbedding a description of the route in a "routing-information field" contained in the information field of the transmitted frame. Bridges interpret the routing-information field to decide when to copy and forward a frame to the adjoining ring. When the frame reaches the ring to which the destination station is attached, this station will recognize its own address in the destination-address field of the frame. The received frame contains a description of the exact route the frame traversed, which the destination station may save and later use to send responses to the originating station.

The mechanism through which a station initially obtains the route to a given station is called "route discovery," a process that allows the station to dynamically discover a route to any other station. Route discovery is accomplished in the following manner.

When a station needs to transmit a frame to a destination for which it does not yet know a route, it transmits the frame as a "single-route broadcast" such that one copy of the frame appears on each ring in the network. After receiving this message, the destination station in turn transmits a "route-explorer" message which traverses all possible paths between local station and the initiating station. This will allow the initiating station to select a route between the two stations. The selected route is communicated to the remote station in subsequent exchanges.

In case of multiple routes, the originating station will receive multiple route-exploring frames, from which it must select one to obtain the routing information to be used. Examples of possible route selection criteria are: 1) Using the routing information from the first route-exploring frame received; 2) the first route-exploring frame with fewer than a specified number of hops, or 3) within a certain time window, the route-exploring frame with the shortest route.

The first bridge to forward a route-exploring frame from one ring to another will insert into its routing-information field the ring number of the ring from which the frame was forwarded, its own bridge number, and the number of the ring on which the frame is to be transmitted. Other bridges forwarding that frame append their bridge number and the ring number of the ring on which the frame is to be transmitted to the routing-information field. When the frame reaches the destination station, the routing information in the frame indicates the path the frame travelled through the network. This provides a mechanism to dynamically determine a route from one station to another. Special measures are taken to prevent establishing routes that could cause a frame to continuously circulate, see [74].

A recent, extensive simulation study [90] showed that the path-finding algorithm which selects the route based on the first reception of a route-exploring frame generally performs well with respect to delay and throughput of the newly established routes and the total throughput of the already existing routes. Under certain conditions, the delay of the existing routes may be adversely affected by the particular route-selection criterion.

B. Logical-Link Control

The interconnection architecture described above assumes that the LLC function is employed end-to-end between the communicating user stations. The LLC standard for LANs, IEEE 802.2, defines two types of operation [91]. Type-1 operation, called unacknowledged connectionless service, is a datagram service by which LLC entities in two devices exchange frames without benefit of link-level error recovery or flow control. Type-2 operation, the connection-oriented service, provides recovery from lost, duplicated, or misordered frames, and allows the stations to regulate the flow of data between them so that neither becomes overrun. To furnish this service, type-2 operation depends on the establishment of connections, and thus has a plurality of different types of frames for connection setup, connection takedown, data transfer, acknowledgments, and flow control. A third type of operation, approved by IEEE 802 in 1986, is called acknowledged connectionless service. It provides some reliability but no flow control and requires less state information and a less rigorous form of initialization than connection establishment.

For the following considerations, we assume that the IEEE 802.2 Type-2 Logical-Link Control protocol is used whose major functions can be described as follows.

Data units provided by a user of the layer-2 services are transmitted in Information (I-) frames. The LLC sublayer makes use of so-called Supervisory (S-) frames for control functions, such as connection establishment and termination, positive and negative acknowledgments, flow-control information, etc.

Flow control is realized by a window mechanism, i.e., a sender is permitted to transmit up to W (the window size) I-frames without having to wait for an acknowledgment. The receiver uses Receive Ready (RR-) frames to acknowledge I-frames correctly received and to indicate to the sender that more I-frames can be transmitted. In the original 802.2 Type-2 protocol, the window size W was specified as a constant value to be defined at link-initialization time. Based on the performance considerations described in the next subsection, the standard was amended by introducing an option in which the window size is dynamically adjusted in case of network congestion.

Any I-frame received with an incorrect Frame Check Sequence (FCS) is discarded. When a received I-frame has a correct FCS, but its send sequence number is not equal to the one expected by the receiver, the receiver returns a Reject (REJ-) frame. It then discards all I-frames until the expected I-frame has been correctly received. Upon receiving a REJ-frame, the sender retransmits I-frames starting with the sequence number received within the REJ-frame.

In addition to REJECT recovery, a time-out mechanism is used. At the instant of transmission of an I-frame, a timer will be started if it is not already running. When the sender receives an RR-frame, it restarts the timer if there are still unacknowledged I-frames outstanding. When the timer expires, the station performs a "checkpointing" function by transmitting an RR-frame with a dedicated bit (the "P-bit") set to one. Upon receiving this frame, the receiver must return an RR-frame with the "F-bit" set to one. When this RR-frame has been received by the sender, it either proceeds with transmitting new I-frames or retransmits previous I-frames depending on the sequence number contained in the received RR-frame.

C. Network Performance

Interconnected token-ring networks exhibit a number of characteristic performance properties; of these we subsequently discuss the most important ones following [92]. The network structure underlying our considerations is depicted in Fig. 19. Several IEEE 802.5 token rings are interconnected through bridges and a "backbone" token ring. Note that

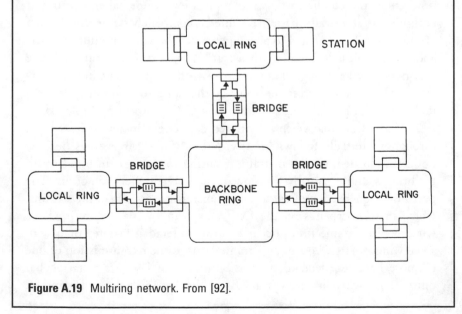

Figure A.19 Multiring network. From [92].

in such a network, the routing problem is almost trivial, since only one path exists between any two stations. User stations are only attached to the local rings but not to the backbone; the latter serves to interconnect bridges.

For all examples in this section, the following parameter values have been used. The transmission rates investigated are 4 Mbps for the local rings and 16 Mbps for the backbone. Bridges are assumed to need 300 & mu.sec to process one frame. The size of each of the two bridge buffer pools (one per data flow direction) is assumed to be 4 kbytes, unless stated otherwise. The maximum I-field length is 0.5 kbyte; the framing overhead and the length of S-frames is 24 bytes, as specified in [19], [90]. The time intervals between generation of messages are assumed to be exponentially distributed. The mean message length is 1 kbyte, the coefficient of variation 1.5. Because of message segmentation and supervisory frames, the resulting overall frame-length distribution resembles the bimodal distribution measured on real systems [93] with a mean of about 250 bytes.

Fig. 20 shows the total network throughput (in Mbps) as a function of the total offered data rate (also in Mbps). Throughput of a connection is defined as the number of bits received at both ends across the layer-2-to-3 interface per unit time. The offered data rate is defined as the number of bits generated by an application for transmission per unit time under the condition that the application is not halted because of backpressure. The traffic pattern assumed is completely symmetrical: Each of the 12 stations attached to a ring generates the same amount of traffic and has a logical link set up to a station on a different ring. I-frame transmission on each logical link is two-way. Bridges operate in nonpriority mode, i.e., they access the rings at the same token-priority level as the normal stations. A fixed-window-size LLC protocol is employed.

We observe that when the offered data rate is increased from zero, throughput initially follows linearly. When the 4 Mbps rings become noticeably loaded, queues of frames waiting to enter them build up in the bridges, and eventually overflow of the bridge buffers will occur. The overflow frequency increases with the transmission activity of the stations, which in turn depends on the fixed LLC window size W. With large window sizes, frame losses occur at lower offered data rates than with small window sizes. Loss of an I-frame leads to the retransmission of one or more I-frames, depending on the number of I-frames a station has outstanding when it receives a REJ or has performed checkpointing.

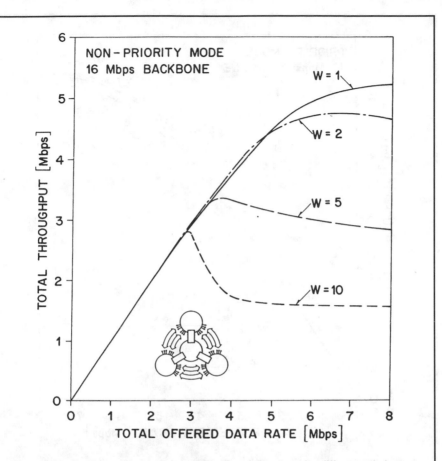

Figure A.20 Total throughput versus total offered data rate for different window sizes W. Non-priority mode. Symmetrical traffic. From [92].

Obviously, the window size sets an upper limit to the number of frames to be retransmitted per lost frame. Therefore, the additional traffic created by retransmissions decreases with smaller window sizes. This explains the significant differences between the throughput values pertaining to different window sizes at high offered data rates. For large window sizes, throughput shows a pronounced maximum. When the offered data rate is increased beyond a certain critical value, throughput drops the more rapidly the bigger the window size and approaches an asymptotic value.

As described in Section IV, the IEEE 802.5 protocol supports multiple levels of access priorities on token rings. In a network of interconnected token rings, such priorities can be used to give better service to bridges as compared to normal ring stations. In Fig. 21, we consider the

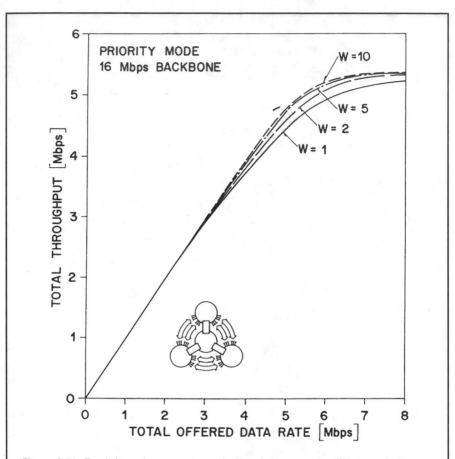

Figure A.21 Total throughput versus total offered data rate for different window sizes W. Priority mode. Symmetrical traffic. From [92].

same scenario as before, however, bridges now operate at a higher ring-access priority level than normal stations and are allowed to transmit continuously until their transmit buffer has been emptied. On the backbone, priority access is not employed and each bridge transmits only one frame per token. We observe substantial improvement, in fact, no or negligibly few frame losses are observed. The explanation of this rather striking effect is the following. Priority access for bridges has two basic consequences. First, it shortens the queues of frames waiting to enter the destination rings and hence contributes to a reduction of buffer overflows. Second, access of stations to their local ring is delayed, which slows down both the injection of new I-frames into the network and the returning of acknowledgments. Delayed acknowledgments, on the other hand, further throttle the transmission of I-frames because of the LLC window flow-control mechanism. The overall effect is similar to flow-control schemes suggested for wide-area networks in which traffic entering the network is handled with lowest priority [94]–[97].

The effectiveness of the priority-mode operation diminishes when the traffic is less symmetric, as assumed in the next example (Fig. 22) where six stations attached to one ring and the same number of stations attached to a second ring transmit I-frames to 12 stations on the third ring. The throughput characteristic observed indicates congestion in the bridge connecting the backbone with the third ring when the offered data rate exceeds a certain value. Again, a small window size is the best choice for this scenario.

As shown in [92], priority access for bridges has a major advantage with respect to fairness It avoids the very pronounced differences in the quality of service among inter-ring and intra-ring traffic that exist when no priorities are used.

Although all results shown so far have indicated that a small window size is an effective means of minimizing frame loss and congestion, it is important to understand that under certain conditions, a small window size can also be a disadvantage. In Fig. 23, we show the throughput

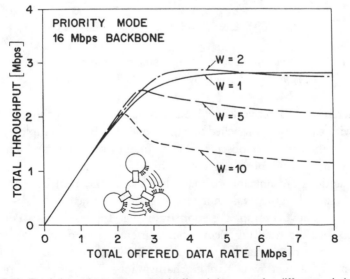

Figure A.22 Total throughput versus total offered data rate for different window sizes W. Priority mode. Asymmetrical traffic. From [92].

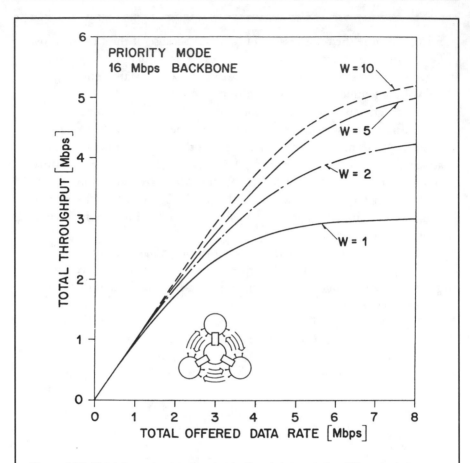

Figure A.23 Total throughput versus total offered data rate for different window sizes W. Priority mode. Symmetrical traffic. From [92].

characteristic of a network in which two stations on each ring communicate with two stations attached to a different ring. Bridges employ priority access to the rings. We observe that the total throughput can be substantially improved by increasing the window size, because at small window sizes, stations are not able to make full use of the available bandwidth since acknowledgments do not return sufficiently fast.

The above examples demonstrate that frames lost in congested bridges and the resulting recovery can severely degrade network performance. This insight has led the IEEE 802.2 Committee to introduce an optional dynamic flow-control mechanism into the Type-2 LLC protocol whose purpose is to guarantee robust and efficient network operation under both normal traffic load and overload. When using this algorithm, stations initially use the window size as defined at link initialization. Whenever a station needs to retransmit an I-frame (either because of a received REJECT frame or after checkpointing), it sets its window size to one. Afterwards, the window size is increased by one (up to the initial value) for every n-th successfully transmitted (i.e., acknowledged) I-frame.

The rationale behind this algorithm is as follows. Under normal conditions, i.e., no congestion, the actual window size used is the one negotiated between the communicating partners. By setting the window size to one, whenever there is an indication of a possible congestion, a high responsiveness of the flow-control mechanism is achieved, because immediate throttling of the network input traffic is performed. Subsequent to reduction, stations again attempt to increase their window sizes. This process is tightly coupled to the reception of acknowledgments, so that the speed by which the window size is increased is controlled by the momentary ability of the network to transport frames successfully.

For the scenario previously studied in Fig. 22, Fig. 24 shows the total network throughput as a function of the total offered data rate for different values of n. The initial window size is ten. We observe substantial improvement in throughput when the window size is dynamically adjusted. The performance gain depends heavily on the choice of parameter n defined above: the larger n, the higher the throughput. However, the incremental gain in throughput decreases as n increases. For small values of n, e.g., one or two, the algorithm works too dynamically in the sense that the window is apparently opened very rapidly and hence the throttling effect does not last long enough. It was generally observed that even under extreme overload, the frame-loss frequency in bridges is never substantially greater than one percent when the dynamic window-size algorithm with $n \geq 8$ is employed. Under this condition, only a very small fraction of the bandwidth is lost for retransmissions, and hence performance is almost ideal.

Finally, Fig. 25 compares the throughput characteristics of the dynamic window-size LLC and the fixed window-size LLC for three different bridge-buffer sizes and demonstrates that for larger *fixed* window sizes, the overload behavior is not improved by bigger bridge buffers. On the other hand, the dynamic window-size algorithm yields a stable throughput behavior even for relatively small bridge-buffer sizes in the sense that throughput never decreases with increasing offered data rate. With the *dynamic* window, throughput is improved by bigger bridge buffers, however, increasing the buffer size beyond the two times 8 kbytes shown in the figure does not yield any noticeable improvement.

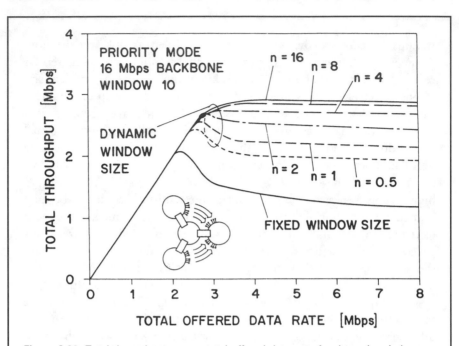

Figure A.24 Total throughput versus total offered data rate for dynamic window-size LLC with initial window size ten and different values of n. Priority mode. (Note: n = 0.5 means that the window size is increased by two per acknowledgment.) From [92].

In all cases where the fixed window-size protocol works without congestion problems, the dynamic flow-control algorithm yields equally good performance since the window size is never or only very rarely reduced. A typical example is that of Fig. 21, for which the throughput characteristic of the dynamic-window LLC with an initial window size of ten is identical to the result for the fixed window of ten, namely, almost ideal.

VII. Concluding Remarks

During the past few years, substantial research activity has been devoted to studying the various facets of the performance of token-ring-based LANs. This work was of significant help in establishing the token-ring technique as a major LAN technology. Today, we have achieved a satisfactory understanding of most practically relevant performance questions.

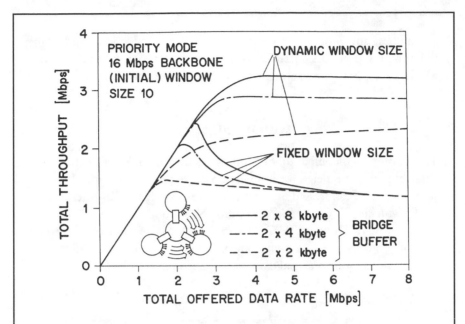

Figure A.25 Total throughput versus total offered data rate for different bridge-buffer sizes. Fixed and dynamic window size LLC with fixed or initial window size ten and n = 8. Priority mode. From [92]. Reprinted with permission from IEEE, Proceedings of the IEEE, Vol 77, No. 2, February 1989.

What is still missing is a better theoretical foundation of a number of problems. A few examples follow.

Analytic models of the basic token-ring operation are typically limited to Poisson arrivals of single frames and very often require symmetry and independence assumptions. Analytically tractable models that lend themselves to numerical evaluation with a more realistic characterization of the traffic would be highly desirable. Equally important would be models yielding more detailed performance measures than just mean delays, e.g., delay distributions.

In the field of more complex protocols, such as the FDDI timed-token protocol, theoretical results are available for the throughput characteristic only; similar results for the delay would be of great help for system design and configuration.

An area where only very little theoretical understanding has so far been achieved is the modeling of multiring local-area networks. Both routing and flow control have only been investigated by simulation techniques; any progress towards a better theoretical understanding of the complex processes in such systems would be extremely useful and would strengthen our confidence in the algorithms used.

Acknowledgment

The author is indebted to his colleagues at the IBM laboratories in Rueschlikon and Raleigh for numerous interactions and technical discussions on issues addressed in this paper, especially to K. Küemmerle, H. R. Müeller, F. Closs, P. Zafiropulo, N. A. Bouroudjian, D. A. Pitt, K.-B. Sy, R. A. Donnan, R. C. Dixon, H. D. Maxey, and N. C. Strole. Special thanks go to D. Dykeman and D. Grillo for their invaluable contributions to the material discussed in Sections V and VI, respectively.

References

[1]   K. Küemmerle and M. Reiser, "Local-Area Networks—Major Technologies and Trends," in: *Frontiers in Communications: Advances in Local Area Networks,* K. Küemmerle, F.A. Tobagi, and J.O. Limb, Eds., New York: IEEE Press, 1987, pp. 2–26.

[2]   R.M. Metcalfe and D.R. Boggs, "Ethernet: Distributed Packet Switching for Local Computer Networks," *Commun. Ass. Comput. Mach.,* vol. 19, pp. 395–404, 1976.

[3]   W.D. Farmer and E.E. Newhall, "An Experimental Distributed Switching System to Handle Bursty Computer Traffic," in: *Proc. ACM Symposium on Problems in the Optimization of Data Communications* (Pine Mountain, GA, 1963), pp. 31–34.

[4]   D.J. Farber er al., "The Distributed Computer System," in: *Proc. 7th IEEE Computer Society Intl. Conf.* (Piscataway, NJ, 1973), pp. 31–34.

[5]   J.R. Pierce, "Network for Block Switching of Data," *Bell Syst. Tech. J.,* vol. 51, pp. 1133–1145, 1972.

[6]   J.H. Saltzer and K.T. Pogran, "A Star-Shaped Ring Network with High Maintainability," *Computer Networks,* vol. 4, pp. 239–244, 1980.

[7]   A. Hopper, "Data Ring at Computer Laboratory, University of Cambridge," in Computer Science and Technology: Local Area Networking. Washington, DC: Nat. Bur. Stand., NBS Special Pub. 500-31, 1977, pp. 11–16.

[8]   W. Bux, F. Closs, P. Janson, K. Kuemmerle and H.R. Mueller, "A Reliable Token-Ring System for Local-Area Communication," in: *Conf. Rec. NTC '81* (Piscataway, NJ, 1981), pp. A2.2.1–A2.2.6.

[9]   E.R. Hafner, Z. Nenadal, and M. Tschanz, "A Digital Loop Communications System," *IEEE Trans. Commun.,* vol. COM-22, pp. 877–881, June 1974.

[10] C.C. Reames and M.T. Liu, "A Loop Network for Simultaneous Transmission of Variable Length Messages," in: *Proc. 2nd Annual Symp. Comput. Architecture* (Houston, TX, 1975), pp. 7–12.

[11] D.E. Huber, W. Steinlin, and P.J. Wild, "SILK: An Implementation of a Buffer Insertion Ring," *IEEE J. Select. Areas Commun.,* vol. SAC-1, pp. 766–783, Nov. 1983.

[12] ANSI/IEEE Standard 802.5-1985, Token Ring Access Method and Physical Layer Specifications.

[13] "FDDI Token Ring Media Access Control (MAC)," American National Standard, X3.139-1987.

[14] "FDDI Physical Layer Protocol (PHY)," Draft Proposed American National Standard, X3.148-198X.

[15] "FDDI Token Ring Physical Layer Medium Dependent (PMD)," Draft Proposed American National Standard, X3.166-198X.

[16] "FDDI Token Ring Station Management (SMT)," Draft Proposed American National Standard, X3T9/85-, X3T9.5/84-49, Rev. 3.0, Aug. 1, 1987.

[17] "An Introduction to the IBM 8100 Information System," IBM Publication order number GA27-2875.

[18] D.A. Pitt, "Standards for the Token Ring," *IEEE Network Magazine,* vol. 1, pp. 19–22, Jan. 1987.

[19] E. Enrico, "The European Approch to LAN Standardization," in: *Proc. IEEE INFOCOM '86* (Miami, FL, 1986), pp. 343–354.

[20] Standard ECMA-89: Local-Area Networks—Token Ring, Sep. 1982.

[21] H.R. Müeller, H. Keller, and H. Meyr, "Transmission in a Synchronous Token Ring," in: *Proc. IFIP TC 6 International Symposium on Local Computer Networks,* P. Ravasio, G. Hopkins, and N. Naffah, Eds., Amsterdam: North-Holland, pp. 125–148, 1988.

[22] R.C. Dixon, N.C. Strole and J.D. Markov, "A Token-Ring Network for Local Data Communications," *IBM Syst. J.,* vol. 22, pp. 47–62, 1983.

[23] N.C. Strole, "A Local Communications Network Based on Interconnected Token-Access Rings: A Tutorial," *IBM J. Res. Develop.,* vol. 27, pp. 481–496, Sept. 1983.

[24] S. Joshi and V. Iyer, "New Standards for Local Networks Push Upper Limits for Lightwave Data," *Data Communications,* pp. 127–138, July 1984.

[25] F.E. Ross, "FDDI— A Tutorial," *IEEE Commun. Magazine,* vol. 24, May 1986.

[26] F.E. Ross, "Rings Are 'Round for Good!," *IEEE Network Magazine,* vol. 1, pp. 31–38, Jan. 1987.

[27] "Draft Addendum to IEEE Standard 802.5, (Token Ring Access Method) for Dual Ring Operation with Wrapback Reconfiguration," IEEE 802.5 Working Document D1.0 (12/01/87), Dec. 1987.

[28] H.J. Keller, H. Meyer, and H.R. Müeller, "Transmission Design Criteria for a Synchronous Token Ring," *IEEE J. Select. Areas Commun.,* vol. SAC-1, pp. 721–733, Nov. 1983.

[29]   W. Bux, F. Closs, K. Küemmerle, H. Keller, and H.R. Müeller, "A Reliable Token Ring for Local Communications," *IEEE J. Select. Areas Commun.,* vol. SAC-1, pp. 756–765, Nov. 1983.

[30]   H. Takagi, "Analysis of Polling Systems," The MIT Press, Cambridge, MA, 1986.

[31]   M. Eisenberg, "Queues with Periodic Service and Changeover Times," *Oper. Res.,* vol. 20, pp. 440–451, 1972.

[32]   O. Hashida, "Analysis of Multiqueue," *Rev. Elect. Commun. Lab.,* vol. 20, pp. 189–199, 1972.

[33]   G.L. Brodetskii and V.P. Vinnitskii, "Some Characteristics of Systems with Cyclic Data Processing," *Cybernetics,* vol. 9, pp. 407–413, May–June 1973.

[34]   P.A. Humblet, "Source Coding for Communication Concentrators," Report ESL-R-798, Electronic Systems Laboratory, MIT, Jan. 1978.

[35]   M.J. Ferguson and Y.J. Aminetzah, "Exact Results for Nonsymmetric Token Ring Systems," *IEEE Trans. Commun.,* vol. COM-33, pp. 223–231, March 1985.

[36]   A.G. Konheim and B. Meister, "Waiting Lines and Times in a System with Polling," *J. ACM,* vol. 21, pp. 470–490, July 1974.

[37]   G.B. Swartz, "Polling in a Loop System," *J. ACM,* vol. 27, pp. 42–59, 1980.

[38]   L.F.M. de Moreas, "Message Queueing Delays in Polling Systems with Applications to Data Communication Networks," Dept. of Systems Science, School of Engineering and Applied Science, University of California, Los Angeles, CA, UCLA-ENG-8106, May 1981.

[39]   I. Rubin and L.F.M. DeMoreas, "Message Delay Analysis for Polling and Token Multiple-Access Schemes for Local Communication Networks," *IEEE J. Select. Areas Commun.,* vol. SAC-1, pp. 935–947, Nov. 1983.

[40]   B. Avi-Itzhak, W.L. Maxwell and L.W. Miller, "Queueing with Alternating Priorities," *Oper. Res.,* vol. 13, pp. 306–318, 1965.

[41]   M.F. Neuts and M. Yadin, "The Transient Behavior of the Queue with Alternating Priorities with Special Reference to the Waiting Times," Mimeo Series No. 136, Dept. of Statistics, Purdue University, Jan. 1968.

[42]   L. Takacs, "Two Queues Attended by a Single Server," *Oper. Res.,* vol. 16, pp. 639–650, 1968.

[43]   R.B. Cooper and G. Murray, "Queues Served in Cyclic Order," *Bell Syst. Tech. J.,* vol. 48, pp. 675–689, 1969.

[44]   R.B. Cooper, "Queues Served in Cyclic Order: Waiting Times," *Bell Syst. Tech. J.,* vol. 49, pp. 399–413, 1970.

[45]   J.S. Sykes, "Simplified Analysis of an Alternating-Priority Queueing Model with Setup Times," *Oper. Res.,* vol. 18, pp. 1182–1192, 1970.

[46]   M. Eisenberg, "Two Queues with Changeover Times," *Oper. Res.,* vol. 19, pp. 386–401, 1971.

[47]   W. Bux and H.L. Truong, "Mean-Delay Approximation for Cyclic-Service Queueing Systems," *Performance Evaluation,* vol. 3, pp. 187–196, Aug. 1983.

[48] K.S. Watson, "Performance Evaluation of Cyclic Service Strategies—A Survey," in *Performance '84*, E. Gelenbe, Ed., Amsterdam: Elsevier Science Publ. B.V. (North Holland), 1984, pp. 521–533.

[49] O.J. Boxma and W.P. Groenendijk, "Pseudo-Conservation Laws in Cyclic-Service Systems," Report OS-R8606, Centre for Mathematics and Computer Science, Amsterdam, June 1986.

[50] D. Everitt, "Simple Approximations for Token Rings," *IEEE Trans. Commun.*, vol. COM-34, pp. 719–721, July 1986.

[51] M. Eisenberg, "Two Queues with Alternating Service," *SIAM J. Appl. Math.*, vol. 36, pp. 287–303, April 1979.

[52] J.W. Cohen and O.J. Boxma, "The M/G/1 Queue with Alternating Service Formulated as a Riemann-Hilbert Problem," in *Performance '81*, F.J. Kylstra, Ed., Amsterdam: North-Holland, 1981, pp. 181–199.

[53] J.W. Cohen and O.J. Boxma, "Boundary Value Problems in Queueing System Analysis," in *Mathematics Studies 79*, Amsterdam: North-Holland, 1983.

[54] S. Iisaku, N. Miki, N. Nagai and K. Hatori, "Two Queues with Alternating Service and Walking Time," *Trans. Inst. Electron. Commun. Eng. Jpn.*, vol. J64-B, pp. 342–343, April 1981.

[55] O.J. Boxma, "Two Symmetric Queues with Alternating Service and Switching Times," in *Performance '84*, E. Gelenbe, Ed., Amsterdam: Elsevier Science Publ. B.V. (North-Holland), 1984, pp. 409–431.

[56] M. Nomura and K. Tsukamoto, "Traffic Analysis on Polling Systems," *Trans. Inst. Electron. Commun. Eng. Jpn.*, vol. J61-B, pp. 600–607, July 1978.

[57] H. Takagi, "Mean Message Waiting Times in Symmetric Multi-Queue Systems with Cyclic Service," in: *Proc. IEEE Int. Conf. on Commun.*, (Chicago, MI, June 1985), pp. 1154–1157.

[58] O.J. Boxma and B. Meister, "Waiting-Time Approximations for Cyclic-Service Systems with Switch-Over Times," Proc. Performance '86 and ACM SIGME-TRICS 1986, *ACM Perform. Eval. Rev.*, vol. 14, pp. 254–262, May 1986.

[59] O. Hashida and K. Ohara, "Line Accommodation Capacity of a Communication Control Unit," *Rev. Elec. Commun. Lab.*, vol. 20, pp. 245–253, March–April 1981.

[60] P.J. Kuehn, "Multiqueue Systems with Nonexhaustive Cyclic Service," *Bell Syst. Tech. J.*, vol. 58, pp. 671–698, 1979.

[61] K. Kurosawa and S. Tsujii, "Analysis of Asymmetric Polling Systems with Limiting Service," in: *Proc. Nat. Telecommunications Conference 1981* (New Orleans, LA, Nov.–Dec. 1981), pp. G.4.2.1–G4.2.5.

[62] K. Arndt and H. Sulanke, "A Queueing System with Relative and Cyclic Priorities," *Elektronische Informationsverarbeitung und Kybernetik*, vol. 20, pp. 423–425, Sept. 1984.

[63] M.M. Srinivasan, "An approximation for mean waiting times in cyclic server systems with non-exhaustive service," to appear in *Performance Evaluation*, 1989.

[64] W. Bux, P. Janson, H.R. Müeller, and D.T.W. Sze, "Method of Transmitting Information between Stations Attached to a Unidirectional Transmission Ring," European Patent Application, Appl. No. 80107706.6, Date of Publ.: 23.06.82, Bulletin 82/25.

[65] W. Bux, "Local-Area Subnetworks: A Performance Comparison," *IEEE Trans. Commun.,* vol. COM-29, pp. 1465–1473, Oct. 1981.

[66] R.M. Grow, "A Timed Token Protocol for Local Area Networks," Electro/82, Token Access Protocols (17/3), May 1982.

[67] P. Zafiropulo, H.R. Müeller, and F. Closs, "Data/Voice Integration Based on the IEEE 802.5 Token-Ring LAN," in: *Proc. EFOC/LAN 86* (Amsterdam, The Netherlands, June 1986), Boston: (IGI Europe) Information Gatekeepers Inc., pp. 67–76.

[68] K.C. Sevcik and M.J. Johnson, "Cycle Time Properties of the FDDI Token Ring Protocol," *IEEE Trans. Software Eng.,* vol. SE-13, pp. 376–385, March 1987.

[69] M.J. Johnson, "Proof that Timing Requirements of the FDDI Token Ring Protocol are Satisfied," *IEEE Trans. Commun.,* vol. COM-35, pp. 620–625, June 1987.

[70] M.J. Johnson, "Reliability Mechanisms of the FDDI High Bandwidth Token Ring Protocol," *Computer Networks and ISDN Systems,* vol. 11, pp. 121–131, Feb. 1986.

[71] J.M. Ulm, "A Timed Token Ring Local Area Network and Its Performance Characteristics," in: *Proc. 7th Conf. on Local Computer Networks* (Minneapolis, MS, Feb. 1982), pp. 50–56.

[72] D. Dykeman and W. Bux, "An Investigation of the FDDI Media Access Control Protocol," in: *Proc. EFOC/LAN 87* (Basel, Switzerland, June, 1987), Boston: (IGI Europe) Information Gatekeepers, Inc., pp. 229–236.

[73] D. Dykeman and W. Bux, "Analysis and Tuning of the FDDI Media-Access Control Protocol," *IEEE J. Sel. Areas in Commun.,* vol. SAC-6, pp. 997–1010, July 1988.

[74] A. Goyal and D. Dias, "Performance of Priority Protocols on High-Speed Token Ring Networks," in: *Proc. 3rd Int. Conf. on Data Communication Systems and Their Performance,* (Rio de Janeiro, Brazil, June 1987), L.F.M. de Moreas, E. de Souza e Silva, and L.F.G. Soares, Eds., Amsterdam: Elsevier Science Publishers B.V. (North-Holland) IFIP 1988, pp. 25–34.

[75] M. Frontini and G. Watson, "An Investigation of Packetised Voice on the FDDI Token Ring," in: *Proc. 1988 Int. Zurich Sem. on Digital Commun.,* B. Plattner and P. Guenzburger, Eds., pp. 171–178.

[76] "ANSI/IEEE Standard 802.4-1984, Token-Passing Bus Access Method and Physical Layer Specifications."

[77] M.A. Colvin and A.C. Weaver, "Performance of Single Access Classes on the IEEE 802.4 Token Bus," *IEEE Trans. Commun.,* vol. COM-34, pp. 1253–1256, Dec. 1986.

[78] A. P. Jayasumana, "Performance Analysis of a Token Bus Priority Scheme," in *Proc. INFOCOM '87,* (San Francisco, CA, 1987), pp. 46–54.

[79]   J.W.M. Pang and F.A. Tobagi, "Throughput Analysis of a Timer-Controlled Token-Passing Protocol under Heavy Load," in *Proc. INFOCOM '88*, (New Orleans, LA, 1988) pp. 8B.1.1–8B.1.9.

[80]   D. Dykeman and W. Bux, "An Investigation of the FDDI Media Access Control Protocol," IBM Zurich Research Laboratory, Research Report, RZ 1591, May 1987.

[81]   "Draft Addendum to ANSI/IEEE Standard 802.5-1985, (Token Ring MAC & PHY Specification) Enhancement for Multi-Ring Networks," IEEE 802.5 Working Document 802.5D/D6 Nov. 1987.

[82]   W. Hawe, A. Kirby, and R. Stewart, "Transparent interconnection of local area networks with bridges," *J. Telecommun. Networks*, vol. 3, pp. 116–130, Summer 1984, reprinted in: *Advances in Local Area Networks*, K. Kuemmerle, J. Limb, and F. Tobagi, Eds., New York: IEEE Press, 1987, pp. 482–495.

[83]   IEEE Network Magazine, vol. 2, No. 1, January 1988.

[84]   D.J. Farber, and J.J. Vittal, "Extendibility considerations in the design of the distributed computer system (DCS)," in: *Proc. Nat. Telecommun. Conf.*, Nov. 1973.

[85]   B. Forss, E. Hafner, and M. Tschanz, "Distribution of logical processes in telephone systems," in: *Proc. Nat. Telecommun. Conf.*, Nov. 1977.

[86]   C.A. Sunshine, "Source routing in computer Networks," *ACM Comput. Commun. Rev.*, vol. 7, Jan. 1977.

[87]   J.H. Saltzer, D.P. Reed, and D.D. Clark, "Source routing for campus-wide internet transport," in: *Local Networks for Computer Communications*, A. West and P. Janson, Eds. Amsterdam, The Netherlands: North-Holland, 1981.

[88]   K-B. Sy, D.A. Pitt, and R.A. Donnan, "Source routing for local area networks," in *Proc. Global Telecommun. Conf.* (New Orleans, LA, 1985), pp. 1019–1023.

[89]   D.A. Pitt and J.L. Winkler, "Table-Free Bridging," *IEEE J. Sel. Areas in Commun.*, vol. SAC-5, pp. 1454–1462, Dec. 1987.

[90]   J.W. Wong, A.J. Vernon, and J.A. Field, "Evaluation of Path-Finding Algorithm for Interconnected Local Area Networks," *IEEE J. Sel. Areas in Commun.*, vol. SAC-5, pp. 1463–1472, Dec. 1987.

[91]   ANSI/IEEE Standard 802.2-1985, Logical Link Control.

[92]   W. Bux and D. Grillo, "Flow Control in Local-Area Networks of Interconnected Token Rings," *IEEE Trans. Commun.*, vol. COM-33, pp. 1058–1066, Oct. 1985.

[93]   J.F. Shoch and J.A. Hupp, "Measured Performance of an Ethernet Local Network," *Comm. ACM*, vol. 23, pp. 711–721, 1980.

[94]   M. Gerla and L. Kleinrock, "Flow Control: A Comparative Survey," *IEEE Trans. Commun.*, vol. COM-28, pp. 553–574, April 1980.

[95]   A. Giessler, A. Jaegemann, E. Maeser and J.O. Haenle, "Flow Control Based on Buffer Classes," *IEEE Trans. Commun.*, vol. COM-29, pp. 436–443, April 1981.

[96]   S.S. Lam and M. Reiser, "Congestion Control of Store-and-Forward Networks by Input Buffer Limits—An Analysis," *IEEE Trans. Commun.*, vol. COM-27, pp. 127–134, Jan. 1979.

[97]   M. Schwartz, "Telecommunications Networks: Protocols, Modeling and Analysis," Reading, MA: Addison-Wesley, 1987.

[98]   W. Bux, "Performance Issues," in *Lecture Notes in Computer Science 184: Local Area Networks: An Advanced Course,* G. Goos and J. Hartmanis, Eds., Berlin Heidelberg New York Tokyo: Springer-Verlag, 1985, pp. 108–161.

[99]   W. Bux, "Modeling Token Ring Networks—A Survey," in: *Proc. 4. GI/ITG-Fachtagung Messung, Modellierung und Bewertung von Rechensystemen* (Erlangen, Sept.–Oct. 1987), U. Herzog und M. Paterok, Eds., Berlin, Heidelberg, New York: Springer-Verlag, pp. 192–221.

Werner Bux (Member, IEEE) received the M.S. and Ph.D. degrees in electrical engineering from the University of Stuttgart, Stuttgart, West Germany, in 1974 and 1980, respectively.

From 1974 to 1979 he was with the Institute of Switching Techniques and Data Processing, University of Stuttgart, where he worked primarily in the field of performance analysis of data communication networks and computer systems. He joined the IBM Zurich Research Laboratory, Rüschlikon, Switzerland, in 1979 and spent 1984 on assignment at the IBM T.J. Watson Research Center, Yorktown Heights, NY where he served on the Technical Staff of the IBM Director of Research. He is currently Manager of the Communications and Computer Science Department at the IBM Zurich Research Laboratory. His work for the past several years has been in the area of architecture and performance evaluation of local and wide area networks. He has also been involved in local network standardization through his work in TC24 and TC32 of the European Computer Manufacturers Association (ECMA) and IEEE Project 802.

Dr. Bux is the Managing Editor of *Performance Evaluation* and was a guest editor of a recent issue on LAN interconnection of the *IEEE Journal on Selected Areas in Communications.* He was the co-recipient of the 1981 Stephen O. Rice Prize Paper Award in the field of communication theory of the IEE Communications Society and has received two IBM Outstanding Innovation Awards and a Corporate Award for his contributions to the IBM Token Ring network.

Reprinted with permission from IEEE, *Proceedings of the IEEE,* Vol. 77, No. 2, February 1989.

B

Investing in Token Ring: A Guide for Network Managers

(The following is a paper by John Morency originally published on the Alliance for Token Ring Advancement and Leadership (ASTRAL) Web site, www.ASTRAL.org, in 1995.)

Table of Contents

CHANGE MANAGEMENT CONSIDERATIONS
PRODUCT LIFECYCLE VALUE
SUMMARY & CONCLUSIONS

Executive Summary

Given the current levels of industry activity, network managers are continually dealing with an unprecedented volume of technology, product, vendor, and service options for the growth and expansion of their enterprise internetworks. On the one hand, new applications that take advantage of PC and LAN technology are being developed at a faster pace, and are delivering higher levels of productivity than ever before. The down side, however, is that decisions concerning which technologies, vendors, products and services to implement are often made suboptimally.

Many network managers are now dealing with this exact issue in regards to continued long term investment in Token Ring technology and products. Our strong recommendations to users performing this evaluation is that by using the principals of network cost of ownership, network complexity, cost of network downtime, performance and manageability considerations as well as overall lifecycle value, he or she will be able to direct a far more complete and thorough decision making process. Our belief is that, in many cases, superficial cost advantages for alternative LAN technologies may well be negated in light of increased support complexity which may lead to greater network downtime and reduced business effectiveness.

By use of the methodologies and principles contained in this white paper, the reader will be able to view the fundamental strengths and advantages of Token Ring technology in a whole new light. By use of these principles as well as Token Ring offerings for switching, ATM coexistence and full duplex services, we believe that the great majority of users will find a sound justification for continued investment over the next three to five years, if not well into the 21st century.

The Impact of Internetworking Complexity

Given the current levels of industry activity, network managers are continually dealing with an unprecedented volume of technology, product, vendor, and service options for the growth and expansion of their enterprise internetworks. On the one hand, new applications that take advantage of PC and LAN technology are being developed at a faster pace, and are delivering higher levels of productivity than ever before. The down side, however, is that decisions concerning which technologies, vendors, products and services to implement are often made suboptimally.

It is often be the case, for example, that claims will be made regarding a product's ability to provide the "least cost of ownership" simply based upon product price. Faced with a lack of information on what real cost of network ownership means, network managers often have no choice but to accept these claims on faith. The result can too often be rising support costs, unhappy end users and lost business opportunity due to a disappointing implementation. Many network managers are now dealing with this exact issue in regards to continued long term investment in Token Ring technology and products.

Research sponsored by ASTRAL shows that one of the most significant detriments to the long term health and well being of enterprise internetworks is unmanaged network complexity. Network complexity is characterized by the addition of new network protocols, capital equipment, software and network management infrastructure necessary for the effective implementation of applications that support both the existing operations and long term growth of the business. In today's highly competitive corporate environment, it is difficult if not impossible to find an internetwork that is not experiencing increased complexity in at least one of these dimensions. The growth of internetwork complexity juxtaposed against a backdrop of a fixed set of support resources can create a phenomenon known as *complexity inflation.*

Conceptually, complexity inflation is very similar to economic inflation. In the case of economic inflation, we have the case of too many dollars chasing too few goods. The result is diminished buying power for a fixed set of dollars. In the case of complexity inflation, the situation is too much network infrastructure being chased by too few network wizards. The result here is diminished support power or quality of network service that can be delivered by a fixed size support staff. A comparison of economic and network complexity is provided in Figure 1.

Network Complexity Management by Network Value Assessment

Rather than avoiding this problem in the hope that it will resolve itself, many leading edge users are taking a proactive approach towards limiting

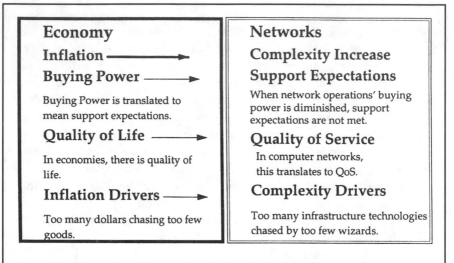

Figure B.1 Economic to Network Complexity Inflation Mapping.

complexity while still achieving the benefits of open networking through healthy inter-vendor competition. This approach often takes the form of a proactive strategy whose objective is to specify the key technologies, topologies, products, and services that will make up the network over the course of a particular planning period (typically eighteen months to three years). One key side effect is that technologies or vendors that do not make the short list will either have their support outsourced in some way or be completely phased out of the network altogether.

The experience of those users covered in our research shows that implementation of this type of proactive plan can help control long term catastrophic network complexity. The approach taken by these companies has been to prioritize the support provided for individual protocols, vendors, and products in a way that reflects the value of their contribution and the timeframe for the delivery of that value. By implementing a proactive planning process of this type, network managers at these companies have in some cases been able to effectively support network desktop and application growth rates ranging from 25% to 35% per year while still meeting stringent budget objectives.

Experience in the implementation, managing, and evolution of existing internetworks strongly suggests that two independent, yet related, analyses provide the most effective means of determining the real short and long term value of any internetwork investment. These analyses involve the assessment of *Network Cost of Ownership* and the determination of the impact of complexity upon network service quality. A Network Cost of Ownership analysis provides the means by which network managers can determine how to best optimize short and long term network investment. A network complexity analysis provides a means to project the support impact of particular implementation decisions.

Taken together, these analyses can provide an effective value assessment that guides short and long term network investment by defining the actual business benefit delivered by a particular technology, product, or service. In short, it provides the means of determining effective value propositions that will guide both short and long term decision making.

Token Ring Value Assessment Objectives

This white paper provides guidance for investment decisions concerning Token Ring products and services using cost of ownership and network complexity as key criteria. As a result of understanding these criteria and cost factors, network managers will be in a far better position to make the most optimal decision regarding longer term Token Ring investment. In addition, they will be able to gain better insight regarding how future functionality such as switching, Dedicated Token Ring (DTR) and ATM coexistence can best benefit the long term transition of their networks.

Cost of Network Ownership—The Framework

One of the key issues associated with effectively evolving an existing internetwork lies in determining exactly where the majority of its costs are incurred. A related question concerns how those costs can be more effectively managed in the future while still maintaining network quality of service. In general, network costs are incurred in one of three cost categories—the costs for the hardware and software capital equipment that make up the network, the salaries and fringe benefits paid to the technical staff responsible for designing and maintaining the network, and the recurring non-people costs associated with ongoing network operation including circuit costs, cabling costs, and hardware and software maintenance costs. A time perspective can be added by the addition of cost assessment across three network lifecycle phases:

	Startup	Operations	Change
Capital	Cost of Network Capital in Startup Phase	Cost of Network Capital in Operations Phase	Cost of Network Capital in Change Phase
Staff	Cost of Network Staff in Startup Phase	Cost of Network Staff in Operations Phase	Cost of Network Staff in Change Phase
Facilities	Cost of Network Facilities in Startup Phase	Cost of Network Facilities in Operations Phase	Cost of Network Facilities in Change Phase

Figure B.2 Network Cost of Ownership Assessment Matrix.

- the *Initial Startup* phase in which the network is designed, configured, and initialized;
- the *Operations* phase which reflects the state of the network on a day by day basis after initial startup and configuration; and
- the *Change* phase in which modifications are made to the network to reflect both business and end user requirements. These include the addition of new capital equipment; reconfiguring a hub or router to support new protocols or operating parameters; or a Move, Add, or Change operation performed to add a new desktop device or server to the network.

A structure for assessing network cost of ownership by cost category across the three individual lifecycle phases is provided in Figure 2.

LAN Cost of Network Ownership—Industry Results

Studies conducted by Strategic Networks Consulting and The Registry show that the great majority of cost allocation (over 80%) within a local area network environment is incurred for Capital Equipment and Support Staff in the Operations and Change lifecycle phases. The largest cost allocation generally occurs in the Capital Equipment Change category which represents approximately 40% of the total cost of network ownership reflecting the dominating role of physical infrastructure costs. The next highest allocation is for Support Staff in the Operations Phase with an allocation of approximately 25%. Overall results are shown in Figure 3.

The implication of this cost allocation is significant for both short and long term planning purposes. First, it means that internetwork changes must involve those technologies, products, and vendors that have either the best track record or strongest potential to most significantly optimize or add the greatest value to *where the greatest percentage of cost of ownership exists.* For a local area network environment, this means the existing network infrastructure as well as the technical support staff. The key challenge is to achieve the best balance in most optimally leveraging existing investments (which applies to both capital equipment and support staff) and selectively making new ones. A related corollary is that change management is a significant enough activity to warrant its consideration as part of a longer term network plan, particularly with respect to streamlining the staff costs incurred for performing Moves, Adds, and Changes.

Dominant Cost Allocations (over 80%)

	Startup	Operations	Change
Capital	1 %	2 %	4 0 %
Staff	2 %	2 5 %	1 4 %
Facilities	2 %	6 %	8 %

Figure B.3 LAN Cost of Ownership Results & Key Hot Spots.

Cost of Network Complexity—The Framework & The Methodology

Because capital costs constitute such a significant portion of total cost of network ownership in LAN environments, many network managers responsible for Token Ring environments are faced with a very significant issue. That issue involves the use of alternative LAN technologies in order to effectively grow the network while minimizing the growth in overall capital costs.

These changes can vary by scope. On the one hand, deployment can be limited to a small workgroup implementation of between ten and twenty nodes. At the other extreme, the scope can be substantially larger involving hundreds or perhaps thousands of end systems in a large building or campus environment.

A course of action commonly taken is to assume that the best choice for new LAN technology is the one that offers the least cost of capital equipment for implementation, primarily for both network interface cards (NICs) as well as corresponding network hub ports. However, given that LAN Cost of Ownership findings show that both capital equipment *and* support staff costs constitute the greatest critical mass of overall cost of ownership, a far more effective basis for implementation decision making involves fully assessing the implementation impact of a particular product and technology approach on *both* support costs *and* capital costs. By considering them together, a more complete and accurate assessment of both implementation cost and short and long term quality of service can more reliably be made.

In those cases where an alternative LAN technology choice is a viable possibility, a quantitative assessment of the impact of complexity inflation upon the support staff must be performed, particularly in those cases where minimal growth in the size of the support staff is forecasted for the foreseeable future. The next few sections of this white paper provide a step-by-step approach on how to calculate a complexity inflation index, the meaning of that index and how to apply it to the decision making process for new LAN deployment.

Determining Complexity Inflation

As previously defined, the metaphor of complexity inflation is too much network infrastructure being chased by too few network support staff.

The major impacts of network complexity inflation are the additional time requirements which are imposed in order to provide continued quality of service for infrastructure changes. These changes generally take the form of:

- new protocols such as TCP/IP, IPX, Appletalk, Data Link Switching, Fast Ethernet or ATM

- new equipment types such as stackable hubs, CSU/DSUs, or LAN switches

- new vendor relationships which may result from the acquisition of new equipment types

- new network management agents which manipulate Management Information Base (MIB) data to provide monitoring and control data for a given piece of network equipment

- additional network management host software such as HP Open-View, Sun Solstice Enterprise Manager, or Remedy Action Request System

- new base computing platforms such as SunSoft's Solaris, Novell's Netware, or Microsoft's Windows NT

- locally developed or shrink-wrapped client-server applications which require network-wide deployment

Each of these can be thought of as a network *complexity driver* which will require incremental support effort due to its addition, and be measured in any unit of support effort such as person-months. Person-months are often the best choice since they can be easily quantified in units of budget dollars so that both the human and cost impacts of making network changes can be more effectively understood.

Network complexity drivers can be quantified in two ways—the *initial effort* required for a support staff to come up to speed in providing high service quality for it and the *steady state effort* which is required to maintain ongoing support (such as installing new releases, learning new functionality, providing incremental documentation to the end user community and so on). Study data shows that typical values for initial effort can range from one to six person-months for any complexity driver. Similarly, steady state effort most often ranges from one to three person months reflecting the reduced effort required to maintain ongoing competency.

The actual effort that is required is determined by the support staff itself as a function of their skill sets, network size, available support tools and other related factors. Once these efforts are quantified, complexity inflation is then measured as a function of the additional support effort impact of both the initial and steady state efforts for each relevant complexity driver.

Research performed across a large number of multinational networks shows that the seven identified complexity drivers offer an effective means of characterizing any network change that is implemented. Some changes may involve an impact in all seven driver categories while others may affect only one or two. In cases where alternative LAN technologies are deployed, at least five drivers are generally affected. These drivers include new protocols, equipment types, vendors, management agents and management applications. The following example will show exactly how complexity inflation can be calculated for this type of deployment.

Complexity Impact—Token Ring to Ethernet—Industry Example

A large money center bank has made a decision to supplement their existing Token Ring investment with select deployment of shared and switched Ethernet equipment in select remote branches. This deployment will be both within small remote branches supporting less than 25 networked desktops as well as in relatively large branches which can support up to 150 networked desktops. In total, 350 networked desktops will initially be removed from Token rings and connected to either shared or switched Ethernet hubs. This transition will occur within five different branches with longer term plans to transition an additional 500 desktops over the coming year.

All remote branches are currently connected to the main corporate data center via branch routers and this will continue to be the case for the foreseeable future. A new vendor that offers excellent low end Ethernet shared and switched stackable products is being used for the new implementation. The modular hub vendor of choice will remain the same. The bank currently has 20 support staff—eight of whom are Help Desk Staff, eight of whom are technical backup support people, three of whom perform LAN administration people and one manager. The staff is responsible for the support of a total of 1,500 networked desktops within the bank. A block diagram of the network structure is shown in Figure 4.

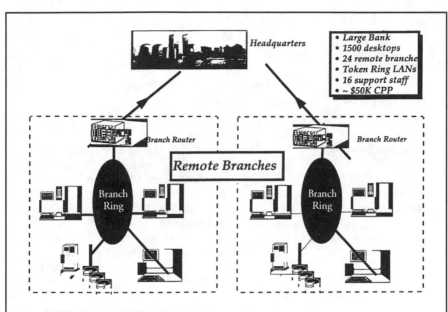

Figure B.4 Sample Network Block Diagram.

The proposed implementation affects five of the seven complexity drivers—protocols (Ethernet), equipment types (shared Ethernet hub modules, shared Ethernet stackables, switched Ethernet hub modules and switched Ethernet stackables), vendors (the new vendor for shared and switched stackables), management agents (the vendor-specific SNMP management data necessary to manage each of the four new product types) and the vendor-specific management applications necessary to manage the new modules and hubs atop an industry standard network management platform (such as HP OpenView, IBM SystemView or Sun's Enterprise Manager). The actual number of additions for each complexity driver as well as the supporting basis for each number is shown below in Table 1.

Now that the amount of each complexity driver change is known, the next step is to determine the amount of incremental person effort required to support the proposed change. This requires estimates of the staff ramp time or *initial effort* required to achieve product quality competency and support effort with respect to each individual driver. While this effort will vary and is best determined by each individual shop, the use of average values from studies conducted by Strategic Networks Consulting and The Registry Inc. provides an effective starting point. These values are shown below in Table 2.

Table 1
Complexity Driver Impact—Token Ring to Ethernet Transition

Complexity Driver	Additional Number	What
Protocols	1	Ethernet Protocol
Equipment Types	5	Switched & Shard Ethernet stackable hubs; shared and switched Ethernet hub modules; Ethernet NIC cards
Vendors	1	Stackable Vendor
Management Agents	5	New SNMP MIB data for each individual equipment type
Management Applications	3	Individual device manager for shared and switched hub modules; new management application for stackable products
Total	15	

Table 2
Complexity Driver Effort—Industry Study Averages

Complexity Driver	Initial Effort (Person-Months)	Steady State Effort (Person-Months)
Protocols	4.2	2.0
Equipment Types	2.3	0.8
Vendors	4	0.9
Management Agents	2.2	1.7
Management Applications	2.2	1.2
Network Applications	2.1	2.5
Computing Platforms	2.8	1.2

It is important to note that these person-effort averages reflect the *total* effort for the support staff as a whole. Therefore this effort will typically be spread across the entire support organization with some individuals incurring more effort than others due to their more in-depth support responsibilities.

Once the incremental number of each complexity driver and corresponding support effort is known, the total incremental staff effort necessary to support the proposed change is obtained by summing their products together. The results of our user example are provided in Table 3.

Now that we have the total incremental effort to support deployment of the new LAN technology (37.3 person months in Year 1 plus 19 person months per year starting in Year 2), we can then determine the rate of complexity inflation over the course of a calendar year by dividing that incremental effort by the total support effort that can be provided by the staff. In this example, delivery of backup support for the Ethernet and Token Ring-attached nodes will be provided by Help Desk (eight people) and technical backup staff (another eight people).

Using the results of industry studies which show that on average each person is performing support tasks only 80% of the time (allowing a 20% overhead factor for holidays, vacation, meetings, writing reports, vendor briefings, training and so on), the complexity inflation rate for our example network is

(37.3 additional person months)/(11*16*.8 total person-months)=
37.3 additional person months /140.8 total person-months=
26% complexity inflation rate for Year 1 and
(19 additional person months)/(11*16*.8 total person-months)=
19 additional person months /140.8 total person-months=
13% for Years 2 and beyond

On the surface, this figure seems quite high. Certainly, economic inflation rate of 26% and 13% per year would be considered disastrous in any developed nation.

Table 3
Total Incremental Effort from Increased Complexity

Complexity Driver	First Year Effort (Person-Months)	Subsequent Year Effort (Person-Months)
Protocols	4.2*1= 4.2	2*1= 2
Equipment Types	2.3*5 = 11.5	.8*5 = 4
Vendors	4*1 = 4	.9*1 = .9
Management Agents	2.2*5 = 11	1.7*5 = 8.5
Management Applications	2.2*3 = 6.6	1.2*3 = 3.6
Total	37.3 Person Months	19 Person Months

Assessing Complexity Inflation & the Real Quality of Service Impact

A perspective gained from this research is that the actual rate of complexity inflation, while important, is really secondary to the incremental amount of person time it represents. In our example, over three person years of time investment were necessary to support the full LAN transition within the span of a single calendar year. Given the fact that no staff growth was planned, it was clear that the network manager was placed in a quagmire in that he would either have to postpone the transition in order to maintain the existing quality of service or de-commit on planned projects and/or support responsibilities in order to enable the staff to perform the transition. Due to this and other reasons the planned transition was placed on indefinite hold.

A closer look at the actual person cost impact of the complexity inflation shows why this is the case. Given an approximate $50,000 cost per support person run rate (including salary and fringe benefits), the person cost of 37.3 person months is approximately $186,500. This cost impact needs to be assessed against the capital cost difference (on a cost per port basis) between continued Token Ring investment and the alternative technology. Using list prices obtained from a major internetworking vendor for Ethernet and Token Ring products, we have capital costs of

- $125 for a 16 bit ISA Ethernet Network Interface Card
- $73 per port for a 12 port unmanaged Ethernet stackable
- $299 for a 16 bit ISA 4/16 Token Ring Network Interface Card
- $169 per port for a 16 port unmanaged Token Ring stackable

This results in an average list price per port difference of **$270** between the Ethernet and Token Ring implementations. Note that street prices will probably be lower and future purchases can take advantage of further cost reductions.

In this example, we find that the **$186,500** of support costs represents approximately **80%** of the list price cost difference for capital equipment in Year 1 with the **break even point being reached midway through Year 2.** This is graphically shown in Figure 5. Given a three year cost of ownership lifecycle perspective coupled with the need to

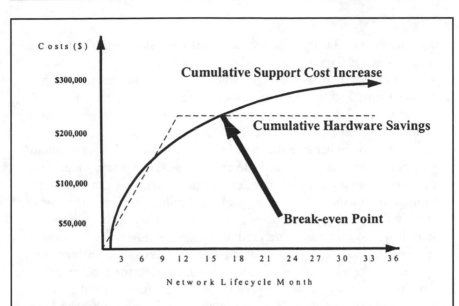

Figure B.5 Capital Equipment Cost—Support Cost Payback Period Graph.

deliver other critical projects, a much sounder business decision was reached than would otherwise have been the case.

Some further considerations are worth noting. Since the first year learning investment is a fixed time allocation, it is independent of the number of affected desktops. In other words, the time impact is the same independently of whether the transition affects eight or eight hundred desktops. The relative cost per desktop becomes more important in subsequent years when the ongoing support effort can be more effectively prorated across the impacted end users.

Network Downtime Impact

An additional perspective to consider when assessing the investment question concerns the price difference between Token Ring products and the alternative technology being considered. In the case of 10 megabit Ethernet, over two-thirds of the product pricing difference is consumed in the cost of silicon that supports the configuration and fault management services necessary for the effective self-operation and self-healing properties of the Token Ring protocol. Using representative product pricing information, this results in approximately a $270 price per port difference between Ethernet and Token Ring products when both NIC and hub port prices are considered. Quality of service shortfall costs shown in the previous section often make up some or all of this difference.

For those shops where this may not be the case, however, the impact of down time and lost productivity of professional staff may also merit consideration. For example, given a range of annual salary and fringe benefits for most professional staff ranging from $200 to $300 per day, a break even point between the cost of Ethernet and Token Ring technologies can be reached when as little as a single day's worth of network down time per professional staff desktop in a given year can be avoided through use of the self managing and self healing properties of most of today's Token Ring products. As the cost per person of the professional staff increases, a proportionately reduced degree of down time is required. If more than one day's worth of down time is avoided, an increasing cost avoidance advantage is afforded to Token Ring for each successive day, independently of the per person cost for professional staff.

Our background research has shown a consistently higher reliability and manageability rating for Token Ring products and services by those network managers who were jointly responsible for managing Token Ring, Ethernet, FDDI, and, in some cases, ATM networks. Surveyed network managers consistently cited performance, reliability, and manageability as three of the strongest assets for the Token Ring portions of their network. This is significant in light of the fact that nearly two thirds of the surveyed managers were also responsible for managing Ethernet subnets. Numerous anecdotal quotes cited the strong reliability and self-healing properties present in the current generation of products as one of the primary reasons why these users (responsible for managing well over 100,000 Token Ring desktops) planned to continue to invest in Token Ring technology.

Performance Considerations

If a network complexity analysis leads to the conclusion that a significant quality of service shortfall may occur as the result of a new product, technology or service implementation, it behooves the network manager to more seriously assess the advantages of continuing to invest in the particular product, technology or service that is already in place. In the case of Local Area Network technologies, another important item to consider in this analysis is the effective performance that will be provided to end users.

Interviews conducted with hundreds of internetworking users and all major internetworking vendors have led us to the following conclusion. We strongly believe that desktop application workloads for the vast majority of users over the next two years will be more than adequately sustained by LAN capacities which support packet bursting in the range of two to four megabits per second per end station.

These workloads not only include those of mainstream application utilities such as X-Windows, file transfer, and electronic mail but also more sophisticated multimedia usage that supports more cutting edge applications such as desktop video, document imaging, shared whiteboard, and World Wide Web access. For these workload classes, either Ethernet, Token Ring, or FDDI technology more than suffice purely on the basis of offered capacity alone. The key question for users to consider is which of these technologies can offer the best cost performance option for delivering the required capacity in a way that still provides future performance headroom.

Performance testing results and perspective from the Tolly Group offer useful insight into answering this exact question. Consistently, test results have shown that the peak offered throughput of an Ethernet segment lies in the range of approximately six megabits per segment with a mixture of packet sizes that approximate what most users are likely to see in actual production usage. By contrast, additional results from the Tolly Group have shown that effective per segment throughput for Token Ring technology is over 99% of the theoretical wire speed at both the four and sixteen megabit per second data rates.

In those cases where possible transition from Ethernet to Token Ring is being considered for either a portion or the great majority of the existing network infrastructure, it behooves the network manager to really assess what offered capacity will be required per user, the short and long term solutions that will provide that capacity while minimizing resultant network complexity and the growth paths available from those solutions to the eventual goal of end-to-end ATM internetworking. Given a representative burst capacity range of between two and four megabits per user, shared Token Ring segments of between eight and sixteen users per segment will generally be more than adequate while corresponding Ethernet implementation would likely require individual switched segments.

For many of Token Ring shops, it will generally be far more effective to continue to invest in Token Ring technology since, for the same number of supported users with the same application workload requirements, at least **60% more effective capacity** will be delivered per user with Token Ring technology. In addition, the implementation of both Token Ring Switching products and availability of Dedicated Token Ring (DTR) will provide the additional performance headroom that many users require for their evolving client-server applications with far less severe complexity inflation implications.

For example, use of many Token Ring Switching products would only require an increase of one for the equipment, management agent, and management application drivers. Using industry averages for person effort, this would result in an incremental person effort of 6.7 person months which is substantially less than the 37.3 person months calculated in our Ethernet transition example by a factor of nearly five times. A similar argument would apply for the implementation of Dedicated Token Ring, although the scope of implementation for the new functionality could be far more limited (thus more effectively limiting the effective quality of service shortfall) due to the coexistence features with existing NICs and hub ports that will be a fundamental architectural feature of DTR products as required by the emerging standard.

Change Management Considerations

In addition to performance reasons, interviewed managers also cited potential change management cost reductions through deployment of switching. In contrast to any of today's shared media technologies, most of today's switched LAN products also provide services for the establishment and maintenance of Virtual LAN services. Some Virtual LAN services can obviate the need to perform network layer address remapping when Moves, Adds, and Changes are performed. Although Virtual LANs are vendor-specific today, experience with their implementation and management may well provide some opportunity for addressing the 10–15% of total network ownership cost for change management while providing valuable experience for the implementation of ATM LAN Emulation which will become practically mandatory in any significant ATM implementation over the coming few years.

Other users interviewed opted for direct implementation of ATM for strategic reasons. In general, risk was bound by an initial implementation scope that was fairly small in size. None of the users interviewed chose to effect major transitions to any other protocols for the primary reason that the complexity, manageability, and quality of service risks would be too high in proportion to the questionable benefit provided by any of the mainstream LAN technologies.

Product Lifecycle Value

An additional point to consider when evaluating use of a particular product or technology over a given time period is the extent to which that product or technology can continue to support both functional and performance requirements over the target time period as a base requirement and even to exceed that time period for those technologies that truly deliver long term value. Perspective worth considering regarding Token Ring is that the technology has fundamentally met the performance, reliability, and manageability requirements of its users for the past ten years without requiring major changes or extensions in order to do so.

The software, interface, and cabling infrastructure that supported the technology at its initial inception still remain compatible and functional today. Given the increasing availability of Token Ring switching products coupled with the expected availability of Dedicated (or Full Duplex) Token Ring (DTR) in the latter half of 1995, it is likely that the technology will continue to effectively address user performance and reliability requirements while minimizing infrastructure disruption for the same four year performance lifecycle period. The net result for existing users of Token Ring is that they have achieved an extremely effective thirteen year return on their initial technology investment.

Summary & Conclusions

Given current networking industry activity, users are currently faced with an unprecedented volume of technology, product, vendor, and service options for the growth and expansion of their enterprise internetworks. This phenomenon coupled with greater autonomy for departments and business units within an enterprise has resulted in many networks being particularly susceptible to the onset of complexity inflation, a form of high-tech cancer that, left unchecked, can seriously undercut the quality of network service and significantly hamper the effective realization of new business opportunity.

Given that open networking is likely to remain the operational norm for some time, network managers responsible for the growth and development of their company's enterprise internetwork need to become increasingly proactive in regards to determining the key network technologies, products, vendors, and services they will and will not support in the future. This involves decisions as to what new technologies they will embrace as well as what existing technologies they will choose to carry forward.

In order to deliver real business value and not be unnecessarily subjected to unwanted complexity inflation, network managers are well advised to assess the cost of ownership for their own network environments; develop an understanding of where the major cost allocations are incurred; and prioritize technology, product, and vendor support as a function of the real business value delivered in return for the price that is paid. Only by doing so will effective critical support mass be attained and an effective balance struck between increasing user expectations and what can realistically be delivered by the support operation.

Within this white paper, we have focused upon the key cost categories of local area network cost of ownership—capital equipment and support staff—and how network managers should plan both short and longer term deployment of Token Ring technology based upon key properties of the technology, key features of its products, and the actual working experience of its nearly 20 million users worldwide. It is clear that for those areas that users value the most—performance, reliability and manageability—Token Ring technology and products have a very compelling story to tell today.

Our strong recommendations to users evaluating continued Token Ring investment is use the principals of network cost of ownership, network complexity, cost of network downtime, performance advantages, manageability advantages and lifecycle value to guide a complete and thorough decision making process. Our belief is that, in many cases, superficial cost advantages for alternative LAN technologies will be negated in light of support complexity impact and the cost of network downtime. In addition, this will cause some of the fundamental strengths and advantages of Token Ring technology to be perceived in a whole new light. Coupled with current and future Token Ring offerings for switching, ATM coexistence and full duplex services, we believe that the great majority of users will find a sound justification for continued investment over the next three to five years, if not well into the 21st century.

Reprinted by permission of ALLIANCE FOR STRATEGIC TOKEN RING ADVANCEMENT AND LEADERSHIP,
Prepared By: John Morency of The Registry, Inc., October 1995

C

Migration Issues and Strategies for Token Ring

(The following is an article by Bengt Beyer-Ebbesen, Mark Cowtan, Sharam Hakimi, and Robert D. Love originally published in the International Journal of Network Management, *Spring 1997.)*

C.1 Growth of a Token Ring LAN

Token Ring LANs have become an indispensable part of corporate infra-structures. In the mid 1980's 4 megabit per second Token Ring LANs were first being deployed on a trial basis with typical LAN utilizations running at a few percent. As the benefits of networking became better understood and better appreciated, LANs grew in number of attaching stations, in use, and in topological complexity, spanning large campuses. Soon 4 Mbit/s bandwidth on the LANs was not enough, and migration to 16 Mbit/s Token Ring began. Because of the predictable guaranteed access provided by Token Ring networks, collision problems which limited the useful bandwidth of Ethernet were eliminated, and Token Ring's full 16 Mbit/s capacity was exploited. Networks continued to grow along with a dependency on them. Given the new higher bandwidth available on the LANs, applications have been developed which exploit this new capability. As a result, LANs have again become clogged with traffic. The need for increased LAN capacity has expanded and will continue to expand. How will this need be met?

Since wholesale replacement of a LAN is financially unacceptable, the only solution is to migrate to higher performance gradually. In doing so, it pays to acknowledge the strengths and weaknesses of existing solutions. Let's look at each of these areas: Token Ring's primary strengths derive from its deterministic access to the ring, its high reliability, and its superb network management capabilities. Its primary weakness is common to all shared LANs, that is the requirement that the LAN bandwidth be shared among all attached stations. On average, only a fraction of Token Ring's usable information carrying capacity is available to any one station. A solution is needed which will preserve Token Ring's strengths while eliminating this bandwidth limitation.

It is important to recognize that projected bandwidth requirements for stations that will have full-motion video plus voice capability, while simultaneously sending and receiving data is well contained within Token Ring's 16 Mbit/s bandwidth. For example, full-motion video using MPEG2 compression only requires about 1.5 Mbit/s dedicated bandwidth, which would leave over 14 Mbit/s available for voice and data if this bandwidth were fully dedicated to the single attaching station.

The channel from the end station to its concentrator or hub port is more than adequate for most projected applications but the channel is not where the problem lies. Many users are seeing congestion on the LAN backbone. As networks grow, and more stations are added to individual rings, there will soon be congestion on the local rings, and unacceptable delays going through bridges and, especially, through routers. So if 16 Mbit/s is enough for most foreseeable applications, there must be a way forward that will beef up the backbone infrastructure to allow delivery of voice and video data to the desktop.

With these considerations in mind, the IEEE 802.5 Token Ring Working Group has developed a new standard, Dedicated Token Ring (DTR), to enable migration to Switched Token Ring while preserving users' significant investment in Classic Token Ring (today's shared Token Ring as defined by the 1995 IEEE 802.5 Standard). Full Duplex Switched Token Ring provides a high performance capability using Token Ring framing, but is not fundamentally compatible with Classic Token Ring, today's solution. DTR encompasses both Switched Token Ring and Classic Token Ring, providing both the required compatibility and the full range of intermediate solutions enabling the network performance to be gradually and economically increased as required. Dedicated Token Ring defines a DTR Concentrator which is a Token Ring switch to provide both switching capability needed for higher performance, and backward compatibility to enable smooth migration.

The following sections describe DTR requirements for both the workgroup and backbone switching environments. They also provide scenarios and a migration strategy for evolving networks using DTR. The strategy includes potential introduction of ATM into the network backbone, and important considerations in making this technology step.

C.2 Matching Technology with Network Needs

To satisfy differing application requirements, many vendors have recognized the need to offer a choice of switching solutions, ranging from low-cost, fixed-configuration, desktop, and segment switches through to scaleable, fully featured backbone switches.

The key switch features required in the workgroup, and backbone switching environments are outlined next, highlighting features that may be required as the switched internetwork grows and evolves.

C.2.1 Workgroup Switching Environment

Workgroups with local servers, AS/400 systems, network printers, or other shared resources have evolved as a LAN topology designed to provide high performance by eliminating the need to send most traffic over bridges and a possibly congested backbone. However, network demands continue to grow, so that today even these configurations suffer from bandwidth starvation at times. In many Workgroups, the biggest limitation is server performance. Other servers and peer-to-peer applications consuming some of the available bandwidth limit the server's access to the ring. Likewise, in Workgroups that rely on backbone resources for e-mail, corporate databases, and 3270 access, only part of the 16 Mbit/s bandwidth is available to the local server. The result is client/server performance that fluctuates according to network usage.

Workgroup switching can alleviate these problems. The workgroup ring is segmented into multiple shared segments, each supporting fewer users, with servers and selected power users each getting a dedicated switch port. The minimum features that should be considered when choosing a switch for segmenting a congested workgroup are discussed.

Low Latency: When migrating from shared-LAN Workgroups to a switching environment, with servers placed on dedicated ports, the switch delay is introduced into the network. This delay makes local switching performance a top selection criterion. When server utilization is low, and few users are sharing the server's 16 Mbit/s pipe, most of the transmission delay is the inherent latency of the switch. It is in this environment that cut through switching provides its biggest advantage, keeping the latency between local devices to a minimum.

Low cost: Cost is always an important consideration for Workgroups. Workgroup switches were priced at $600–$1500 per port in mid-1996. Prices are often based on a fully populated switch and don't include high-speed uplink costs. If the uplink is ATM, there may be added costs for the LAN Emulation required to integrate LANs with ATM devices.

High-speed server option: For dedicated Workgroups, full-duplex 16 Mbit/s connectivity will provide adequate bandwidth for all but the most demanding applications today. For those few applications that may require still more bandwidth, multiple attachments from a server to the switch can be used.

High-speed backbone uplink: Users in segmented Workgroups often need access to resources on the backbone. A high-speed uplink should provide a data path that can match the peak demands of the LAN ports on the switch.

Traffic management: When servers are moved to dedicated ports on a workgroup switch, they are no longer local to their clients. For this network configuration, clients need to generate additional broadcast and source-route explorer traffic to locate the servers. Broadcast and explorer traffic consumes some of the bandwidth on all segments. In the case of power users on a dedicated port, the impact of unwelcome broadcast traffic may be trivial given the additional bandwidth now available. However, on shared segments the bandwidth is still relatively scarce and any significant increase in broadcast traffic can affect performance. MAC address filtering should be adequate for preserving bandwidth on selected segments. Filtering beyond the MAC address usually involves buffering the frame which adds latency. Virtual LANs (VLANs) may also help, especially if they do not introduce additional delay.

Remote Management: Most workgroup switches will be attached to backbone switches, and a few will be stand-alone, acting as the LAN backbone. Some, however, will be located in branch offices at the end of an SNA network, and therefore can't be managed easily over the network. For these switches, remote management may be essential.

Future Growth: Many of today's Token Ring workgroup switching requirements are fairly straightforward and easily accommodated. However, the switch needed for tomorrow's micro-segmented workgroup may require the robustness and full features found in the high-end switches employed in backbone applications. Workgroup switch selection focuses on price/performance, whereas backbone switch selection focuses on reliability, salability, and future-proofing. Capacity planning for the longer term requirements of a switched internetwork is one of the best weapons in selecting the switches that will grow to match evolving needs.

C.2.2 Backbone Switching Environment

The greatest need for Token Ring switching may well be in the backbone. Switches can replace congested and unscaleable bridge-laden 16 Mbit/s backbones, and relieve overused, collapsed backbone routers. Many shared 16 Mbit/s backbone rings are oversubscribed today, causing frame delivery delays. Bridges and routers connecting workgroup rings to the backbone add additional latency to these networks, especially with high traffic volume. Because all devices attached to the backbone must share its available 16 Mbit/s bandwidth, each device is forced to operate at a small fraction of its potential capacity.

To overcome this congestion, some network managers have already dispensed with Token Ring as a backbone technology, opting for FDDI in the backbone with FDDI/Token Ring translational bridging or routing to workgroup rings, while others have settled for routers in a collapsed backbone architecture. Although these approaches alleviate some of the bandwidth problem, they increase latency and add complexity.

Using a Token Ring switch as a collapsed backbone, and replacing bridges or front-end routers, reduces end-to-end latency and boosts available bandwidth for clients and shared resources alike. When choosing a switch for this application, it pays to understand the nature of the traffic that will go through the switch. Unlike workgroup traffic which is predominantly many-to-one, backbone traffic tends to be any-to-any. The backbone supports shared resources such as servers IBM cluster controllers and front end processors, all of which may be accessed simultaneously from any of the attached workgroup rings. Since attached Workgroups support many users, there is usually a high bi-directional load on every switch port.

Discussed below are features that should be considered when choosing a switch for the backbone.

Aggregate throughput: To accommodate the backbone's traffic conditions, backbone switches should offer high aggregate capacity and be able to handle constant loads on all ports as well as bursty client/server traffic. The backplane capacity should be adequate to cope with the potential demands that could result from all ports running in full-duplex mode. Some switch designs have a non-blocking architecture. For these switches, congestion on one output port does not result in loss or delay of other incoming frames. Blocking can occur in some switches if incoming frames destined for other output ports are held waiting until a frame intended for a congested port is cleared from the input buffers.

Reduced latency: Switching can significantly reduce a network's latency by providing dedicated bandwidth, a high-speed backplane interconnect, and efficient forwarding of frames. Some switches offer cut-through forwarding, some offer store and forward, and some offer both (adaptive cut-through). Cut-through switches offer the largest advantage where there is no high-speed backbone, and where a high percentage of the messages forwarded are single frame. For Token Ring switches, cut through is not plagued by forwarding of invalid frames, since the Token Ring protocol results in very few of these. The latencies associated with transmitting a message using store and forwarding and cut-through switches are similar when (a) the message must traverse a high-speed backbone so forwarding requires data rate conversion, (b) when most forwarded messages are multiple frames and (c) when frames are sent from a 16 Mbit/s ring and destined for a 4 Mbit/s ring.

Congestion control: Servers on dedicated ports can potentially deliver a full 16 Mbit/s to their users, but since there is no longer a shared medium governing access to the server, simultaneous access from multiple clients can result in spikes of demand that exceed 16 Mbit/s. Similarly, on shared segments where many clients are communicating with multiple servers across the switch, response by more than one server at the same time can result in excess load on the workgroup port. Whenever there is many-to-one traffic, a variety of conditions can give rise to port congestion. Ample buffer memory and a buffer management scheme are required to deal with bursty traffic efficiently without losing frames. A switch with a high port density requires more buffers than a low-density switch, given the higher probability of simultaneous, many-to-one accesses.

Priority queuing, allowing high-priority packets to be forwarded in preference to low-priority packets, may become useful in the future to ensure that time-sensitive applications such as multimedia are delivered within their transmission delay and skew requirements even when a switch port becomes congested during traffic bursts. Token Ring's built-in priority levels are available for applications which require this capability. In addition, the Token Ring IEEE 802.5 standard contains recommended guidelines for the use of multimedia traffic classes based on latency requirements.

Bridging methods: Not all Token Ring sites use source-route bridging. In routed networks, mixed Token Ring and Ethernet environments and those with IP or Novell IPX traffic, many users employ source-route transparent (SRT) or transparent bridging. Support for all three bridging methods eases implementation and facilitates network evolution. Support for IBM or IEEE 802.1d Spanning Tree to allow redundant paths maintains compatibility with other network components.

Dedicated connections: To support servers cost-effectively on a dedicated port, the switch must allow a device to be connected directly without needing a hub. DTR provides this first step toward high-speed servers, providing dedicated Token Ring bandwidth to servers with Classic Token Ring adapters, and providing full, duplex bandwidth to servers with DTR adapters. (Note that many of today's modern Classic Token Ring adapters are software-upgradeable to DTR.) The gain in effective server bandwidth for dedicated 16 Mbit/s attachment as well as for full duplex attachment, results in significant improvements in client/server throughput.

High-speed server option: When a 16 Mbit/s dedicated connection becomes saturated, converting it to full duplex provides a significant increase in capacity. Some switches also allow for multiple attachments. For example, two, three or even four full-duplex Token Ring cards can be used to directly attach a server to a switch. If the network is configured with a high-speed backbone, the servers could be directly attached to that backbone.

High-speed backbone links: As the number of switches grows it becomes necessary to provide for high-speed interconnection. Some networks can achieve the required performance level when configured in a collapsed backbone configuration, using one of the DTR Concentrators as a "master" switch. Its Data Transfer Unit (DTU) provides the high-speed backbone for the LAN. Some hardware will allow attachment of the switches via multiple full-duplex Token Ring ports, thereby providing an interconnection of two, three, or even four times the data transfer capacity of a single 16 Mbit/s Full Duplex connection. Some switches provide an FDDI uplink for the high-speed connection. However, many networks will require still higher speed interconnect, or a distributed backbone. For these networks, ATM provides the most scaleable solution with the potential to grow with increasing networking demands.

Network virtualization: These days, more and more users who work together are not necessarily collocated. Some estimates suggest that as many as 20% of desktops are relocated every year. This high worker mobility has fueled a growing interest in network virtualization. By assigning groups of ports or MAC addresses into a logical broadcast domain, traffic can be contained within a logical ring or group of bridged rings that can span many switches, allowing desktops to be moved easily with little regard for physical location.

Although many vendors offer VLANs that span multiple switches, for all devices in a multi-vendor switched internetwork to participate in common VLANs a standardized VLAN solution is needed. IEEE 802 is currently working on developing a standard for Virtual LANs (IEEE 802.1q). This standard will provide VLAN operations on all 802 MACs including Token Ring, Ethernet, and FDDI. However, ATM already provides a standardized Virtual LAN capability for ATM adapters that support LAN Emulation. This capability will allow native ATM devices and LAN devices to belong to the same VLAN (or ELAN, to be precise, see side bar).

Traffic management: The initial replacement of an existing backbone with a Token Ring switch normally entails little change in the flow of traffic from workgroup rings onto the backbone. However, as power users, servers and other backbone devices are moved from shared rings to dedicated ports, there is a potential increase in protocol broadcast and source route explorer traffic, from devices trying to find each other on the network. Network managers must explicitly address this new source of backbone traffic, employing broadcast control, filtering, enhanced route discovery and segregating the network with VLANs or using routers as fire-walls to minimize the effect of these new frames. With the implementation of advanced applications including intranet, multicast video, video conferencing and other time-sensitive traffic flows, it will become increasingly critical to manage the traffic, eliminating unnecessary frames and minimizing broadcast traffic.

Filters can be an invaluable traffic management tool whose effectiveness is dependent on the switch architecture. Therefore, it is important to understand the answers and impact to questions such as: Can you filter both transmit and receive? How many bytes deep into the frame, and how many bytes wide is the filter mask? Is filter processing centralized or distributed? and Does the switch provide automatic broadcast control features, thereby reducing administration and support overhead?

One important application for filtering is protocol troubleshooting and diagnosis. With shared LANs we employed LAN analyzers which capitalized on the shared LANs' principle weakness, the fact that bandwidth must be shared by all stations, because all stations receive every frame. That weakness and resultant capability does not exist with switched LANs, but we still have need for LAN analyzers in our networks. This requirement is addressed in a number of ways. One way is to design a switch to provide mirroring or redirection to an analyzer port. Another is building in RMON support to allow full traffic analysis and packet decode on the fly. The capabilities and flexibility of the different vendor offerings should be carefully examined.

Solution cost: In backbone switching, the cost of each port can be spread across many users, so cost per port is often less important than it is for desktop equipment. For example, when using switches to micro-segment existing Token Ring LANs, each port may support a dozen or more stations. For these LANs the switch may provide the network's high-speed backbone. For larger LANs, with many switches linked to a high-speed backbone, there will be additional costs of both the implementation of the high-speed backbone and the costs of connecting to it. High-speed backbone link prices vary enormously. In estimating switch cost/port supported, one must include the switch ports that will be used to uplink to the backbone.

Upgradability: When both the number of users and the applications required to run on the LAN are static, equipment purchase decisions can be made strictly on the ability to handle the present work load. However, most application environments are far more dynamic with increasing numbers of users and evolving applications. For these environments, purchase decisions must include an evaluation of the types of applications planned, the level of network management that will be required, traffic volumes and latency requirements.

An example of the type of concern to be addressed is planning for multicast video. Most video server applications today rely on multicast IP. Unfortunately, few LAN protocol stacks are able to set a frame's destination MAC address to a functional address. This deficiency results in multicast IP traffic appearing as general broadcasts with switches designed to copy such frames to every port. If there are no devices on a switched segment that want to receive this traffic, the bandwidth it consumes is effectively wasted. How will this unwanted traffic be filtered? At what layer? Does the switch have the processing capacity for important functionality such as Layer 3 switching to be added successfully?

Switch Management: Switch management is another area that needs to be examined in terms of both present requirements and future plans. What is the extent of your LAN? If the switch is the collapsed backbone, a network management scheme that stops at the switch may be adequate. If the switch is one of many, effective network management must address the whole virtual network. There may be requirements to see both the logical and physical views of the switched network and do path tracing or fault diagnosis end-to-end.

Summary: In meeting future application needs, a simple collapsed backbone implementation can evolve to become a multi-switch inter-network. When this happens, two issues that are less critical today will become paramount. These are: how to provide more backbone bandwidth, and how to avoid wasting what you've got. When evaluating backbone switches, performance and port cost are not the principle issues, scaleability and future-proofing are. Server consolidation, network virtualization and new applications will place ever greater demands on the backbone bandwidth. A close look at your traffic flow today and in the future will help in assessing your backbone scaleability requirements.

C.3 Migration of Token Ring Networks

Driven by the ever growing demand for network bandwidth, network managers are increasingly finding their shared Token Ring LANs becoming saturated by today's demanding users. At last there is a selection of Token Ring switches on the market, available to solve present and projected bandwidth shortages. The networking industry with three years of experience in switched technologies employed in the older, slower LAN technologies has recognized that segment switching alone will address only short term bandwidth requirements. An array of new applications, in particular multi-media and intranets, will drive demand for more bandwidth at a faster pace than ever before.

C.3.1 Migration Steps: (1) Backbone Switching

Consider the migration path from a small Classic Token Ring network consisting of a number of concentrators interconnected in a single wiring closet on a single ring as shown in the Figure C.1. For this configuration, the average available bandwidth per station is the ring data rate (4 or 16 Mbit/s) divided by the number of stations on a ring. For a typical large ring of 100 stations operating at 16 Mbit/s, with 10 users simultaneously accessing the network, the average available bandwidth per station is 400 kbit/s. This bandwidth is more than sufficient for most of today's legacy applications. Over a short interval such as the time it takes for a large file transfer, the average bandwidth available to a station is 16 Mbit/s divided by the average number of stations demanding bandwidth during that interval. New bandwidth intensive applications and high-speed server capabilities are now starting to outstrip this capacity.

Using Classic Token Ring, the network can be subdivided, typically, as shown in Figure C.2, into smaller more manageable groups, interconnecting these groups via bridges to a backbone ring. The new configuration significantly increases network bandwidth capacity, especially if there is substantial traffic that is limited to each of the local rings. The potential drawbacks to this configuration are based on the latency associated with the bridges, which will delay the packet transport across the network compared to a single ring configuration, and the creation of a new potential bottleneck, the backbone ring.

Using dedicated Token Ring, the DTR Concentrator (the Token Ring switch defined by the IEEE 802.5 DTR Standard) is used to

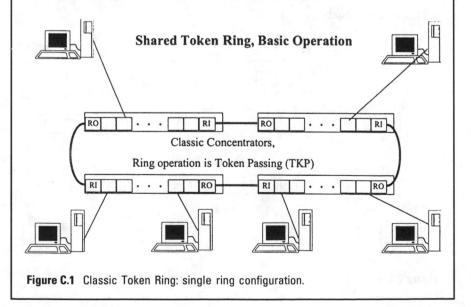

Figure C.1 Classic Token Ring: single ring configuration.

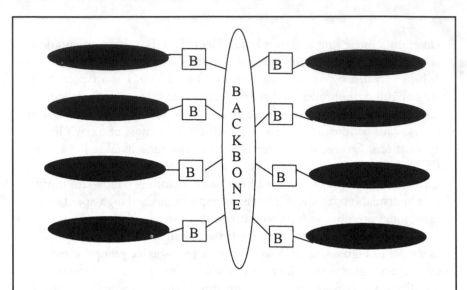

Figure C.2 Classic Token Ring: bridged backbone configuration.

interconnect micro-segmented LANs as shown in Figure C.3. Used this way, all existing shared Token Ring stations and concentrators are retained. The DTR concentrator serves as a high-speed collapsed backbone for data transfer between the micro-segmented LANs. This configuration provides much greater capacity than the single ring, and delivers data with lower latency and at lower cost than by bridging the micro-segmented LANs onto a backbone.

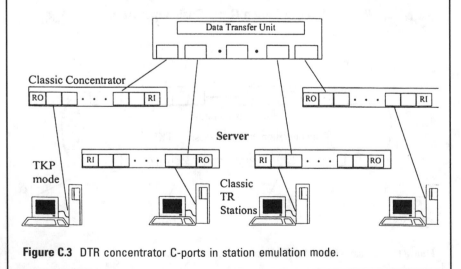

Figure C.3 DTR concentrator C-ports in station emulation mode.

Therefore, beginning the migration process with a bridge-laden backbone, the first step will be to replace the backbone and the bridges with a DTR concentrator. With a large network, it is likely that key resources will need to be directly attached to the switch ports. Switch selection may include consideration of port density, because as soon as port capacity for the first switch is exceeded, multiple, interconnected switches will be required.

If you've come from the IBM mainframe world, you may have witnessed the introduction of a new CPU where the smart mainframe administrator does not release all of the power of the new processor on day 1. Instead he turns the tap on gradually over a 3–5 year period. Using this approach he provides a consistent response over the life of the machine, rather than have users see a giant leap in performance and then a gradual deterioration over time. If cost is a concern, a similar approach might provide a more gradual migration to a switched Token Ring environment. The investment in backbone switch ports is staggered by clustering a few servers together on the same switch port, and then moving them to a dedicated port as demand increases. Even 2 or 4 servers on one port will give users better performance than when all the servers are competing for bandwidth on the backbone ring.

Dedicated 16 Mbit/s for servers and key resources makes a big difference to the users of these devices. But eventually even 16 Mbit/s may not be enough to satisfy some users. At that time, full-duplex Token Ring can be used to provide an additional boost in bandwidth. Many switches and adapters today provide full duplex capability. You can expect most vendors that do not already offer this capability to have software upgrades for it by early 1997.

C.3.2   Migration Steps: (2) Workgroup Segmentation

As the network grows, and especially as the balance between client/server and peer-to-peer traffic changes, it may become necessary to split up heavily populated rings into multiple segments each carrying fewer users. The main drivers for workgroup switching are power-users with local servers. When segmenting a workgroup ring with local servers, a work-group switch cascaded from the backbone switch is usually the most cost-effective approach, allowing the ring to be split into many segments and minimizing the ports consumed on the backbone switch. However, if the workgroup does not have local servers it may not need much segmentation.

Thus, depending on the extent of peer-to-peer traffic and workgroup server deployment in the network, some of this demand is best met with additional backbone ports and some with workgroup switches.

C.3.3   Migration Steps: (3) Building a Multi-Switch Backbone

One obvious way to build a multi-switch backbone is to use full-duplex Token Ring as the inter-switch connection. After all, this is a Token Ring environment, and Token Ring in the backbone eliminates any requirement for frame reformatting. One key advantage to this network architecture for small networks is that it is possible to build an entirely cut-through switching architecture from backbone to desktop. Virtual LANs are used with this configuration as a means of segregating network traffic. While this approach has its attraction in very low end-to-end latency, it ultimately has scaleability limitations for large networks. Therefore, many large system network managers who have taken this approach, regard it as a stop-gap measure while they plan their backbone and high-speed network migration. These managers will eventually move the switches first deployed in the backbone out to workgroups as they build a high-speed backbone infrastructure to replace them. With this in mind, it is important to look closely at the scaleability of the inter-switch connectivity options. The inter-switch links should provide adequate capacity to switch the maximum aggregate inputs of all the LAN ports on the switch.

Although at first glance, there appear to be several high-speed LAN technology choices, on closer inspection it becomes clear that there are vast differences between their long term suitability. In fact, although each has unique merits, only ATM offers the scaleability and low latency to satisfy the ever-increasing bandwidth demands and stringent quality-of-service (QoS) requirements of tomorrow's multimedia applications.

One of the main attractions of ATM is its scaleability. With ATM it is possible to connect switches to one another over multiple links. In addition, with a choice of link speeds, if a link between two switches becomes saturated, it is easily scaled to a higher connection speed. These capabilities provide levels of scaleability, load sharing and redundancy that cannot be matched by any other technology. Building an inter-switch backbone around ATM kills two birds with one stone: First, it creates the backbone infrastructure to support the expected demands of Token Ring desktop switching and second, it provides the core building block from which high-speed servers and high-speed networking to the desktop can evolve. Furthermore, ATM provides the only standardized method for creating VLANs.

Some LAN/ATM backbone switches require that an intermediate ATM switch is used for inter-connecting switches, whereas others allow a simpler back-to-back or meshed configuration. If you choose to do without an ATM switch at the network center, be careful to ensure that the LAN/ATM switches can support the ATM port density that may be needed for high-speed servers and cascaded ATM workgroup switches as well as the high-speed switch inter-connections.

ATM can be used in the backbone in a number of ways, with varying degrees of complexity. ATM Forum LAN Emulation (LANE), allows servers to be migrated from the local LANs to an ATM backbone for greater network access speeds, and to prepare the network for further migration to ATM. However, implementing LANE is a significant commitment so some network managers are implementing solutions that allow for a more incremental migration path. Solutions offered by some vendors have done away with the requirement to run LANE for networks that only use ATM for interconnecting switches. Others require only ATM signaling support.

C.3.4 Migration Steps: (4) High-Speed Server Migration Using ATM

In a multiple backbone switch configuration, one of the issues is where to put servers. If there is a delay when traffic crosses the inter-switch connection, and all the servers are on one switch, then depending on where users are located, they will get a different level of performance. Therein lies the attraction of cut-through switching end-to-end. The ideal configuration would be an inter-switch connection method that exhibits zero latency. Then it wouldn't matter which backbone switch the servers were on. Unlike frame-based networks, which can add variable store-and-forward delays at each hop, ATM ensures low and consistent latency, since the forwarded traffic is always in 53 byte cells. But when switching Token Ring frames there is also a frame-to-cell conversion at each end of the ATM link. On a well designed switch the additional latency this introduces can be kept to a minimum, resulting in very little additional delay for clients crossing an ATM inter-switch link to reach a Token Ring server.

For servers that require more bandwidth there is merit in migrating them to the same high-speed technology used for inter-connecting the switches. This strategy minimizes latency by reducing the frame formatting operations involved in getting data over the backbone to workgroup segments, while providing a bigger pipe into the shared resource. Offering a range of link speeds from 25 Mbit/s through to 622 Mbit/s, ATM is an ideal platform for high-speed server migration.

C.3.5 Migration Steps: (5) Network Virtualization

One thing that has become clear in large switched environments is the need for virtualization. More and more network managers are taking advantage of the capabilities of Virtual LANs to be able to link geographically dispersed groups who need to share the same resources. Network virtualization allows these physically dispersed groups to be treated as though they are on one shared media LAN segment (See VLAN sidebar).

    With continued micro-segmenting of workgroups and migration of servers and power-users to dedicated and high-speed ports, the need for network virtualization becomes ever more pressing. Most vendors allow virtual LANs to be configured on their Token Ring switches. But in an integrated LAN / ATM environment, a common VLAN implementation is required to enable both LAN and ATM devices to participate in the same Virtual LAN. ATM LAN Emulation provides the solution by allowing ATM devices and LAN devices to be members of the same Emulated LAN (ELAN). In addition, ATM devices can participate in multiple ELANs concurrently. ATM LANE is a standardized virtual LAN implementation that will enable VLANs to span a multi-vendor switched internetwork.

C.3.6 Migration Steps: (6) High-Speed Desktop Migration

With falling switched port costs, one can envision a high percentage of LAN users eventually being upgraded to a dedicated switched port. For most imaging and even video applications dedicated and full duplex Token Ring connections will provide ample performance for average desktop users. The key to success for desktop switching will be ensuring there is enough backbone bandwidth to allow these desktops to reach shared resources without encountering bottlenecks. This assurance will require that a high-speed backbone and high-speed server connections are in place first. If still higher speeds to the desktop are required, then making the last step, ATM to the desktop, becomes a small manageable change.

C.3.7 Summary

Switched collapsed backbones, desktop switching and LAN segmentation are widely recognized as the ways to maintain high performance at the desktop while maximizing available bandwidth. Beyond LAN switching, a scaleable high-speed backbone is needed that will enable migration to the next generation of high-speed networking. For the Token Ring user, ATM is the high-speed backbone technology of choice.

A typical Token Ring to ATM migration strategy comprises several phases:

1. Collapse an existing backbone into a backbone switch and provide dedicated ports for key resources.
2. Microsegment heavily used workgroup rings, using workgroup switches where applicable.
3. Interconnect multiple backbone LAN switches, via ATM if needed.
4. Migrate key shared resources to high-speed ATM connections.
5. Implement network-wide VLANs to improve management and ease further microsegmentation.
6. Selectively upgrade power users to direct attached full duplex connections.

End-to-end network performance will be one of the most telling indicators of a successful implementation. To this end, high-speed backbone scaleability is a key network design criterion. Besides ensuring adequate bandwidth, minimizing end-to-end latency is also crucial.

C.4 The Bottom Line Benefits of Migrating to Dedicated Token Ring

As described previously, the migration from Classic Shared Token Ring to Dedicated Token Ring:

- Provides the additional bandwidth required;
- Preserves the use of existing end station network interface cards;
- Continues to use existing classic concentrators.

All this is done while minimizing administration costs and retraining.

By continuing to use classic adapters in the end stations, and to support classic concentrators, attaching them to the Dedicated Token Ring Concentrator Switches, DTR allows for a network migration which minimizes system upheaval. The desktop user never has to be aware of any network change, aside from the better performance he receives. Because he is never directly involved with the migration, and because the small steps preserve operation of existing applications and protocols, no productivity is lost.

For large networks we have seen that implementing an ATM backbone may well be an appropriate step in the migration process. As we examine the implications of adding ATM to our network, we see that there are some equipment and training costs involved. However, to achieve the higher performance required for these large networks, there is a strategic need to make this transition. Choosing DTR as the migration path, and ATM backbone as its first implementation provides benefits here. The initial ATM introduction is limited to a small number of connections, providing the LAN administrators time to learn and deal with the new technology in a bounded environment. Typically, with any major technology change there is more retraining required than can be funded at the outset of the transition. Planning on the gradual introduction of ATM in the network, allows for a fundable training program over time that keeps pace with the requirement for the new expertise. Best of all, during the transition period, DTR provides the needed network performance enhancements.

In conclusion, the bulk of your existing Token Ring LAN investment can be carried forward into the world of switched internetworks. How long your network remains suitable for addressing mission-critical applications, rests largely upon how carefully you select the LAN/ATM switching components, especially with regard to capacity and scaleability. Before you start evaluating components or prototyping solutions, give serious attention to capacity planning. If you can answer the question: "How much data may need to be simultaneously piped from where, to where?" and apply the considerations described above, you'll be well on the way to a successful migration.

C.5 Sidebar: IEEE802.5 Dedicated (Switched) Token Ring

Dedicated Token Ring, as defined by the emerging IEEE 802.5 standard, provides the framework for Token Ring Switching, and for the interoperation of Switching Concentrators and Full Duplex Adapters with Classic, shared Token Ring concentrators and adapters.

C.5.1 What is (Switched) Dedicated Token Ring?

Classic Token Ring is a shared LAN architecture where all data is received by each station attached to the ring. Consequently, the ring bandwidth must be shared by all the attaching stations. Dedicated Token Ring (DTR) begins with a different concept. Using the same physical topology as shared Token Ring, and the same Token Ring physical layer to preserve operational capability over the same cabling, DTR stations are connected to DTR concentrators as shown in Figure C.4. Each DTR station attached to the C-Port of the DTR concentrator forms with it a two node ring. Within this ring each node takes advantage of the fact that there are no other stations by dropping the shared Token Ring requirement to repeat received frames and tokens, and by transmitting at will, rather than only when a token is received. This new transmission protocol is called Transmit Immediate (TXI). Conceptually, the station signals are "switched" onto the DTR Concentrators Data Transfer Unit (DTU) or back plane, hence the name Switched Token Ring. In practice, the DTR Concentrator is like a switch or multi-port bridge with the station data transferred between ports via the DTU's high-speed internal bus.

There is a substantial performance advantage in using DTR as shown here. Each station has access to a dedicated full duplex 16 Mbit/s of bandwidth, rather than only a few percent of that value in

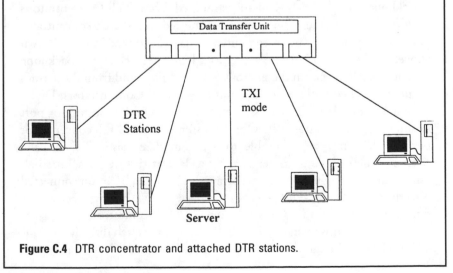

Figure C.4 DTR concentrator and attached DTR stations.

a typical shared Token Ring environment. However, the issues of wholesale migration from shared to dedicated Token Ring are significant. Each concentrator or hub port must be replaced by a DTR Port, and each station adapter must be capable of operating in TXI mode, or must be replaced. Often a far more modest network performance increase would suffice. To address these important issues, DTR has built in capability to operate with Classic (shared) Token Ring adapters and concentrators.

The full flexibility of DTR, as shown in Figure C.5, includes the following devices and interconnections:

- Direct attachment of Dedicated Token Ring (DTR) Stations to DTR Concentrators;
- Direct attachment of Classic Token Ring Stations to DTR Concentrators;
- Attachment of DTR Concentrator C-Ports directly to Classic Concentrator ports;
- Attachment of DTR Concentrator C-Ports to each other;
- Accommodation of a high-speed interconnect from the DTR Concentrator.

The DTR Standard defines an optional Station Emulation Mode for the C-Ports. Used this way, all existing Shared Token Ring Stations and Concentrators are retained. Instead of interconnecting the concentrators in a single large ring, or in separate rings connected via bridges to a backbone ring, the LAN is micro-segmented. Each DTR Concentrator's C-Port in station emulation mode is connected to a shared concentrator port in each of the micro-segmented rings. In connection (1) shown above, the DTR concentrator serves as a high-speed collapsed backbone for data transfer between the micro-segmented rings and from those rings to direct attached stations, servers, and other DTR Concentrators. Direct attachment of Classic Token Ring stations to a DTR concentrator port is shown by connection (2). One requirement on DTR concentrator ports is that they must be able to accommodate classic Token Ring stations. These stations are serviced by a dedicated 4 or 16 Mbit/s half duplex channel to the DTR Concentrator's DTU. Each station connected directly to a DTR C-Port selects the speed it will run at exactly as it does when attached to a Classic Token Ring concentrator.

DTR Stations operating in full duplex mode attach directly to DTR C-Ports (3). Although the normal case is to have each DTR Station run

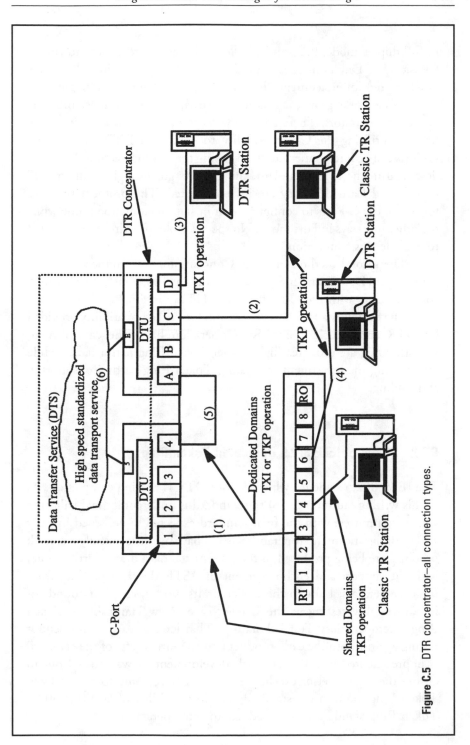

Figure C.5 DTR concentrator–all connection types.

in full duplex mode, the standard allows the station to choose to run in Classic Half Duplex mode. DTR Stations also operate in half duplex (TKP mode) for attachment directly to a Classic Concentrator port (4).

As networks grows, it becomes necessary to interconnect multiple DTR Concentrators. There are two ways to make this connection. The first, by utilizing the C-Port's ability to emulate a DTR station, two C-Ports can be interconnected (5). By using the "Autodetection Protocol" described in the DTR Standard, the C-Ports go through a "discovery" process and learn that they are interconnected. The protocol provides for one of the C-Ports to configure itself in station emulation mode while the other one stays in Port mode. No special cables or switches are needed to provide this connection.

The second method of DTR Concentrator interconnection is via a high-speed Data Transfer Service port, such as 155 Mbit/s ATM, which communicates with the C-Ports via the DTU's internal backplane (6).

The flexibility and full backward and forward compatibility provided by DTR and Classic Token Ring operation bring the capability to smoothly migrate to increasingly higher Token Ring performance while maintaining the investment in existing Token Ring hardware, software, and cabling.

C.5.2   Status of IEEE 802.5 Dedicated Token Ring

The IEEE 802.5 standard which defines DTR is nearing completion. As of this writing (autumn 1996) the standard is undergoing final balloting within committee, with the final standard expected to be issued in early 1997. All significant interoperability issues have been successfully resolved; products are being produced to the present standard draft. Furthermore, with the support and encouragement of ASTRAL, early product from multiple vendors, built to the earlier drafts, underwent four rounds of interoperability testing at the University of New Hampshire's Token Ring Interoperability Test Laboratory. That level of early mutli-vendor testing against both the early product and the early drafts of the standard is unprecedented in LAN standards development. It was carried out to ensure the high level of excellence that you have come to expect from Token Ring, both in the standard, and in the ability of all IEEE 802.5 Token Ring standard based products to interoperate.

C.6 Sidebar: Asynchronous Transfer Mode

ATM is a worldwide recognized standard supported by many vendors for transmission over both copper and fiber media. It operates at standardized data rates from 25 to 155 Mbit/s with near future support for 622 Mbit/s cell transfer rates. ATM is intended to support a wide variety of concurrent services and applications that include data, voice, and video with negotiable quality of service (QoS) and performance. In addition to the negotiable QoS and performance, ATM has been architected as

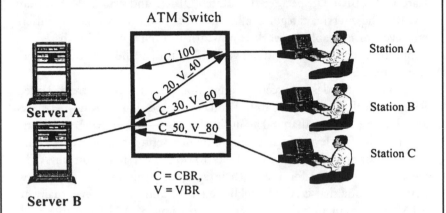

Figure C.6 Providing quality of service with ATM switching.

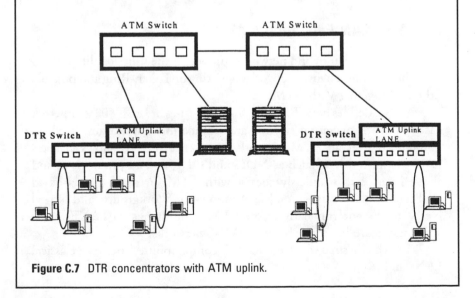

Figure C.7 DTR concentrators with ATM uplink.

a scaleable transport layer to allow you to increase capacity as demand grows. These fundamental capabilities will enable ATM to meet the demands of even the largest campus and enterprise networks for years to come.

The constant cell size of 53 bytes allows for silicon based switching, management, and real time control thereby enabling network monitoring and management. To achieve and provide different levels of performance the ATM standard specifies five service categories: Constant Bit Rate (CBR), real-time Variable Bit Rate (rt-VBR), non-real-time Variable Bit Rate (nrt-VBR), Unspecified Bit Rate (UBR), and Available Bit Rate (ABR). These service categories allow network managers to allocate different levels of available network bandwidth to each application based on application requirements and network utilization at the time the connection is made.

ATM provides an excellent migration path for the current Token Ring networks of today. Many of today's shared Token Ring networks are running at less than 80% utilization and Token Ring switching provides the first level of bandwidth enhancement. These switches could then be connected via ATM uplinks to an ATM switch on an ATM backbone. Providing for a smooth integration from an all Token Ring network to one that is ATM backbone based, and possibly eventually all ATM, will result in hybrid networks consisting of ATM together with Token Ring. These networks will be the norm in the coming years.

C.7 ATM LAN Emulation (LANE)

LANE is the glue in hybrid networks, providing the functionality needed for implementing a hybrid network consisting of Token-Ring and possibly Ethernet, together with ATM.

LANE defines how IEEE 802.5 Token-Ring or IEEE 802.3 Ethernet based network operating systems and compatible applications can run unmodified over an ATM network. In essence, LANE allows classical LAN adapters, drivers such as NDIS and ODI, and all protocols at and above Layer 2 to seamlessly coexist with ATM. Presently, LAN-based applications like Lotus Notes, MS Windows for Workgroups, and Novell NetWare do not run natively on ATM. But thanks to LANE, these applications are interoperable with ATM networks.

LANE is desired to solve several "incompatibilities" between classical LANs and ATM:

- Mapping of connection-less service onto connection-oriented ATM;
- Implementation of broadcast;
- Mapping of addresses.

To achieve these goals, LANE functionality consists of the following components:

- LAN Emulation Clients (LECs);
- LAN Emulation Server (LES);
- Broadcast and Unknown Server (BUS);
- LAN Emulation Configuration Servers (LECS).

LAN Emulation Clients: LECs must be configured in the different parts of the network where LAN to ATM conversion needs to be carried out. Examples are (1) in the LAN switches, and (2) in all servers or end stations directly attached to ATM. LECs may represent multiple stations (MACs) on a LAN switch, but a LEC may only be associated with one ELAN.

LAN Emulation Server: LES and BUS are essential for emulating the connectionless service and address mapping mechanisms that are used on Token-Ring and Ethernet. A minimum of one LAN Emulation Server must be present in each emulated LAN to provide address resolution, and the discovery and announcement of appearing and disappearing stations on the emulated LANs.

Broadcast and Unknown Server: Broadcasting is a fundamental principle used in device discovery and service announcement. LAN Emulation must support this functionality that is inherent in classical LANs. The BUS forwards broadcasts to all LECs in the same ELAN. The BUS is also a part of the address resolution scheme for unknown LAN stations.

The LES and the BUS can be configured upon a dedicated station, be integrated in a switch, or provided as software for a PC or UNIX workstation. Implemented correctly LANE causes no performance degradation in the network although special hardware is needed in situation with heavy data traffic, such as LAN switch uplinks.

Token Ring and Ethernet can coexist on the same physical ATM network, but there is no bridging between the two LAN technologies.

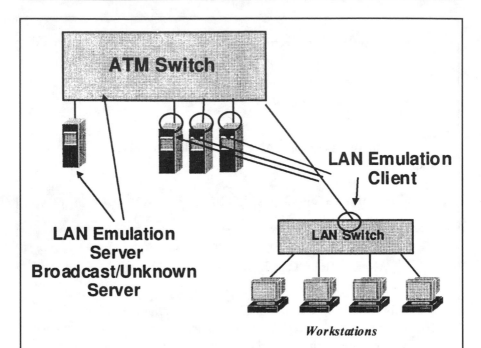

Figure C.8 LAN emulation services deployment (Example 1).

Signaling between the Token Ring and Ethernet segments must be done with routers.

C.7.1   LAN Emulation Client Servers (LECS)

The LECS is used by LANE Clients to obtain information about ELANs before actually joining an ELAN. A configuration request sent by a LEC to the LECS can ask for the following characteristics: ELAN Name, ELAN Type and Maximum Frame Size. The LECS will then try to find the ELAN that best matches the requested configuration based on configuration parameters.

C.7.2   Applications for LAN Emulation

LAN Emulation Services works in three distinct connectivity environments:

- All-ATM;
- Classical-LAN-to-ATM;
- Classical-LAN-to-classical-LAN.

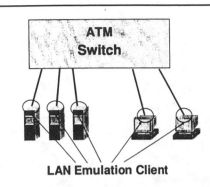

LAN Emulation Client

Figure C.9 LAN emulation services deployment (Example 2).

All-ATM connectivity: Both a workstation and a file server, for example, are physically located on the ATM network (with ATM adapters, etc.). Using LANE, the workstation and file server can run, for example, Novell NetWare and take advantage of the bandwidth of ATM. Note that you only need LANE because conventional NOS are still architected around shared media topologies. Eventually, it will be possible to dispense with LANE and talk directly to AAL5. In fact, some multimedia applications do this today.

Classical-LAN-to-ATM connectivity: A workstation physically located on a Token-Ring can access a file server physically located on an ATM network. The workstation on the classical LAN contains a classical adapter, NDIS or ODI driver, etc., and does not need any LANE software.

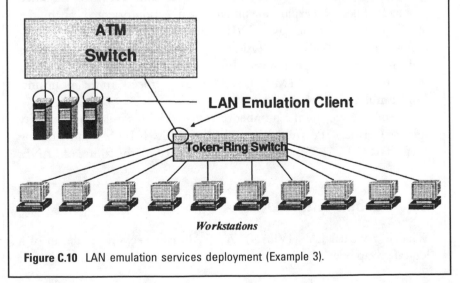

Figure C.10 LAN emulation services deployment (Example 3).

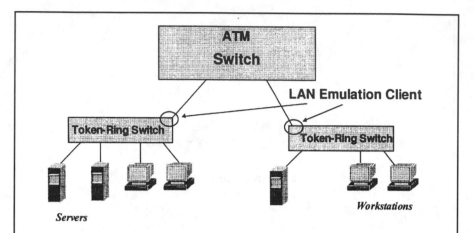

Figure C.11 LAN emulation services deployment (Example 4).

Classical-LAN-to-classical-LAN connectivity via ATM: Token-Ring workstations on two different Token-Rings can communicate using ATM as the backbone,. Using LANE, ATM-capable LAN switches enable connectivity between the Token-Ring and ATM networks.

C.7.3 The Future for LANE

A number of different trends are influencing the future of LANE:

It is expected that within a few years, current network operating systems will define native ATM driver interfaces and will include modified protocols. These will exploit the inherent connection-oriented capabilities of ATM. Protocols such as TCP/IP which were originally designed for point-to-point links, are the easiest to move to ATM, since ATM uses end-to-end, point-to-point links. However, modifying all existing applications to run on native ATM APIs will be a long process. In the meantime, LAN Emulation is essential for migration.

Also the success of a number of concepts with similar functionality such as Classical IP, Multi Protocol Over ATM (MPOA) and Routing over ATM (RFC1483) will have some influence on the future of LANE.

C.8 Sidebar: Virtual LANs

What is a virtual LAN (VLAN)? A VLAN is a network made up of a logical group of end stations that are physically attached to different ports of a switch or to ports on different switches. Using network management software, end stations in such a logical group can be managed with increased flexibility.

C.8.1 Example of a VLAN

Traditional LAN protocols are based on broadcasts of services. As LAN usage grows in a corporation, this broadcast traffic will occupy an increasing amount of available bandwidth ultimately resulting in bandwidth shortages. Virtual LANs may be introduced to overcome the potential problems that occur when switched LANs grow into large, enterprise wide networks. They help address the issues of broadcast flooding, security, and moves, adds and changes.

Broadcast flooding. Network protocols designed for classical LANs are based on broadcasting of services. These broadcasts can use a large portion of the bandwidth in large networks. Virtual LANs are used to filter out much of the unneeded broadcast traffic, and to restrict the domain of messages intended for the Virtual LANs.

Security: Connecting all users on one LAN may raise issues about security and potential access to company-confidential information. Virtual LANs can limit the distribution of messages, and therefore, the security exposure. However, a comprehensive LAN security plan is still required.

Moves, adds, and changes: When users are moved within the organization, some LAN switches will automatically discover the new position of the station and reconfigure the VLAN layout, accordingly. Therefore, the user will be a part of the VLAN regardless of physical location.

Virtual LANs are configured to allow direct communication between stations dispersed throughout the network. A station normally communicates with other stations belonging to the same VLAN. However, stations such as servers can be configured to belong to multiple VLANs. Some implementations allow the network administrator to configure VLANs based on the station's address, in which case the network will automatically recognize if the station is moved to another port.

The simplest implementations of a Virtual LANs are for topologies where all the ports in the same logical domain are attached to switches joined in a single logical domain. For these networks, all of the attached stations will be a part of the configured VLAN. The implementation of VLANs at Layer 2 is based upon the MAC address of the attached stations and will use the standard spanning tree algorithm or a ring design for their switched topologies. Due to the limitation of scaling to large networks, switch vendors have made proprietary enhancements for the implementation of VLANs.

Specific to Token-Ring networks, VLAN can also be based upon Source Route information whereas a logical network with one ring number can span multiple ports and switches.

Layer 3 Virtual LANs: More sophisticated implementations of VLANs at Layer 3 are based on LAN switches inspecting the subnet fields of IP and other network protocols. A VLAN is configured by assigning ports to the subnet of the stations attached. This assignment provides functionality related to routing, but will require heavy processing power in the switch, normally using the management CPU in associated switches. Benefits of using this implementation are the possibilities of creating multiple VLAN for each network protocol and for one station to be part of different VLANs depending upon the protocol used. This implementation will however also create an extensive burden for the network administrator to configure and maintain the multidimensional VLAN.

C.8.2   What Do Virtual LANs Cost?

VLANs add direct and indirect costs to a switched network.

- Increased overhead in data frames is introduced by "Routing information." All packets sent between two stations in a VLAN

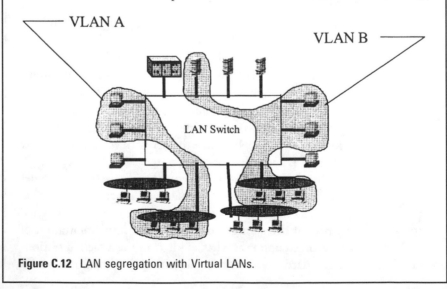

Figure C.12 LAN segregation with Virtual LANs.

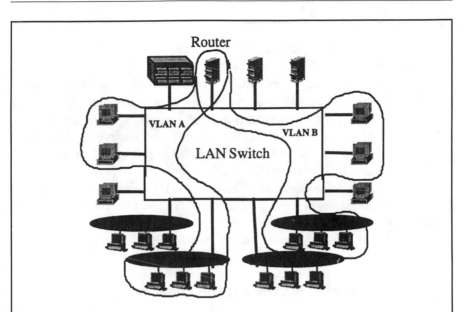

Figure C.13 Router in a virtual network.

must be extended with an additional "Identification TAG" to indicate the VLAN of which the stations are a part of. Depending upon the implementation this may add 4 to 20 bytes to the original LAN packets.

- There is an increase in management traffic used for signaling between switches to be keep them updated with information of VLAN changes.

- The more sophisticated the implementation of a VLAN, the more processing of LAN frames are required resulting in a lower performance of the switches.

- Increased sophistication in configuration and network management will become necessary to provide the network managers with the capability to easily add and users from different VLANs.

C.8.3 Routers and VLAN

Port-based and Layer 2 VLANs can be regarded as pure bridged networks. They do not use any information on the Network layer, and can not provide any service based upon router information. Even though some implementations of Layer 3 VLAN use the information from the network layer, a true router function is required in order to provide VLAN-to-VLAN connectivity.

In an environment where VLANs are configured based upon network addresses, e.g. IP addresses, the VLAN is normally limited to one subnet. The standard implementation of IP requires a router function (Gateway) to be identified prior to connection to stations located on another subnet. This router function must be implemented within the switches or as an external device.

The prerequisite for interoperability of LAN switches and Routers in a VLAN environment is compatibility between the different implementations of VLAN. Although a number of strategies have been announced, only a limited number of vendors can show any real interoperability.

C.8.4 Interoperability of VLAN and LAN Emulation

Because LAN Switches regard LAN Emulation as just another LAN connection, VLANs can be extended to switches connected via LAN Emulation and an ATM backbone. This configuration can be considered as a multi level VLAN where the first level, the VLAN within the ATM Network, is the LAN configured via LAN Emulation and the second level, is the logical network configured via the VLAN functionality of the LAN switches.

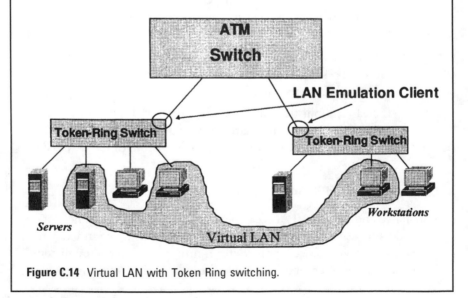

Figure C.14 Virtual LAN with Token Ring switching.

C.9 International Standardization of VLANs: A Note of Caution

VLANs promise to make physical and logical moves adds and changes easier and to reduce unnecessary broadcast traffic. These functions are very appealing to network administrators because of the recent increases in workgroup dynamics and information broadcasting. However, in the absence of standards for VLAN transmission, almost all of today's VLAN solutions are proprietary. The IEEE 802 is in the process of developing a VLAN standard for Token Ring, Ethernet, and FDDI. Although some of the fundamental aspects of VLANs are being nailed down, the IEEE 802.1 working group is still wrestling with many of the details. Only ATM LANE offers a standardized VLAN solution today. Therefore, if you are implementing a VLAN solution for your network in the near future, you may have to depend on a single vendor, non-standard solution. For long range plans, look to multi-vendor interoperability evaluations of the ATM and emerging IEEE 802 VLAN standards to see if they will deliver on their promise.

C.10 About ASTRAL–Alliance for Token Ring Advancement and Leadership

ASTRAL is an alliance of leading Token Ring technology providers dedicated to supporting Token Ring users as they prepare for the future of high-demand networking. ASTRAL provides vendor-independent education and information about Token Ring and new technology developments that will be important to the future of Token Ring customers. To receive more information about ASTRAL, visit our web site at www.astral.com or call us at 1 (415) 328-5555.

C.11 About the Authors

Bengt Beyer-Ebessen is Product Manager for LAN Switch Products at Olicom. He has a Ph.D from the technical University of Copenhagen and has more than 10 years experience in the networking industry.

Mark Cowtan is a Product Marketing Manager for the Enterprise Business Unit of Bay Networks, Inc.. Mark has over 10 years networking experience in sales support, product management and product marketing with several networking vendors. Before joining Bay Networks, Mark was a Product Manager with 3Com's Network Adapter Division, following 5 years in various sales and marketing positions at Madge Networks. Mark has also been involved in network implementation and support for several corporate users in traditional IBM SNA environments. Mark Cowtan holds a Bachelor of Science degree in Psychology from Hull University in the United Kingdom.

Sharam Hakimi is director of Engineering at UB Networks.

Robert D. Love, a Senior Engineer at IBM's Network Hardware Division in Research Triangle Park, North Carolina, is a physical layer expert on Token Ring, involved in Token Ring development and standardization since 1981. He currently serves as Chairman of IEEE 802.5. Bob published numerous articles, presented papers and taught seminars on Token Ring and LAN Cabling, both nationally and internationally. He received his Bachelor's degree in Electrical Engineering from Columbia University and his Masters in Electro-Physics from the Brooklyn Polytechnic Institute. His E-Mail address is: rdlove@us.ibm.com.

D

Vector and Subvector Definitions

This appendix contains details for the vectors and subvectors used by the TKP Access protocol defined in Chapter 4. The MAC frame format, specified in Section 4.3.4.2, is repeated in Figure D.1.

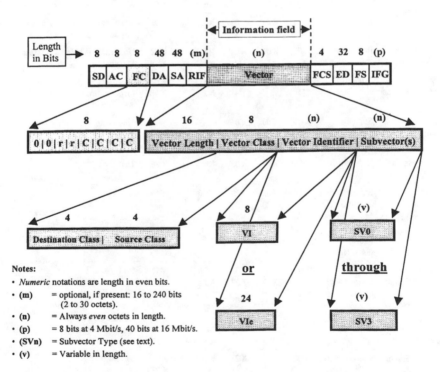

Notes:
- *Numeric* notations are length in even bits.
- **(m)** = optional, if present: 16 to 240 bits (2 to 30 octets).
- **(n)** = Always *even* octets in length.
- **(p)** = 8 bits at 4 Mbit/s, 40 bits at 16 Mbit/s.
- **(SVn)** = Subvector Type (see text).
- **(v)** = Variable in length.

Figure D.1 MAC frame format.

D.1 Vector Definitions

Table D.1 defines the vectors that the station is required to transmit. See Section D.2 for a definition of the subvector names.

Table D.1
MAC Frame Transmit Definitions

Vector Name	Destination Address	FC, Pm	VC value	VI	Optional?	SVI	Subvectors Subvector(s) Name **9
Response	Source address of frame received	X'00', **2	X'*0'	X'00'	No No	X'09' X'20'	Correlator **5 Response Code **7
Re-IMPL	**11	**11	**11	X'01'	**11	**11	**11
Beacon	Broadcast	X'02', **3	X'00'	X'02'	No No No	X'01' X'02' X'0B'	Beacon Type Upstream Neighbor's Address Physical Drop Number
Claim Token	Broadcast	X'03', **3	X'00'	X'03'	No No	X'02' X'0B'	Upstream Neighbor's Address Physical Drop Number
Ring Purge	Broadcast	X'04', **3	X'00'	X'04'	No No	X'02' X'0B'	Upstream Neighbor's Address Physical Drop Number
Active Monitor Present (AMP)	Broadcast	X'05', **1	X'00'	X'05'	No No	X'02' X'0B'	Upstream Neighbor's Address Physical Drop Number
Standby Monitor Present (SMP)	Broadcast	X'06', **1	X'00'	X'06'	No No	X'02' X'0B'	Upstream Neighbor's Address Physical Drop Number
Duplicate Address Test	Station's address (DA = MA)	X'01', **1	X'00'	X'07'			This vector has no subvectors
Lobe Media Test	Null address	X'00', **1	X'00'	X'08'	No	X'26'	Test Data **8
Transmit Forward	Station address or Network Management's functional address (FA = NM)	X'00'	X'00' or X'80'	X'00' or X'09'	**10	**10	**10
Resolve	**11	**11	**11	X'0A'	**11	**11	**11
Report Station Addresses	Source address of Requester	X'00', **2	X'*0'	X'22'	No No Yes Yes No No	X'02' X'09' X'0B' X'21' X'2B' X'2C'	Upstream Neighbor's Address Correlator **5 Physical Drop Number Individual Address Count Group Address(es) Functional Address(es)
Report Station State	Source address of Requester	X'00', **2	X'*0'	X'23'	No Yes Yes No	X'09' X'23' X'28' X'29'	Correlator **5 Ring Station Version number Station Identifier Ring Station Status

Table D.1 *(continued)*
MAC Frame Transmit Definitions

Vector Name	Destination Address	FC, Pm	VC value	VI	Optional?	SVI	Subvectors Subvector(s) Name **9
Report Station Attachments	Source address of Requester	X'00', **2	X'*0'	X'24'	No	X'06'	Authorized Function Class
					No	X'07'	Authorized Access Priority
					No	X'09'	Correlator **5
					Yes	X'21'	Individual Address Count
					Yes	X'22'	Product Instance ID
					No	X'2C'	Functional Address(es)
Report New Active Monitor	Configuration report server's functional address (FA = CRS)	X'00', **1	X'40'	X'25'	No	X'02'	Upstream Neighbor's Address
					Yes	X'0B'	Physical Drop Number
					Yes	X'22'	Product Instance ID
Report SUA Change	Configuration report server's functional address (FA = CRS)	X'00', **1	X'40'	X'26'	No	X'02'	Upstream Neighbor's Address
					Yes	X'0B'	Physical Drop Number
Report Neighbor Notification Incomplete	Ring error monitor's functional address (FA = REM)	X'00', **1	X'60'	X'27	No	X'0A'	Source address of last "Active Monitor present" or "Standby Monitor present" MAC frame received
Report Active Monitor Error	Ring error monitor's functional address (FA = REM)	X'00', **1	X'60'	X'28'	No	X'30'	Error code
					No	X'02'	Upstream Neighbor's Address
					Yes	X'0B'	Physical Drop Number
Report Error	Ring error monitor's functional address (FA = REM)	X'00', **1	X'60'	X'29'	No	X'02'	Upstream Neighbor's Address
					Yes	X'0B'	Physical Drop Number
					No	X'2D'	Isolating Error Counts
					No	X'2E'	Nonisolating Error Counts
Report Transmit Forward	Network management's functional address (FA = NM)	X'00'	X'80'	X'2A'	**10	**10	**10

VC='*0' indicates the destination class of the transmitted frame is equal to the source class of the *received* Frame.

Optional transmit (Yes) indicates the subvector is *optional* and may be *omitted*.

**1 indicates the frame's priority (Pm) is equal to 7 (preferred; see Section 4.3.3.2) *or* 3.

**2 indicates the frame's priority (Pm) is equal to 7 (preferred; see Section 4.3.3.2) *or* 0.

**3 indicates the frame is transmitted without waiting for a Token (ring recovery; see Section 4.4.8).

**4 indicates the frame is *never* transmitted with a RIF.

**5 indicates the correlator *will* contain an unpredictable value if received frame does not contain a correlator subvector.

**6 indicates that during the Join process, a station can transmit *multiple* request initialization MAC frames (VI = X'20') to assured delivery of the frame. The first request frame is *always* transmitted with a FC value of X'00'. If the ring parameter server sets the A-bits and it does not respond to the station's request, then the station may transmit a request initialization MAC frame with a FC value of X'01'.

**7 indicates the response codes are defined in Section D.2 by subvector X'20'.
**8 indicates this subvector may have the SV0 or SV1 form defined in Section 4.3.3.2.
**9 indicates subvectors are listed by their numeric values and may be transmitted in any order.
**10 indicates reserved by ISO/IEC 8802-5:1995 (see Appendix E for definition).
**11 indicates reserved by ISO/IEC 8802-5:1995 (VI defined; but DA, FC, VC, and SVI not defined).

Table D.2 defines the vectors that the station is required to receive. See Section D.2 for definitions of the subvector names.

Table D.2
MAC Frame Receive Definitions

Vector Name	Source Address	FC	VC Value	VI	Optional?	SVI	Subvectors Subvector(s) Name **6
Beacon	Station detecting Beacon condition	X'02'	X'00'	X'02' **4	No No **3	X'01' X'02' X'0B'	Beacon Type Upstream Neighbor's Address Physical Drop Number
Claim Token	Station detecting Claim Token condition	X'03	X'00'	X'03' **4	No **3	X'02' X'0B'	Upstream Neighbor's Address Physical Drop Number
Ring Purge	Active Monitor	X'04'	X'00'	X'04' **4	No **3	X'02' X'0B'	Upstream Neighbor's Address Physical Drop Number
Active Monitor Present (AMP)	Active Monitor	X'05'	X'00'	X'05' **4	No **3	X'02' X'0B'	Upstream Neighbor's Address Physical Drop Number
Standby Monitor Present (SMP)	Any Standby Monitor	X'06'	X'00'	X'06' **4	No **3	X'02' X'0B'	Upstream Neighbor's Address Physical Drop Number
Duplicate Address Test	Station executing DAT test	X'01'	X'00'	X'07' **4			This vector has no subvectors
Transmit Forward	Station executing Transmit Forward (SC = 0) or Network Management (SC = 8)	X'00'	X'00' X'08'	X'09'	**7	**7	**7
Remove Ring Station	Configuration Report Server (SC = 4) or Network Management (SC = 8)	X'01'	X'04' or X'08' **2	X'0B'			This vector has no subvectors
Change Parameters	Configuration Report Server (SC = 4) or Network Management (SC = 8)	X'00'	X'04' or X'08' **2	X'0C'	Yes Yes Yes Yes Yes Yes	X'03' X'04' X'05' X'06' X'07' X'09'	Local Ring Number Assign Physical Drop Number Error Timer Value Authorized Function Classes Authorized Access Priority Correlator **5

Table D.2 *(continued)*
MAC Frame Receive Definitions

Vector Name	Source Address	FC	VC Value	VI	Optional?	SVI	Subvectors Subvector(s) Name **6
Initialize Station **1	Ring Parameter Server		X'00' or X'01'	X'0D'	Yes Yes Yes Yes	X'03' X'04' X'05' X'09'	Local Ring Number Assign Physical Drop Number Error Time Value Correlator **5
Request Station Address	Requester's Source Address	X'00'	X'0*'	X'0E'	Yes	X'09'	Correlator **5
Request Station State	Requester's Source Address	X'00'	X'0*'	X'0F'	Yes	X'09'	Correlator **5
Request Station Attachments	Requester's Source Address	X'00'	X'0*'	X'10'	Yes	X'09'	Correlator **5

VC='0*' indicates the received frame can have any source class except for the ring station (0).
Optional receive (No) indicates the subvector must be present for the frame to be considered valid.
Optional receive (Yes) indicates the subvector is *optional* and may not be present.
**1 indicates a station may optionally receive this frame with a FC value of X'01'.
**2 indicates a station may optionally receive this frame with a VC value of X'08' (network management).
**3 indicates a station may optionally require this subvector to be present for the frame to be considered valid.
**4 indicates frame is *invalid* if received with a RIF.
**5 indicates the response MAC frame's correlator subvector (if transmitted by the receiving station) *will* contain an unpredictable value if this correlator is not present.
**6 indicates subvectors are listed by their numeric values and may be received in *any* order.
**7 reserved by ISO/IEC 8802-5:1995: see Appendix E for definition.

D.2 Subvector Definitions

Table D.3 defines the subvectors defined by ISO/IEC 8802-5.2:1995. The format of these subvectors is described in Section 4.3.4.2.

Table D.3
Subvector Definitions

Subvector Name	SVI	SVL/SVLe	SVV (Length in Octets—Meaning)
Beacon Type	X'01'	SVL = X'04'	*2 octets:* Reason for Beaconing: 1. X'0001': Recovery Mode (used only by IEEE 802.5c: Dual-Ring Operation) 2. X'0002': Signal Loss Detected 3. X'0003': Claim Token failure, no Claim Token MAC Frames received 4. X'0004': Claim Token failure, Claim Token MAC Frames received

Table D.3 *(continued)*
Subvector Definitions

Subvector Name	SVI	SVL/SVLe	SVV (Length in Octets—Meaning)
Upstream Neighbor's Address (UNA)	X'02'	SVL = X'08'	*6 octets:* Upstream neighbor's specific address. If this address is unknown, then value is zero.
Local Ring Number	X'03'	SVL = X'04'	*2 octets:* Local ring number of the reporting station
Assign Physical Drop Number	X'04'	SVL = X'06'	*4 octets:* Physical address of the reporting station. The value of this SVV is not defined by ISO/IEC 8802-5.2:1995.
Error Report Timer Value	X'05'	SVL = X'04'	*2 octets:* States the value of the timer, error report in 10-ms intervals.
Authorized Function Classes	X'06'	SVL = X'04'	*2 octets:* Identifies the functional classes allowed to be active in the station. Each bit 0 through 15 represents a functional class as follows. X'xxxx 1xxx xxxx xxxx' is CRS, Class Value X'4' X'xxxx x1xx xxxx xxxx' is RPS, Class Value X'5' X'xxxx xx1x xxxx xxxx' is REM, Class Value X'6' X'xxxx xxxx 1xxx xxxx' is Network Management, Class Value X'8' A station may have more than one active functional class.
Authorized Access Priority	X'07'	SVL = X'04'	*2 octets:* Maximum Access Priority (Pm) allowed for the station transmit of LLC frames.
Correlator	X'09'	SVL = X'04'	*2 octets:* Value of the correlator received on a request received by the station. If the request does not contain a correlator, then this subvector may be omitted or have an unpredictable value.
SA of last AMP or SMP Frame	X'0A'	SVL = X'08'	*6 octets:* An indication given when the station transmits the neighbor notification incomplete MAC frame as follows: 1. A null address means no AMP MAC frame has been received (only occurs prior to Join complete). 2. An address equal to the reporting station's specific address means that no AMP or SMP MAC frame was received. 3. The address of the last received AMP or SMP address.
Physical Drop Number	X'0B'	SVL = X'06'	*4 octets:* The physical drop number of the reporting station.
Response Code[1]	X'20'	SVL = X'06'	*6 octets:* Used to indicate why the response code is being sent. It includes a 2-octet response code, and from the received MAC frame being responded to, the 1-octet VC field and the 1-octet vector identifier. The response codes defined include: 1. X'0001': Positive response

Table D.3 *(continued)*
Subvector Definitions

Subvector Name	SVI	SVL/SVLe	SVV (Length in Octets—Meaning)
			2. X'8001': MAC frame data field incomplete 3. X'8002': Vector length error 4. X'8003': Unrecognized vector 5. X'8004': Inappropriate source class 6. X'8005': Subvector length error 7. X'8006': Not used-reserved 8. X'8007': Missing subvector 9. X'8008': Subvector unknow 10. X'8009': MAC frame too long 11. X'800A': Function requested is disabled
Response Code[2]	X'20'	SVL = X'08'	*8 octets:* Used to indicate why the response code is being sent. It includes a 2-octet response code, and from the received MAC frame being responded to: the 1-octet VC field and the 3 SVIe field.
Individual Address Count	X'21'	SVL = X'04'	*2 octets:* The number of individual addresses supported by the reporting station as follows. 1. X'0000': One individual address is supported. 2. Nonzero: The number of individual addresses supported.
Product Instance ID	X'22'	SVL = not defined	*Length not defined:* Contains a manufacturer's identification of station characteristics. It is recommended by ISO/IEC 8802-5.2:1995 this SVV be equal to the "ResourceTypeID" managed object defined by ISO/IEC 10742:1994.
Ring Station Version Number	X'23'	SVL = not defined	*Length not defined:* Contains the station's version number assigned by the manufacturer. The data content of this SVV is not defined by ISO/IEC 8802-5.2:1995.
Wrap Data[3]	X'26'	SVL = X'02' through X'FE'	*2 through 255 octets:* Data for the Lobe Media test MAC frame. The data contents of this SVV is not defined by ISO/IEC 8802-5.2:1995.
Wrap Data[4]	X'26'	SVL = X'FF' SVLe = X'00 FF' through maximum length allowed	*256 through the number of octets of the longest supported information field size (see Section 4.3.3.4):* Data for the Lobe Media test MAC frame. The data contents of this SVV is not defined by ISO/IEC 8802-5.2:1995.
Frame Forward[5]	X'27'	SVL = not defined	*Reserved (see Appendix E for definition)*
Station Identifier	*X'28'*	*SVL = X'08'*	*6 octets:* Used to identify the station, it is recommended by ISO/IEC 8802-5.2:1995 to be the station's universal address.

Table D.3 *(continued)*
Subvector Definitions

Subvector Name	SVI	SVL/SVLe	SVV (Length in Octets—Meaning)
Ring Station Status[6]	X'29'	SVL = not defined	*Length not defined:* Contains the station's status number assigned by the manufacturer; the data content of this SVV is not defined by ISO/IEC 8802-5.2:1995.
Transmit Forward Status Code[7]	X'2A'	SVL = not defined	Reserved (see Appendix E for definition)
Group Address	X'2B'	SVL = X'06'	*4 octets:* This SVV contains the 4-octet version of the group address (4 low-order octets) being used by the station. 1. If the value is zero, no group address is supported. 2. If the value is nonzero, no assumptions can be made about the 2 high-order octets of the group address.
Group Address	X'2B'	SVL = X'08'	*6 octets:* This SVV contains the group address being used by the station. 1. If the value is zero, no group address is supported. 2. If the value is nonzero, the group address being supported by the station is reported. If the station supports more than one group address, the SVV can be any of the supported group addresses.
Functional Address	X'2C'	SVL = X'06'	*4 octets:* Contains the functional address supported by the reporting station.
Isolating Error Counters	X'2D'	SVL = X'08'	*6 octets:* Contains the isolating error counters as follows: 1. Octet 0—Line error (CLE) 2. Octet 1—Internal error (CIE) 3. Octet 2—Burst error (CBE) 4. Octet 3—AC error (CACE) 5. Octet 4—Abort Sequence transmitted (CABE) 6. Octet 5—Reserved (always zero) See Section 4.3.7.2 for definitions.
Nonisolating Error Counters	X'2E'	SVL = X'08'	*6 octets:* Contains the nonisolating error counters as follows: 1. Octet 0—Lost Frame error (CLFE) 2. Octet 1—Receive Congestion error (CRCE) 3. Octet 2—Frame-copied error (CFCE) 4. Octet 3—Frequency error (CFE) 5. Octet 4—Token Error (CTE) 6. Octet 5—Reserved (always zero) See Section 4.3.7.3 for definitions.

Table D.3 *(continued)*
Subvector Definitions

Subvector Name	SVI	SVL/SVLe	SVV (Length in Octets—Meaning)
Function Request ID[8] X'2F'		SVL = not defined	Reserved (not defined)
Error Code	X'30'	SVL = X'04'	*2 octets:* Contains the following error codes. 1. X'0001': Active Monitor Error 2. X'0002': Duplicate Active Monitor 3. X'0003': Duplicate Address Error

[1]Even though a response code could be sent, if the Source Address of the MAC Frame being responded to is zero (MAC Sublayer), a Response MAC Frame is *not* sent.
[2]This form is not used by ISO/IEC 8802-5.2:1995, but is reserved for future expansion (SVIe).
[3]Form 1: Data field less than 256 bytes.
[4]Form 2: Data Field 256 bytes or greater. This is the only Subvector known to use SVLe.
[5]This code point is used by early station's and is not defined by ISO/IEC 8802-5.2:1995, but is reserved to prevent older stations from being nonconformant.
[6]The receiver of this subvector may be able to identify the format of the data content of this SVV by examination of the product instance ID subvector (X'22').
[7]This code point is used by early station's and is not defined by ISO/IEC 8802-5.2:1995, but is reserved to prevent older stations from being nonconformant.
[8]This code point is used by early station's and is not defined by ISO/IEC 8802-5.2:1995, but is reserved to prevent older stations from being nonconformant.

E

Transmit Forward Process

This appendix details the Transmit Forward process referenced by Section 4.2.4.3 but *not* specified by the ISO/IEC 8802-5 Standard. This process is a network management function originally used to test the data path between multiple stations on one ring (*on-ring*) or multiple rings (*off-ring*) using the station's MAC layer Transmit Forward function. The Transmit Forward process is also known as the "Path Trace process" in some implementations.

The Transmit Forward process defined in this appendix is a MAC layer function. Because of a late IEEE 802.1d (Bridging Standard) ruling to prevent MAC frames from being exported *off-ring*, this MAC layer function is now only an *on-ring* test function. The *off-ring* MAC layer Transmit Forward function has been replaced in some networks by a similar function operating at the LLC layer, but the frame flow test logic of the Transmit Forward process remains as originally designed.

This appendix is offered as a guide for the *intent* of the Transmit Forward process and not as a specification of any implementation. It is hoped that the Transmit Forward process with its powerful diagnostic capabilities will be used in other LANs to assist in the diagnosis of failing paths between stations.

E.1 Process

Each station that receives a Transmit Forward MAC frame uses its MAC layer transmit forward function to:

- Transmit a Transmit Forward MAC frame built from the data contained in the Information field of the received transmit forward MAC frame;

- Transmit a report Transmit Forward MAC frame upon successful transmission of the preceding Transmit Forward MAC frame.

The intent of the Transmit Forward process is to allow checking the connectivity (path testing) between one or more stations on the ring, as illustrated in Figure E.1.

Figure E.1 illustrates network management sending the *first* Transmit Forward MAC frame "1" to test the path between itself and stations 1 to 5. Normal operation of the Transmit Forward process terminates when network management receives frames "e" and "A."

Any failure of the path under test causes premature termination of the Transmit Forward process. For example, it would appear station 3 failed to forward frame "4" if network management failed to receive frames "d," "e," and "A." Using this information coupled with frame "c," network management can determine that either station 3 failed to forward frame 4 (frame "c" is *not* received by network management) or station 4 failed to receive frame 4 (frame "c" *is* received by network management).

Transmit Forward Setup Information

In practice, setting up the preceding Transmit Forward process is difficult because of the complexity of the *first* Transmit Forward MAC frame. For this reason, network management limits its connectivity test to itself and one or two other stations, as explained in the example in Section E.4.

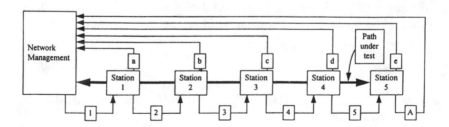

Figure E.1 Overview of Transmit Forward. Frames "1" to "5" are transmit forward MAC frames sent to Destination Class 0 (MAC layer). Frame "A" is a transmit forward MAC frame sent to Destination Class 8 (network management). Frames "a" to "e" are report transmit forward MAC frame sent to Destination Class 8.

E.2 Vector and Subvector Definitions

The definition of the Transmit Forward MAC frame's vectors and subvectors have been *reserved* by the ISO/IEC 8802-5.2:1995 Standard to prevent older implementations manufactured *prior* to the acceptance of this Standard from being noncompliant.

E.2.1 Vectors

The Transmit Forward process uses the MAC frame vectors defined by Table E.1.

Table E.1
Transmit Forward Vectors

Vector Name	Destination Address	FC, Pm	VC Value	VI	Subvectors (see Table E.2) Optional?	SVI	Subvector(s) Name
Transmit Forward (MAC frame sent to a station)	Source address of Target	X'00', *	X'00' or X'08'	X'09'	No	X'27'	Frame Forward Data
Report Transmit Forward	FA = NM	X'00', *	X'80'	X'2A'	No	X'2A'	Transmit Forward Status
Transmit Forward (MAC frame sent to network management)	Network Management's functional address (FA = NM)	X'00', *	X'80'	X'09'	No	X'mm'	Transmit Forward Response

* indicates a Pm (transmit priority) of 0 or 7.
X'mm' is a SVI value specified by network management.

E.2.2 Subvectors

The Transmit Forward process uses the MAC frame subvectors defined by Table E.2.

Table E.2
Transmit Forward Subvectors

Subvector Name	SVI	SVL	SVV (Meaning)
Frame Forward	X'27'	X'**'	Contains the "AC field through the last I-field octet" of the frame to be forward
Transmit Forward Status Code	X'2A'	X'04'	Two octets of product unique status code (1)
Transmit Forward Response	X'mm'	X'nn'	Value is uniquely defined by network management

X'**' denotes that the subvector length (SVL) is variable with a *maximum* of X'FE'.
X'mm', a subvector identifier (SVI), is defined by network management.
X'nn', a subvector length (SVL), is defined by network management.
(1) is a status code usually equal to the station's transmit completion status.

E.3   Transmit Forward MAC Frame Utilization

Figure E.2 illustrates an overview of how the station's MAC layer Transmit Forward function uses the data contained *within* the Transmit Forward MAC frame "1" to build transmit forward MAC frame "2".

E.4   Transmit Forward Frame Flow Example

Figure E.3 expands on Figure E.2 to illustrate an example of the data flow between stations A, B, and C. This flow occurs when network management (a higher layer function) in station A causes the transmission of transmit forward MAC frames between these stations.

Before explaining the operation illustrated in Figure E.3, the following points are made.

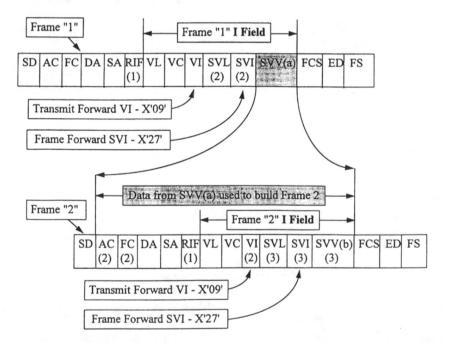

Figure E.2   Transmit Forward MAC frame. (1) means the RIF is optional. The RIF is *not* present for the *on-ring* testing but was present in early *off-ring* testing (no longer supported). (2) means this entity *is* syntax checked as defined in Chapter E.5. (3) means this entity *is not* syntax checked by the station building frame "2." Even though the structure of the data within SVV(b) *must* have the same structure as the data within SVV(a), this is not assured by the station's Transmit Forward function.

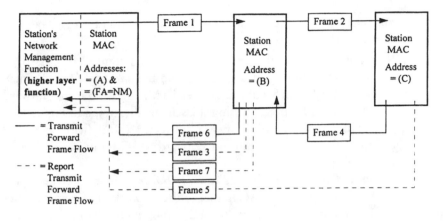

Figure E.3 Transmit Forward data flow-overview.

1. Frame transmit order is logically "1," "2," "3," "4," "5," "6," and "7" but, because of the queuing process, frames "3," "4," "5," "6," and "7" may be out of order. For the purposes of explanation, assume these frames are transmitted in order.

2. Frame "1" is the *original* Transmit Forward MAC frame containing the data from which stations B and C generate their:
   a. Transmit Forward MAC frames "2" (from station B) and "4" (from station C) containing a Transmit Forward subvector used to build another Transmit Forward MAC frame;
   b. Report Transmit Forward MAC frames "3" (from station B), "5" (from station C), and "7" (from station B) used to indicate the successful reception *and* transmission of the Transmit Forward MAC frame;
   c. Transmit Forward MAC frame "6" (from station B) containing a transmit response subvector used to indicate to network management the end of the transmit forward process.

3. Even though not shown in Figure E.3, if stations B or C detect a syntax error in frames "1," "2," or "4," then they would send a Response MAC frame as specified in Appendix E.5 and terminate the Transmit Forward process.

Referring to Figure E.3, a step-by-step definition of the Transmit Forward process follows.

1. The network management function in station A *starts* the Transmit Forward process when station A transmits frame "1" (a Transmit

Forward MAC frame containing data used to build other Transmit Forward MAC frames) to station B.

2. If station B successfully copies frame 1 (i.e., frame "1" passed its *normal* frame receive syntax checking and the *additional* frame syntax checking specified in Section E.5), then its MAC layer transmit forward function builds frame "2" (using the scheme illustrated in Figure E.2) from frame "1's" Frame Forward subvector (SVV field) and then transmits frame "2" (a Transmit Forward MAC frame) to station C. If station B:

   a. *Successfully* strips frame "2," then it *transmits* frame "3" (a Report Transmit Forward MAC frame) to the network management functional address (in station A) and terminates its Transmit Forward function;

   b. *Unsuccessfully* strips frame "2," then it *terminates* its Transmit Forward function *without* transmitting frame "3."

3. If station C successfully copies frame "2," then its MAC layer transmit forward function builds frame "4" from frame "2's" frame forward subvector (SVV Field) and then transmits frame "4" (a Transmit Forward MAC frame) to station B. If station C:

   a. *Successfully* strips frame "4," then it *transmits* frame "5" (a Report Transmit Forward MAC frame) to the network management functional address and terminates its Transmit Forward function.

   b. *Unsuccessfully* strips frame "4," then it *terminates* its Transmit Forward function *without* transmitting frame "5."

4. If station B successfully copies frame "4," then its MAC layer Transmit Forward function builds frame "6" from frame "4's" frame forward subvector (SVV Field) and then transmits frame "6" (a Transmit Forward MAC frame containing a "transmit response" subvector) to station A. If station B:

   a. *Successfully* strips frame "6," then it *transmits* frame "7" (a Report Transmit Forward MAC frame) to the network management functional address and terminates its Transmit Forward function.

   b. *Unsuccessfully* strips frame "6," then it *terminates* its Transmit Forward function without transmitting frame "7."

5. Network management in station A terminates its Transmit Forward process when it receives frame "6" (a Transmit Forward MAC frame containing a "Transmit Response" subvector).

Close examination of this process reveals that significant data about the paths between stations A, B, and C can be collected about the success or failure of the data flow between these stations. Two examples follow.

- If station B is having stripping problems, then the Transmit Forward process continues but there will be no Report Transmit Forward MAC frame from station B (e.g., frames "3" and/or "7"), indicating station B failed to strip its Transmit Forward MAC frame(frame "2" may or may not have been received by station C).

- If station C does not copy frame "2," then the Transmit Forward process fails because the network management function does not receive all of its expected frames.

E.5 Transmit Forward Function Syntax Checking

The station's Transmit Forward function performs the following *additional* syntax checking on the Transmit Forward MAC frame (after *normal* frame reception syntax checking has completed successfully).

1. The *first* subvector in the Transmit Forward MAC frame *must* be a Frame Forward subvector (SVI = X'27') and have a length (SVL) of *less than* X'FF'.

2. The Frame Forward subvector data (SVV), containing the frame to be forwarded, is syntax checked as follows:
   a. All AC field bits *must be* equal to zero.
   b. All FC field bits except the reserved bits *must be* equal to zero (bits 2 and 3 are not checked).
   c. The vector identifier (VI) *must be* transmit forward (X'09').

The Transmit Forward MAC frame is acceptable for transmission *if* the frame passes the preceding syntax checking.

A Response MAC frame with an error code of X'8006'[1] (invalid Transmit Forward MAC frame) is transmitted to the network management functional address *if* the frame fails *any* of these checks. The network management function must correct the data within the Transmit Forward MAC frame based on the data within the *original* Transmit Forward MAC frame and the *address* of the station reporting the invalid Transmit Forward MAC frame error code.

---

1. This response code is *reserved* in ISO/IEC 8802-5.2:1995 because of its use by the Transmit Forward process.

List of Acronyms and Abbreviations

A-bit	Address-recognized bit
AC	Access control
AM	Active Monitor
AMP	Active Monitor Present
ANSI	American National Standards Institute
AP	Access Protocol
API	Application programming interfaces
APS	Accumulated phase slope
ARE	All routes explorer
ARI	Address recognized indicators
ARL	Adjusted ring length
ASTRAL	Alliance for Strategic Token Ring Advancement and Leadership
ATM	Asynchronous transfer mode
BER	Bit error rate
BRF	Bridging relay function
BUD	Bring-up diagnostics
BUS	Broadcast and unknown server
BW	Bandwidth
CAM	Content addressable memory
C-bit	Frame-copied bit
CPU	Central processing unit
CRC	Cyclic redundancy check

CRF	Concentrator relay function
CRFP	CRF Port
CRS	Configuration Report Server
CSMA/CD	Carrier Sense Multiple Access with Collision Detect
DA	Destination address
DAC	Duplicate Address Check
dB	Decibels
dBm	Decibels above a milliwatt
DC	Destination class
DLC	Data link control
DMA	Direct memory access
DRD	Destination route descriptor
DTR	Dedicated Token Ring
DTU	Data Transfer Unit
E-bit	Error-detect bit
ED	End delimiter
EDFI	Error detection/fault isolation
EDI	Error detect indicator
EFS	Ending frame sequence
EIA	Electronic Industries Association
EMC	Electromagnetic interference
ETR	Early token release
FAPS	Filtered accumulated phase slope
FC	Frame control
FCI	Frame copied indicators
FCS	Frame check sequence
FDDI	Fiber distributed data interface
FIFO	First in, first out
FL	Fiber Loss
FOTCU	Fiber optic trunk coupling unit
FS	Frame status
FSM	Finite state machine
HDLC	High-level data link control
HSTR	High-Speed Token Ring
HSTRA	High-Speed Token Ring Alliance
IAC	Individual address count
I-bit	Intermediate frame bit

IBM	International Business Machines Incorporated
IFG	Interframe Gap
I-field	Information field
ISDN	Integrated Services Digital Network
ISO	International Organization for Standardization
JTOL	Jitter tolerance
LAA	Locally administered address
LAN	Local area network
LANE	ATM LAN emulation
LEC	LAN emulation clients
LECS	LAN emulation configuration servers
LED	Light-emitting diode
LES	LAN emulation server
LLC	Logical link control
LMT	Lobe media test
LSB	Least significant bit
m	Meters
MAC	Media access control
M-bit	Monitor bit
mFCJ	Mean filtered correlated jitter
MFID	Manufacturer's ID
MIB	Management information block
MLL	Maximum lobe length
MRI	Management Routing Interface
ms	Milliseconds
MSAU	Multiple station access units
MSB	Most significant bit
NADN	Nearest active downstream neighbor
NAUN	Nearest active upstream neighbor
NDT	Net data delay time
NEXT	Near end cross talk
NIC	Network interface cards
OSI	Open Systems Interconnection
OUI	Organizational unique identifier
P-bits	Priority bits
PDS	Premises distribution system
PDU	Protocol data unit

PH	Protocol handler
PHY	Physical layer
PLL	Phase locked loop
Pm	Priority of frame queued for transmission
PMC	Physical media components
Prx	Power received
PSC	Physical signaling components
Ptx	Power transmitted
QoS	Quality of service
RAM	Random access memory
RAS	Reliability, availability, serviceability
R-bits	Reservation bits
RD	Route descriptor
REM	Ring error monitor
RI	Routing indicator
RIF	Routing information field
RII	Routing information indicator
ROM	Read-only memory
RPS	Ring parameter server
RVV	Ratification, verification, and validation
SA	Source address
SD	Start delimiter
SFS	Starting frame sequence
SM	Standby Monitor
SMP	Standby Monitor Present
SNA	System Network Architecture
SRF	Specifically routed frame
SRT	Source routing transparent
STE	Spanning Tree explorer
STP	Shielded twisted pair, also in Chapter 5 as Spanning Tree protocol
SUA	Stored Upstream Address
SV	Subvector
SVI	Subvector identifier
SVIe	Subvector identifier extended
SVL	Subvector length
SVLe	Subvector length extended

SVV	Subvector value
T-bit	Token bit
TCP/IP	Transmission Control Protocol/Internet Protocol
TCU	Trunk coupling unit
TI	Texas Instruments Incorporated
TIA	Telecommunications Industry Association
TKP	Token passing protocol
TSB	Technical Supplemental Bulletins
TTL	Transistor-transistor logic
TTRT	Target Token-Rotation Time
TXI	Transmit immediate protocol
UAA	Universally administered address
UNA	Upstream neighbor address
UTP	Unshielded twisted pair
UUT	Unit under test
VC	Vector class
VCO	Voltage Controlled Oscillator
VI	Vector identifier
VL	Vector length
VLAN	Virtual local area network
VLSI	Very large scale integration
XTAL	Crystal

About the Authors

Dr. James Thomas Carlo is a TI Fellow of Texas Instruments and serves as the IEEE 802 LAN/MAN Sponsor Committee Chair. At TI, he has participated in Token Ring since 1981, working closely with IBM during initial development and developing TI's customer base for Token Ring and other LAN products. Jim's activities span both the MAC and PHY aspects of Token Ring to enable him to view the entire system configuration. He is currently responsible for Networking Standards in the World Wide Access Business Unit of TI Semiconductor Group.

As Chair of IEEE 802, Jim is responsible for the development and activities of LAN/MAN standards for the Computer Society. The IEEE 802 committee consists of about 500 engineering representatives from industry and education with standards developments in Ethernet, Token Ring, Cable TV Modems, and VLANS. In his prior responsibility, he was chair for the IEEE 802.5 Token Ring Working Group. Jim is also chair of the ISO/IEC/JTC1/SC6 standards committee for the internationalization of LAN standards. He is also on the editorial board of the Network Magazine of the Communications Society and is on the IEEE Standards Board as director for the New Standards Committee.

Jim joined TI in 1973 after graduation from MIT with a Ph.D. in Electrical Engineering. Jim and his wife Muffy have two sons, both currently in college, and live in Dallas.

Robert D. Love, a Senior Engineer at IBM's Network Hardware Division in Research Triangle Park, North Carolina, is a physical layer expert on Token Ring and has been involved in Token Ring and cabling development and standarization since 1981. Bob developed the original Token Ring wiring rules (at 4 and 16 Mbits/s for both 150 ohm STP, and 100 ohm UTP cabling). He developed and taught a class on Token Ring for the IEEE and has published and delivered numerous papers on both Token Ring and telecommunications cabling around the world. Bob, an active participant in the High Speed Token Ring Alliance (HSTRA), is Chair of the IEEE 802.5 Token Ring Working Group and a recipient of the IEEE Standards Medallion for his Token Ring standards leadership and technical contributions.

He received his bachelor's degree in Electrical Engineering from Columbia University, and his Masters in ElectroPhysics from the Brooklyn Polytechnic Institute. He can be reached via e-mail at rdlove@us.ibm.com.

Michael S. Siegel, a Senior Engineer, joined IBM's Network Hardware Division located at Research Triangle Park, North Carolina, where he became involved in the development of Token Ring Switches in 1993. He joined the 802.5 committee in January 1994 and became a co-editor of the DTR Standard in the same year.

Michael continues to be involved in the development of high speed, high capacity, Layer 2 and Layer 3 switch products for IBM. Prior assignments in IBM include IO Channel development and Vector Processor development for IBM mainframes. He joined IBM's Large System Division located in Poughkeepsie, New York in 1977, after receiving his BSEE from Renesslaer Polytechnic Institute.

Michael holds patents in the fields of Local Area Networks, Vector Processor design, and Scalar Processor design.

Kenneth T. Wilson joined IBM in Kingston, New York, on June 7, 1957, as a field engineer responsible for maintaining the Air Force's Sage Air Defense system. In 1960, he went to Poughkeepsie, New York, to work on IBM's mainframes and developed the first known hardware assisted software diagnostics capable of identifying faulty cards for commercial systems. This led to a management position in charge of developing an On-line Test System (OLTs) that operated not only in the standalone environment, but the S/360 OS and DOS environments. In 1967, he worked on an advanced high speed mainframe and developed a process for identifying faults in early LSI systems still being used in today's supercomputers. In 1976, he transferred to Research Triangle Park, North Carolina, to apply his system knowledge to the Networking field. He helped develop the first Network Management functions to be integrated into IBM's System Network Architecture (SNA).

In 1981, Ken joined the group responsible for developing the IBM Token Ring hardware, microcode, and software. Ken was the key inventor of the IBM Token Ring recovery process which is now the basis of the IEEE 802.5 Classic

Token Ring's ring recovery. This work led to his receiving an IBM Corporate Technical Recognition award, IBM's highest technical award, and several other IBM recognition awards including IBM's RTP Inventor of the year.

Since 1994, he has served on the IEEE 802.5 LAN/MAN Committee as a technical editor of the IEEE 802.5 MAC protocol for the Classic Token Ring (Chapter 4), Dedicated Token Ring (Chapter 5), and for the on-going High Speed Token Ring (Chapter 6). Ken is currently responsible for development of the High Speed Token Ring Networking Standards as part of IBM's Token Ring development group.

Ken is a Senior Programmer in IBM's Networking Hardware Division and holds nine patents in the field of Local Area Networking.

Ken lives in Garner, North Carolina, and has five children and four grandchildren, all living in Raleigh, North Carolina. He can be reached via e-mail at ktwilson@us.ibm.com or k_wilson@ieee.org.

Index

For further information on these and other Artech House titles, including previously considered out-of-print books now available through our In-Print-Forever™ (IPF™) program, contact:

Artech House
685 Canton Street
Norwood, MA 02062
781-769-9750
Fax: 781-769-6334
Telex: 951-659
email: artech@artech-house.com

Artech House
Portland House, Stag Place
London SW1E 5XA England
+44 (0) 171-973-8077
Fax: +44 (0) 171-630-0166
Telex: 951-659
email: artech-uk@artech-house.com

Find us on the World Wide Web at: www.artech-house.com